Mixed-Signal Circuits

Devices, Circuits, and Systems

Series Editor
Krzysztof Iniewski
CMOS Emerging Technologies Research Inc.,
Vancouver, British Columbia, Canada

PUBLISHED TITLES:

Atomic Nanoscale Technology in the Nuclear Industry
Taeho Woo

Biological and Medical Sensor Technologies
Krzysztof Iniewski

Building Sensor Networks: From Design to Applications
Ioanis Nikolaidis and Krzysztof Iniewski

Circuits at the Nanoscale: Communications, Imaging, and Sensing
Krzysztof Iniewski

CMOS: Front-End Electronics for Radiation Sensors
Angelo Rivetti

Design of 3D Integrated Circuits and Systems
Rohit Sharma

Electrical Solitons: Theory, Design, and Applications
David Ricketts and Donhee Ham

Electronics for Radiation Detection
Krzysztof Iniewski

**Electrostatic Discharge Protection of Semiconductor Devices
and Integrated Circuits**
Juin J. Liou

**Embedded and Networking Systems:
Design, Software, and Implementation**
Gul N. Khan and Krzysztof Iniewski

Energy Harvesting with Functional Materials and Microsystems
Madhu Bhaskaran, Sharath Sriram, and Krzysztof Iniewski

Gallium Nitride (GaN): Physics, Devices, and Technology
Farid Medjdoub

**Graphene, Carbon Nanotubes, and Nanostuctures:
Techniques and Applications**
James E. Morris and Krzysztof Iniewski

High-Speed Devices and Circuits with THz Applications
Jung Han Choi

PUBLISHED TITLES:

PUBLISHED TITLES:

FORTHCOMING TITLES:

Semiconductor Devices in Harsh Conditions
Kirsten Weide-Zaage and Malgorzata Chrzanowska-Jeske

Smart eHealth and eCare Technologies Handbook
Sari Merilampi, Lars T. Berger, and Andrew Sirkka

Structural Health Monitoring of Composite Structures Using Fiber Optic Methods
Ginu Rajan and Gangadhara Prusty

Terahertz Sensing and Imaging: Technology and Devices
Daryoosh Saeedkia and Wojciech Knap

Tunable RF Components and Circuits: Applications in Mobile Handsets
Jeffrey L. Hilbert

Wireless Medical Systems and Algorithms: Design and Applications
Pietro Salvo and Miguel Hernandez-Silveira

Mixed-Signal Circuits

EDITED BY

Thomas Noulis
Intel Corporation, Munich, Germany

CRC Press
Taylor & Francis Group
Boca Raton London New York

CRC Press is an imprint of the
Taylor & Francis Group, an **informa** business

CRC Press
Taylor & Francis Group
6000 Broken Sound Parkway NW, Suite 300
Boca Raton, FL 33487-2742

First issued in paperback 2021

© 2015 by Taylor & Francis Group, LLC
CRC Press is an imprint of Taylor & Francis Group, an Informa business

No claim to original U.S. Government works

ISBN-13: 978-0-367-77889-7 (pbk)
ISBN-13: 978-1-4822-6062-5 (hbk)

Library of Congress Cataloging-in-Publication Data

Mixed-signal circuits / Thomas Noulis, editor.
    pages cm. -- (Devices, circuits, and systems ; 46)
  "A CRC title."
  Includes bibliographical references and index.
  ISBN 978-1-4822-6062-5 (alk. paper)
  1. Mixed signal circuits. 2. Integrated circuits. I. Noulis, Thomas, editor.

TK7874.55.M59 2016
621.3815--dc23   2015015695

Visit the Taylor & Francis Web site at
http://www.taylorandfrancis.com

and the CRC Press Web site at
http://www.crcpress.com

Contents

Preface

This book addresses mixed-signal integrated circuits using advanced design techniques to enable digital circuits and sensitive analog circuits to co-exist without any compromise. Different related topics are addressed, such as the advanced process technology to address the performance challenges associated with developing these complex mixed-signal circuits, the related blocking points in the industry design flow, and the general validation of the proposed solutions and implementations. Development and implementation of innovative methodologies to move analog into the digital domain quickly, minimizing and eliminating common trade-offs between performance, power consumption, simulation time, verification, size, and cost containment are also discussed.

Specifically, in this book, the state of the art in integrated circuit design in the context of mixed-signal applications is addressed. New, exciting opportunities in different areas like wireless communications, data networking, and simulation and verification techniques are presented. Design concepts for very low-power performance and approaches for high-speed interfaces, PLL, VCOs, ADC converters, and biomedical filters are described. Respective parts of a full system-on-chip (SoC), from the digital parts untill the baseband blocks, the RF circuitries, the ESD structures and the built-in self-test architectures are provided.

Coverage includes advanced crucial topics like signal integrity, large-scale simulation, and verification and testing. Extremely hot modeling topics are also addressed such as reliability, variability, and crosstalk that define pre-silicon design methodology and trends and are the main research items for all industry leading companies involved in wireless applications.

The book is written by a mixture of top industrial experts and key academic professors and researchers. Practical enough to understand how these technologies work, but not a product manual and, at the same time, scientific enough but not pure academic theory.

This book is a must for anyone involved in mixed-signal circuit design for future technologies. The intended audience is engineers with advanced integrated circuit background working in the semiconductor industry. This book can also be used as a recommended reading and supplementary material in a graduate course curriculum and, in general, the intended audience is professionals working in the integrated circuit design field.

I hope you enjoy reading this book as much as we have enjoyed writing it!

Thomas Noulis
Editor
March 30, 2015

MATLAB® and Simulink are registered trademarks of The MathWorks, Inc.
For product information, please contact:

The MathWorks, Inc.
3 Apple Hill Drive
Natick, MA 01760-2098 USA
Tel: 508 647 7000
Fax: 508-647-7001
E-mail: info@mathworks.com
Web: www.mathworks.com

Editor

Thomas Noulis is a staff RFMS engineer at Intel Corporation in the Mobile & Communications Group in Munich, Germany, specializing in circuit design, modeling–characterization, crosstalk, and SoC product active area minimization. Before joining Intel, from May 2008 to March 2012, Dr. Noulis was with HELIC Inc., initially as an analog/RF IC designer, and then as an R&D engineer specializing in substrate coupling, signal and noise integrity, and analog/RFIC design. Thomas Noulis earned a BSc in physics (2003), an MSc in electronics engineering (2005), and a PhD in the Design of Signal Processing Integrated Circuits (2009) from the Aristotle University of Thessaloniki, Greece, and in collaboration with LAAS (Laboratoire d'Analyse et d'Architectures des Systèmes), Toulouse, France. During 2004–2009, he participated as principal researcher in multiple European and national research projects related to space application and nuclear spectroscopy IC design, while between 2004 and 2010, he also collaborated as a visiting-adjunct professor with universities and technical institutes. Dr. Noulis is the author of more than 30 publications, journals, conferences, and scientific book chapters. He holds one French and World patent. His work has received more than 50 citations. He is an active reviewer of multiple international journals and has given multiple invited presentations at European research institutes and international conferences on crosstalk and radiation detection IC design. Dr. Noulis has received awards for his research at conferences and by research organizations and can be reached at t.noulis@gmail.com.

Contributors

Jacob Abraham
Department of Electrical and
  Computer Engineering
The University of Texas
Austin, Texas

Marise Bafleur
Laboratoire d'Analyse et
  d'Architecture des Systèmes
  (LAAS)
Toulouse, France

Sotiris Bantas
Centaur Technologies
Volos, Greece

Manuel Barragán
Laboratoire TIMA
Centre National de la Recherche
  Scientifique
Grenoble, France

Patrice Besse
Freescale Semiconductor Inc.
Toulouse, France

Fabrice Caignet
Laboratoire d'Analyse et
  d'Architecture des Systèmes
  (LAAS)
Toulouse, France

Francis Calmon
Institut des Nanotechnologies de
  Lyon
Université de Lyon
Lyon, France

Abhijit Chatterjee
Electrical and Computer
  Engineering
Georgia Institute of Technology
Atlanta, Georgia

Ilias Chlis
Tyndall National Institute
and
Electrical and Electronic
  Engineering
School of Engineering
University College Cork
Cork, Ireland

Michael G. Dimopoulos
Laboratoire TIMA
Université Grenoble Alpes
Grenoble, France

Ricardo Doldán
ARQUIMEA DEUTSCHLAND
  GmbH
Frankfurt (Oder), Germany

Ikhwana Elfitri
Department of Electrical
  Engineering
Andalas University
Padang, Indonesia

Nestor Evmorfopoulos
Department of Computer
  Science
University of Thessaly
Volos, Greece

Rafaella Fiorelli
Instituto de Microelectrónica de
    Sevilla (IMSE-CNM-CSIC)
Universidad de Sevilla
Seville, Spain

Antonio Ginés
Instituto de Microelectrónica de
    Seville (IMSE-CNM-CSIC)
Universidad de Sevilla
Seville, Spain

Christian Gontrand
Institut des Nanotechnologies de
    Lyon
Université de Lyon
Lyon, France

Alkis Hatzopoulos
Department of Electrical and
    Computer Engineering
Aristotle University of
    Thessaloniki
Thessaloniki, Greece

Farooq A. Khanday
Department of Electronics
    and Instrumentation
    Technology
University of Kashmir
Srinagar, Jammu, and Kashmir,
    India

Jean-Phillppe Laine
Freescale Semiconductor Inc.
Toulouse, France

Jean-Etienne Lorival
Institut des Nanotechnologies de
    Lyon
Université de Lyon, INSA- Lyon,
    CNRS-UMR
Villeurbanne, France

Yiorgos Makris
Department of Electrical
    Engineering
Erik Jonsson School of Engineering
    and Computer Science
University of Texas
Dallas, Texas

Dzmitry Maliuk
Quantlab Financial LLC
Houston, Texas

Lampros Mountrichas
Electronics Laboratory of the
    Physics Department
Aristotle University of Thessaloniki
Thessaloniki, Greece

Nicolas Nolhier
Laboratoire d'Analyse et
    d'Architecture des Systèmes
    (LAAS)
Toulouse, France

Georgios D. Panagopoulos
Intel Mobile Communications
    GmbH
Munich, Germany

Domenico Pepe
Tyndall National Institute
Cork, Ireland

Eduardo Peralías
Instituto de Microelectrónica de
    Sevilla (IMSE-CNM-CSIC)
Universidad de Sevilla
Seville, Spain

Costas Psychalinos
Physics Department
University of Patras
Rio Patras, Greece

Woogeun Rhee
Institute of Microelectronics
Tsinghua University
Beijing, China

Adoración Rueda
Instituto de Microelectrónica de
    Sevilla (IMSE-CNM-CSIC)
Universidad de Sevilla
Seville, Spain

Stylianos Siskos
Electronics Laboratory of the
    Physics Department
Aristotle University of
    Thessaloniki
Thessaloniki, Greece

Mani Soma
Electrical Engineering
    Department
University of Washington
Seattle, Washington

Alexios Spyronasios
Dialog Semiconductor GmbH
Stuttgart, Germany

George Stamoulis
Department of Computer Science
University of Thessaly
Volos, Greece

Haralampos-G. Stratigopoulos
Sorbonne Universités
Paris, France

Fengyuan Sun
Electronics Department
Northwestern Polytechnical
    University
Xi'an, China

Georgia Tsirimokou
Physics Department
University of Patras
Rio Patras, Greece

Olivier Valorge
EASII-IC
Electronics Design Center
Lyon, France

Diego Vázquez
Instituto de Microelectrónica de
    Sevilla (IMSE-CNM-CSIC)
Universidad de Sevilla
Seville, Spain

Alberto Villegas
Innovaciones Microelectrónicas
S.L. (Anafocus, E2V)
Seville, Spain

Zhihua Wang
Institute of Microelectronics
    Tsinghua University
Beijing, China

Liming Xiu
TAF Microelectronics
Dallas, Texas

Ni Xu
Institute of Microelectronics
Tsinghua University
Beijing, China

Domenico Zito
Tyndall National Institute
and
Electrical and Electronic
    Engineering
School of Engineering
University College Cork
Cork, Ireland

1

Ultra-Low Voltage Analog Filters for Biomedical Systems

Costas Psychalinos, Farooq A. Khanday, and Georgia Tsirimokou

CONTENTS

1.1 Biomedical Signal Processing

Bioelectrical potentials/signals carry information about the health state of the living system. Therefore, monitoring and analyzing these signals allow us to diagnose the various diseases pertaining to the living systems, such as cardiac, neurological, and neuromuscular systems. In fact, in this contemporary world, it is difficult to imagine a situation when diseases related to cardiac, neurological, neuromuscular or any other living system is diagnosed without including certain information derived from bioelectrical signals. The information obtained from the analysis of these bioelectrical signals combined with the study of the impact of these signals on each other continues to improve the quality of life of the patients. Besides, these signals need to be researched at a micro scale so that they can be associated with a living subsystem (e.g., cells/tissues/organs) instead of a system (e.g., nervous/cardio-vascular system, etc.) thereby allowing further comfort to the patients. The need for the analysis of these signals becomes even clearer when we consider home-based patients who are to be monitored for longer durations. This is why, despite being recorded and analyzed for several decades, these patients continue to excite physicians and engineers all over the world.

Biomedical signal processing, in general, includes three operations: acquisition, analysis, and communication of bioelectrical signals. Among these three general areas of biomedical signal processing, biomedical signal analysis is the most widely studied one. Biomedical signal analysis involves the identification and investigation of epochs related to specific physiological events. The corresponding waveform representations of the epochs may be segmented and analyzed in terms of amplitude, wave shape (morphology), time duration, time intervals between events, energy distribution, frequency content, and many more.

Because of the complex nature of biological structures and their associated bioelectrical signals, biomedical signal processing is not a simple task. Biomedical signal processing is complicated because most of the time it needs to be performed in an indirect and noninvasive manner. It is further complicated by the fact that apart from the bioelectrical signal being corrupted with external noise, every single bioelectrical signal acts as the noise signal for every other bioelectrical signal, thereby making their extraction difficult. Biomedical signal processing is even more complicated by the fact that most of the time, the information of interest is contained in the features that either occur temporarily or are more diffuse. Keeping the above facts in mind, novel biomedical signal processing techniques are required for effectively and efficiently diagnosing diseases and improving the quality of life of the patients.

1.2 Issues in Biomedical Circuit Design

Recent advances in semiconductor IC technology, as well as innovations in circuit design techniques, have led to systems with processing capabilities compatible with complex biomedical operations. Further, the complete/partial damage of biological organs has given research impetus toward implantable biomedical devices i.e., the pacemaker, cochlear implant, retinal implant, brain stimulator, etc. The major function of these implantable devices is the detection, analysis, and regulation of biomedical signals. While designing hardware for biomedical applications, the smallest possible dimensions together with low-power consumption without compromising the performance, are the demands for a circuit designer.

The implantable biomedical devices should be small and either the battery life should be longer or the device should consume low power or both, in order to decrease the rate of change of the device thereby giving some comfort to the patients.

In order to implement the biological systems in hardware, analog or digital techniques can be used. Moreover, hybrid techniques, where one technique assists the other, are also used for special applications. However, analog VLSI realization is suitable, because it is simpler and smaller than digital systems in general, and operates in real time. In addition, it is often easier for analog implementation to mimic the function of biological systems, as their working mechanism is analogous to the biological systems.

In analog techniques, there is limitation for large and small signal excursions from the perspectives of nonlinear distortion and device noise, respectively. Moreover, in order to save power in modern analog biomedical systems, the supply voltage is decreased to be compatible with the small dimensions of the advanced technologies. However, this severely limits the linear range of analog circuits. Simultaneously, decreasing the supply voltage also decreases the signal-to-noise ratio (SNR) of the analog circuits and thereby decreases the dynamic range. Consequently, novel device biasing and circuit topologies are required to efficiently perform the biomedical analog computations. Moreover, as many bioelectrical signals of interest are at low frequencies, circuits are highly susceptible to $1/f$ and popcorn device noise. Furthermore, time constants corresponding to these low frequencies are large and special techniques need to be explored to incorporate the design of these large time constants in contemporary IC design.

Despite the immense interest among scientists and tremendous work reported on the hardware implementation of biomedical systems, there is still room for improving their designs and exploring their new applications. The primary requirement for implantable biomedical devices is low-power consumption. Therefore, several designs and techniques have been cited in the open literature for this purpose [1–3]. However, there is still opportunity for further reduction of power consumption. The analog front-end processing stage forms an important part of the implantable biomedical devices because it distinguishes between noise and the desired signal. Therefore, there is a need for designing high dynamic range circuits [1]. Finally, the frequency of operation of biomedical signals is in the sub-Hertz range (e.g., 1 Hz for heart signals). The analog design may thus require large values of passive components, which may not be permissible by contemporary technology. Therefore, there are continuous attempts by the researcher to overcome this issue. One of the solutions is to electronically scale the passive components and thereby achieve large time constants [4]. However, these techniques often come with high-power consumption. Therefore, there are various issues and challenges for designing biomedical systems and care must be taken while choosing a specific technique for their implementation.

1.3 Low Voltage Analog Circuit Design Techniques

The low power analog integrated circuit (LPAIC) design has been the focus of the contemporary research, especially in the areas of portable systems where a low voltage single-cell battery with longer lifetime has to be used. Portable and miniaturized system-on-chip (SoC) applications exhibit an increasing demand in the microelectronics market and, particularly, in the biomedical field with products such as hearing aids, pacemakers, or implantable sensors. System portability usually requires battery supply and, unfortunately, battery technologies do not evolve as fast as applications demand. Therefore, the combination of battery supply and miniaturization often turns into a low voltage and/or low current circuit design problem. In particular, these restrictions affect more drastically the analog part of the whole mixed SoC. As a result, specific analog circuit techniques are needed to cope with such power supply limitations. Consequently, many low voltage/current circuit techniques/strategies such as rail-to-rail [5], multistage [6], bulk-driven [7], supply multiplier [8], adaptive biasing [9], subthreshold biasing [10], etc. have been reported in the open literature.

In addition to the techniques mentioned above, LPAIC design has been achieved by substituting traditional voltage-mode techniques by the current-mode techniques, which have the recognized advantage of overcoming the gain-bandwidth product limitation. Therefore, many current-mode techniques came into existence and companding-mode design is one such technique. "Companding" describes an AIC design technique in which the signals are first compressed to an intermediate integration node and then subsequently expanded after being processed. The distinct characteristic of the technique is that, it is the large-signal transfer function of the circuit that is linearized, not the individual transconductance or active resistive elements as would be the case in more classical circuit design techniques. Depending on whether the intermediate compressed value is proportional to the average measure or the instantaneous value of the input signal strength, the companding techniques are reported as syllabic [11] and instantaneous [12] companding, respectively, in the open literature. Instantaneous companding technique has been studied in detail and the three widely used instantaneous companding filtering techniques are log-domain (LD), square-root-domain (SRD), and sinh-domain (SD).

SD filtering is an important technique for realizing analog filters with inherent class-AB nature. This is originated from the fact that the required current splitting is simultaneously realized with the compression of the linear input current and its conversion into a nonlinear voltage. This is not the case in the LD filters, where a pseudo class-AB operation is realized by establishing two identical class-AB signal paths and employing a current splitter at the input of the whole filter. The produced intermediate output currents are then subtracted in order to derive the final output of the filter. In addition to the aforementioned feature, SD filters also offer the capability for electronic adjustment of their frequency characteristics because the realized time constants are controlled by a DC current. Because of the companding nature, SD filters also allow the operation under a low voltage environment. Compared with their corresponding LD and SRD counterparts, SD offer more power efficient filter realizations but the price paid may be an increased circuit complexity [13]. The basic building blocks of the SD filtering are nonlinear transconductor cells, sinh and sinh^{-1} operators, integrator, and algebraic summation/subtraction blocks which will be introduced in the subsequent sections.

1.4 Sinh-Domain Circuits for Biomedical Applications

1.4.1 Realization of Large Time Constants

The fundamental elements for realizing SD circuits are nonlinear transconductor cells [14,15]. Such multiple-output cell is depicted in Figure 1.1. In order to meet the nowadays trend for ultra-low voltage systems with reduced power consumption, MOS transistors operating in the subthreshold region will be employed next. Thus, following this consideration, the expressions of output currents are given by Equations 1.1 and 1.2 as

$$i_{\cosh} = 2I_B \cdot \cosh\left(\frac{\hat{v}_{IN+} - \hat{v}_{IN-}}{nV_T}\right) \tag{1.1}$$

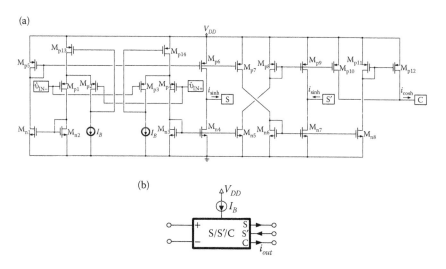

FIGURE 1.1
Multiple-output nonlinear transconductor cell (a) circuitry and (b) associated symbol.

$$i_{cosh} = 2I_B \cdot \cosh\left(\frac{\hat{v}_{IN+} - \hat{v}_{IN-}}{nV_T}\right) \tag{1.2}$$

where I_B is the bias current of the transconductor, n is the subthreshold slope factor $(1 < n < 2)$, V_T is the thermal voltage ($\approx$ 26 mV at 27°C), and $\hat{v}_{IN+}, \hat{v}_{IN-}$ are the voltages at noninverting and inverting inputs, respectively [2].

It should be mentioned at this point that the current at S′ terminal is an inverted replica of that at S terminal and, therefore, it would be also given by the expression in Equation 1.1. Also, additional replicas of the output currents could be derived through the formation of extra current mirrors.

Another useful building block for realizing SD integrators is the two-quadrant divider, which is shown in Figure 1.2.

That topology is realized using appropriately configured S cells as it is demonstrated in Figure 1.2a. The input–output relationship is given by the formula $i_{out} = I_{DIV}(i_1/i_2)$, where i_1 and i_2 are the corresponding input currents, and I_{DIV} is the bias current of divider [15].

Using the aforementioned cells, the general topology of a SD lossy integrator is demonstrated in Figure 1.3a. The realized transfer function is given by Equation 1.3 as

$$H(s) = \frac{1}{\tau s + 1} \tag{1.3}$$

where the time constant (τ) is defined by Equation 1.4 as

$$\tau = \frac{nCV_T}{I_{DIV}} \tag{1.4}$$

It should be mentioned at this point that the electronic adjustment of the gain factor (G) of the transfer function is realized through DC bias current of the corresponding nonlinear transconductor without disturbing the time constant of the filter. This is originated from the fact that the required time constants are still realized through the bias current of the divider (I_{DIV}) [4].

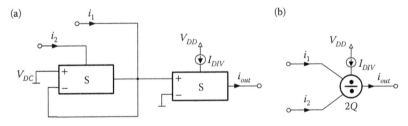

FIGURE 1.2
Two-quadrant divider (a) realization using S cells and (b) associated symbol.

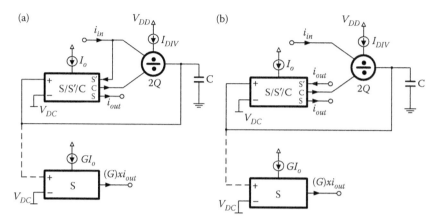

FIGURE 1.3
General schemes of sinh-domain integrators (a) lossy and (b) lossless.

The corresponding lossless integrator could be easily derived from the scheme in Figure 1.3a as it is demonstrated in Figure 1.3b. The realized transfer function will be given by the expression $H(s) = 1/\tau s$, where the time constant is still given by Equation 1.4.

The expression of the time constant in Equation 1.4 indicates that it is dependent on the bias current I_{DIV}, offering electronic adjustment of the frequency characteristics; moreover, it is independent from the value of the bias current I_o, which is employed for biasing the nonlinear S/S'/C cell. Therefore, the range of the input signals which could be handled by the integrator is not limited by the value of the bias current employed for realizing the required time constant. In other words, the topologies in Figure 1.3 offer the capability for realizing large time constants without affecting the level of input currents. Taking into account that the time constant of a conventional SD integrator is given by the formula: $\tau = CnV_T/2I_o$ [2], it could be concluded that they behave as capacitor multipliers with scaling factor equal to $2I_o/I_{DIV}$.

1.4.2 Tinnitus Detection System

Tinnitus is a condition characterized by ringing, hissing, squealing, roaring, or other noises that appear to be originating in the ear or head in the absence of an external stimulation. It is a symptom of an underlying condition, such as age-related hearing loss, explosive to loud noise, earwax blockage, ear, head or neck injuries, and circulatory system disorder [16,17].

Although there is no specific treatment for tinnitus, there are several methods that can eliminate or reduce the severity of symptoms. Those treatments include invasive methods, medication, cognitive and behavioral therapy, and tinnitus retraining therapy. Nevertheless, according to clinical researches there has been introduced several treatments such as repetitive transcranial magnetic stimulation and electrical or magnetic stimulation of brain areas involved in hearing, that seems to present promising results.

In order the patient to avoid the laborious process of programming the applied therapy, an automated electric stimulation is necessary and, thus, a tinnitus detection system is indispensable for this purpose. This system will be suitable for employment in a closed-loop implantable neurodevice in order to electrically stimulate the deceased areas of auditory cortex [18].

It is already known that tinnitus will appear, when the energy level of theta and gamma waves is increased, and that of alpha waves is decreased compared to those obtained from the corresponding healthy regions of the auditory cortex. Consequently, there is a need for a system that will perform the following sequential operations: (a) extraction of the energy of the afore-mentioned waves, (b) comparison of the derived energy levels for waves obtained from both regions (i.e., healthy and deceased) of the auditory cortex, and (c) decision about the occurrence of tinnitus.

Such detection system is depicted in Figure 1.4, which is constructed from three band energy extractors, three current comparators, and an AND gate [17].

The channels 1 and 2 represent the electrodes that records signals from deceased and healthy locations of the auditory cortex, respectively. For each one of the alpha, gamma, and theta waves, there are two energy extractors connected to channels 1 and 2, which form the corresponding band energy extractor used for extracting the energy of the corresponding signals. The outputs of the energy extractors are denoted as a_1 and a_2 in the case of alpha waves, while the corresponding notation for gamma waves is γ_1, γ_2, and θ_1, θ_2 for theta waves.

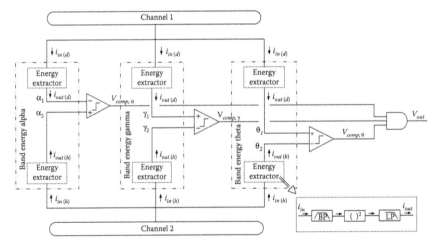

FIGURE 1.4
Tinnitus detection system.

The realization of an energy extractor system is also given in the lower right side of Figure 1.4. It is constructed from a bandpass (BP) filter which is responsible for the selection of the desired band of the electroencephalogram (EEG). The ranges of alpha, gamma, and theta waves are 8–12, 25–100, and 4–7 Hz, respectively. In order to perform the extraction of these components, let us consider the functional block diagram (FBD) of a multifunction second-order filter demonstrated in Figure 1.5. The scaled output current of lossless integrators are achieved through utilization of appropriate biased S cells as demonstrated in Figure 1.3.

The realized second-order BP filter function is given by

$$H(s) = G \cdot \frac{(\omega_o/Q) \cdot s}{s^2 + (\omega_o/Q) \cdot s + \omega_o^2} \tag{1.5}$$

where the resonance frequency ω_o, the Q factor, and the maximum gain G are defined by Equations 1.6 through 1.8, respectively.

$$\omega_o = \frac{1}{\sqrt{\tau_1 \cdot \tau_2}} \tag{1.6}$$

$$Q = \frac{1}{a} \cdot \sqrt{\frac{\tau_1}{\tau_2}} \tag{1.7}$$

$$G = b \tag{1.8}$$

Inspecting the expressions in Equations 1.6 through 1.8, it is readily concluded that the frequency characteristics of the filter (ω_o, Q, and G) could be electronically adjusted without disturbing each other. In other words, the filter offers electronic orthogonal adjustment of the shape of its frequency response and this is an important benefit in the design flexibility point of view. As a result, the same filter topology could be utilized for the extraction of the EEG components [19,20].

The squarer acts as a rectifier and will be realized using an appropriately configured four-quadrant multiplier as shown in Figure 1.6, where

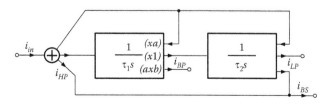

FIGURE 1.5
Functional block diagram (FBD) of a multifunction second-order filter.

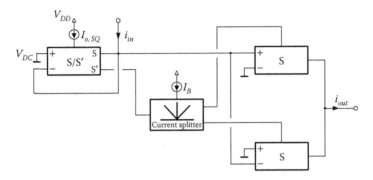

FIGURE 1.6
Two-quadrant class-AB current squarer using nonlinear transconductors.

the required current splitter could be realized as it was demonstrated in References 2, 4. The realized expression for the output current is $i_{OUT} = i_{IN}^2 / I_{o,SQ}$, where $I_{o,SQ}$ is a DC bias current.

The squarer will feed the LP filter, where the cutoff frequency should be much smaller than the maximum frequency of the squarer's output signal; a resonance value could be equal to 1 Hz. This should be fulfilled in order that the output of the LP filter provide information about the energy of the corresponding wave. A second-order LP filter could be easily derived from the FBD in Figure 1.5. The realized transfer function is given by

$$H(s) = G \cdot \frac{\omega_o^2}{s^2 + (\omega_o/Q) \cdot s + \omega_o^2} \tag{1.9}$$

where the resonance frequency ω_o, the Q factor, and the maximum gain G, are still defined by Equations 1.6 through 1.8, respectively.

Having available the required band energy extractors, each pair of signals is applied to the input of a typical comparator, as shown in Figure 1.7a [21]. The

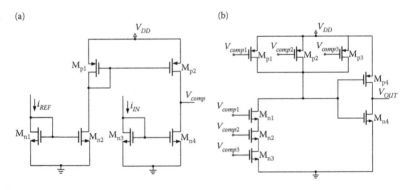

FIGURE 1.7
Topology of (a) current comparator and (b) AND gate.

output voltage of the comparator will be equal to the power supply voltage (V_{DD}) in the case the input current (i_{IN}) is smaller than the reference current (i_{REF}). It should be mentioned at this point that comparators are appropriately configured in order to produce a logical "1" at their outputs in the case that $\alpha_1 < \alpha_2$, $\gamma_1 > \gamma_2$, and $\theta_1 > \theta_2$ in order to indicate the presence of tinnitus.

Owing to the fact that these conditions should be simultaneously fulfilled, the outputs of comparators should feed the input of an AND gate, shown in Figure 1.7b, which extracts the final decision about the presence or absence of tinnitus.

The whole system has been designed using the Cadence IC Design suite as well as MOS transistor models provided by the AMS 0.35 µm C35 CMOS process. The layout design of the tinnitus detection system is demonstrated in Figure 1.8, where the size is 2305.5 µm × 1944.6 µm.

The proper operation of the whole system has been confirmed through schematic and layout simulation results. Considering an EEG signal obtained from Massachusetts Institute of Technology/Beth Israel Hospital (MIT/BIH) database [22], the corresponding output waveforms of the system in the case of tinnitus occurrence, derived using the Analog Design Environment of the Cadence software, are demonstrated in Figure 1.9.

It should be mentioned at this point that the analog part of the system operates in 0.5 V, while the digital part (current comparators, AND gate) operates in 1 V power supply voltage. The corresponding operation voltage for the analog part in Reference 18 was 1 V. Hence, the total power dissipation

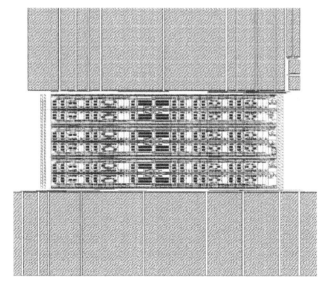

FIGURE 1.8
Layout design of the tinnitus detection system.

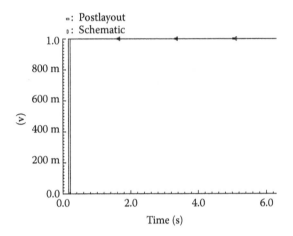

FIGURE 1.9
Postlayout simulation results of the behavior of the system in Figure 1.4 in the presence of tinnitus.

is significantly reduced (40.5 nW) compared to that of the corresponding already published system which was equal to 60 nW.

1.4.3 ECG Signal Acquisition Stage

The main noise components presented in an ECG signal are the following: (a) baseline wander, which is a low-frequency component caused by offset voltages in the electrodes, respiration, and body movement; baseline wander can cause problems to analysis, especially when examining the low-frequency ST segment, (b) interference caused by the 50/60 Hz frequency of the AC power line (as mains) which could significantly destroy the ECG signal, and (c) electromagnetic interference (EMI) often caused from the high-frequency output of the electrosurgical unit (ESU) [23,24].

The FBD of the ECG preprocessor system is depicted in Figure 1.10. It is constructed, from a first-order highpass (HP) filter with cutoff frequency 50 mHz for removing the baseline wander and capturing the ST segment, a fourth-order bandstop (BS) filter with zero frequency at 50/60 Hz for

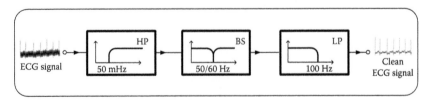

FIGURE 1.10
ECG signal acquisition system.

minimizing the power line interference, and a second-order lowpass (LP) filter with 100 Hz cutoff frequency for minimizing the EMI noise [25].

The BS filter could be readily obtained through a cascade connection of two second-order BS filters which is obtained from the general FBD in Figure 1.5. Also, the LP filter has been realized using the topology in Figure 1.5. The time constants in the BS and LP filters have been realized by choosing appropriate bias currents of the two-quadrant dividers of the corresponding lossless integrators within the filters.

The realization of the HP filter has been performed by employing the FBD depicted in Figure 1.11.

The transfer function of the HP filter is given by

$$H(s) = \frac{\tau s}{\tau s + 1} \tag{1.10}$$

where the time constant (τ) is given by the expression in Equation 1.4.

Although relative large time constants are able to be realized, the current I_{DIV} could not be extremely small in order to achieve sub-Hertz cutoff frequencies. Thus, the required 50 mHz cutoff frequency would be realized by the current division network (CDN) depicted in Figure 1.12. According to Kafe et al. [25], it behaves as a two-quadrant divider with bias current $I_{DIV}/(k_1 \cdot k_2...k_n)$. Hence, the expression of the realized time constant would be given by

$$\tau = \frac{CnV_T}{I_{DIV}} (k_1 \cdot k_2 \cdots k_n) \tag{1.11}$$

The scaling factors k_i, $i = (1,...n)$, offer the capability of realizing an extremely large value of time constant without employing small intermediate bias currents, which is difficult to be performed in practice [26]. Thus, the realization of the 50 mHz cutoff frequency of the HP filter has been achieved by utilizing a CDN constructed from two-quadrant dividers and choosing $I_{DIV} = 50$ pA, $I_{b1} = 400$ pA, $k_1 = 12.5$, and $C = 400$ pF.

The behavior of the system has been evaluated through the utilization of an ECG signal obtained from the Massachusetts Institute of Technology/

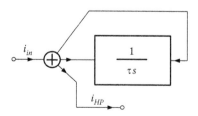

FIGURE 1.11
FBD of a first-order HP filter.

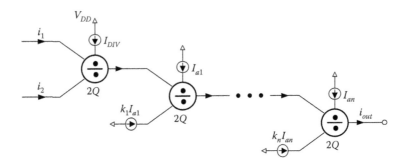

FIGURE 1.12
Current division network (CDN) for realizing extremely large time constants.

Beth Israel Hospital (MIT/BIH) database [22]. The added noise sources were emulated as sinusoidal currents with the following characteristics: 150 pA/20 mHz for the baseline, 50 pA/50 Hz for the AC mains interference, and 50 pA/300 Hz for the EMI interference. The resulted noisy ECG signal is depicted in Figure 1.13a, while the resulted output waveform is demonstrated in Figure 1.13b, where the significant reduction of the noise contents of ECG is evident.

It should also be mentioned at this point that the ECG signal acquisition system in Figure 1.10 operates in a 0.5 V power supply voltage environment, while the DC power dissipation was 59.2 nW. Taking also into account that the required maximum capacitance was 400 pF, the system is fully compatible with the nowadays trend for ultra-low voltage, fully integratable biodevices with enhanced battery life.

1.4.4 Arrhythmia Detector Using RMS-to-DC Converter

Another biomedical application example of SD filter is the design of a system for detecting the heart arrhythmia. This is based on the concept that the average energy between ECGs with normal rhythmus and arrhythmia would be different and, as a consequence, a RMS-to-DC converter topology could be used for detecting the heart arrhythmia [20].

The design of the whole system would be performed using a chain of BP filter and RMS-to-DC converter. The BP filter with passband 5–15 Hz is required for extracting information within the band frequency of interest of an ECG, and suppressing the baseline wander, T-wave interference, muscle noise artifacts, and 60 Hz interference. The output of the filter will be then able to feed the input of RMS-to-DC converter for extracting information about the average energy of the signal. The FBD of that system is depicted in Figure 1.14.

The second-order BP filter could be realized using the FBD in Figure 1.5. Its transfer function is given by Equation 1.5, and the resonance frequency ω_o, Q factor are determined by Equations 1.6 through 1.8. The maximum gain (G) of the filter has been considered with a value equal to one.

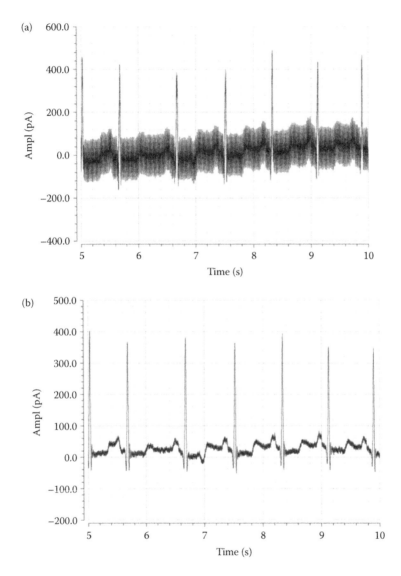

FIGURE 1.13
Demonstration of efficiency of the ECG signal acquisition chain (a) noisy ECG and (b) filtered ECG.

The building blocks required for realizing a RMS-to-DC converter are a squarer and an averaging (LP filter) stage with an appropriate feedback path is established between them. The corresponding FBD is depicted in Figure 1.15.

The current squarer is realized using the topology in Figure 1.6, while the averaging operation could be performed by the topology in Figure 1.3 with a cutoff frequency equal to 100 mHz. The realization of such large time

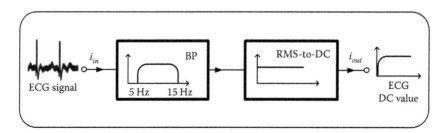

FIGURE 1.14
FBD of a system for measuring the RMS value of an ECG signal.

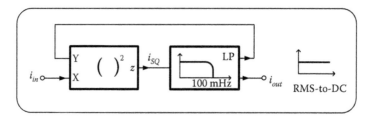

FIGURE 1.15
FBD of RMS-to-DC converter.

constant would be implemented by utilizing the CDN presented in Figure 1.12. Due to the fact that the building blocks of the system in Figure 1.14 are those presented in the previous subsections, it is obvious that it has the capability for operation in a 0.5 V power supply voltage environment.

The correct operation of the system would be verified by stimulating this with ECGs of normal rhythm and arrhythmia, obtained from MIT/BIH database. Such signals are depicted in Figure 1.16.

The output waveforms of the system in Figure 1.14 are simultaneously provided in Figure 1.17, demonstrating the difference of RMS value between ECGs with arrhythmia and normal rhythm.

Taking into account that the proposed RMS-to-DC converter is capable of operation in a 0.5 V environment, it could be considered as an attractive candidate suitable for realizing other high-performance biomedical systems where the extraction of information related to the energy of the signal would be required.

1.4.5 Realization of Wavelet Filters for ECG Signal Analysis

The continuous wavelet transform (WT) of a signal $x(t)$ at scale a and position τ is defined by Equation 1.12 as

$$\mathrm{WT}_x(a, \tau) = \frac{1}{a} \int x(t) \psi\left(\frac{\tau - t}{a}\right) dt = x(t)* \ \psi\left(\frac{t}{a}\right) \qquad (1.12)$$

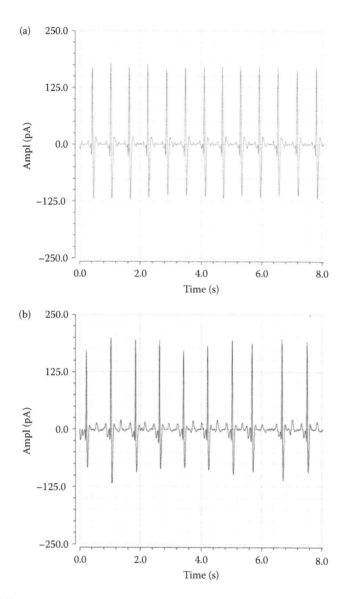

FIGURE 1.16
ECG with (a) normal rhythm and (b) arrhythmia.

where $\psi(t)$ is the mother wavelet, while asterisk denotes the convolution operation. The expression in Equation 1.12 implies that the computation of the WT can be performed through the realization of a filter with impulse response given in the formula: $h(t) = (1/a)\psi(t/a)$.

As it was mentioned in the previous subsections, an ECG signal could be contaminated by background noise and organs of the patient himself.

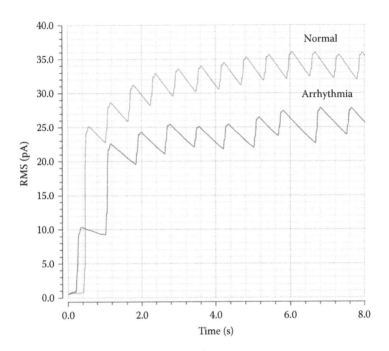

FIGURE 1.17
Output waveforms of the system in Figure 1.14 for ECGs with normal rhythm and arrhythmia.

The design of circuits for such signals is a very difficult procedure, especially in applications where the greatest possible precision in the most important features of the ECG, which are frequency, shape, width, and spacing for each counterpart of the ECG, is required. Those features are utmost important for providing useful information about the health condition of the heart. In order to overcome this, de-noising methods have been introduced in the literature, with the WT being the most successful especially for local analysis of nonstationary and fast transient signals such as cardiac signals [27–29]. This is originated from the fact that WT is a very promising mathematical tool that gives good estimation of time and frequency localization. By decomposing signals into elementary building blocks that are well localized both in time and frequency, distinction of cardiac signal points from severe noise and artifacts could be achieved. In electrocardiogram (ECG) analysis, the most important feature is the QRS complex. Having detected the QRS complex in an accurate manner, all other features of ECG such as P and T waves can be extracted. Thus, WT circuits have been developed for detecting QRS complexes of cardiac signals [29–33].

Owing to the fact that the first-derivative of the Gaussian wavelet is used to analyze the cardiac signal, we will discuss the implementation of a wavelet filter whose impulse response is the first derivative of the Gaussian wavelet. Let us consider the fifth-order Gaussian wavelet filter function

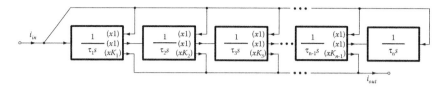

FIGURE 1.18
FBD for realizing a wavelet filter function.

(scale = 1) given by Equation 1.13, which approximates the first-derivative of the Gaussian wavelet

$$H(s) = \frac{8.59s^4 + 1.47 \cdot 10^5 s^3 - 9.69 \cdot 10^7 s^2 + 3.04 \cdot 10^{11} s}{s^5 + 1.28 \cdot 10^3 s^4 + 1.65 \cdot 10^6 s^3 + 9.53 \cdot 10^8 s^2 + 3.86 \cdot 10^{11} s + 6.06 \cdot 10^{13}}$$

(1.13)

Following the theory of Laplace transform, transfer functions of wavelet filters at scales $a = 2^1$ to 2^3 could be directly derived from Equation 1.13.

The implementation of the transfer function in Equation 1.13 will be performed using the FBD in Figure 1.18, where the required multiple-output lossless integrators will be realized using the core given in Figure 1.3b.

To verify the proper functioning of the wavelet filter, an ECG signal obtained from [22] has been applied at its input. The time-domain responses at scales 2, 4, and 8 are demonstrated in Figure 1.19, where the capability of the WT for performing a time scaling of the stimulated signal is evident.

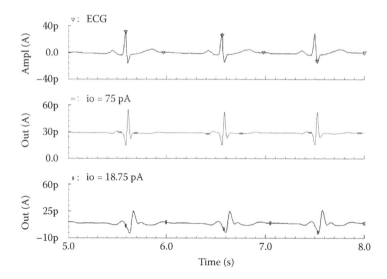

FIGURE 1.19
Time-domain responses of the wavelet filter stimulated by an ECG.

1.5 Fractional-Order Analog Signal Processing

1.5.1 Fractional-Order Filters

Fractional-order filters have recently received increased interest for designing circuits suitable for biomedical applications. The necessity of such devices is their ability of not only describing the electrical behavior of viscoelastic materials and biological tissues but also offering exceptional features that are able to be used in several applications depending on the desired purpose [34–39]. Fractional-order filters, fractional-order differentiators, and integrators constitute a small portion of circuits that will be described next, and an application will be presented too.

The stopband attenuation, which is offered by conventional filters has been limited to be equal to $-6\,n$ dB/octave, and is depended on the integer order of the filter. Integer-order filters are derived from integer-order transfer functions. Fractional-order filters exhibit frequency responses with stopband attenuation equal to $-6(n+a)$dB/octave, when n and a $(0 < a < 1)$ are the integer and fractional parts of the order of the filter, respectively. The advantage of this feature is the more precise control of the attenuation gradient. According to the analysis in [40–41], the transfer function of a LP Butterworth filter with the order $1 + a$ is given by Equation 1.14

$$H_{1+a}^{LP}(s) = \frac{K_1}{s^{1+a} + K_3 s^a + K_2} \tag{1.14}$$

The values of factors K_i are approximated by the following expressions:

$$K_1 = 1 \tag{1.15}$$

$$K_2 = 0.2937a + 0.71216 \tag{1.16}$$

$$K_3 = 1.068a^2 + 0.161a + 0.3324 \tag{1.17}$$

A second-order approximation of the term s^a is given by Equation 1.18.

$$s^a \cong \frac{a_0 s^2 + a_1 s + a_2}{a_2 s^2 + a_1 s + a_0} \tag{1.18}$$

where $a_0 = a^2 + 3a + 2$, $a_1 = 8 - 2a^2$, and $a_2 = a^2 - 3a + 2$.

Having the above approximation available by using Equation 1.14, the transfer function of the Butterworth LP filter is given by Equation 1.4

$$H_{1+a}^{LP}(s) \cong \frac{K_1}{a_0} \frac{a_2 s^2 + a_1 s + a_0}{s^3 + b_2 s^2 + b_1 s + b_0} \tag{1.19}$$

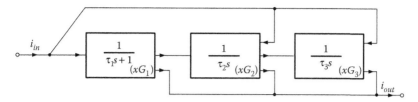

FIGURE 1.20
Functional block diagram (FBD) for approximating a fractional-order filter $(1 + \alpha)$.

where $b_2 = (a_2 + a_0 K_3 + a_2 K_2)/a_0$, $b_1 = (a_1 (K_2 + K_3) + a_2)/a_0$, $b_0 = (a_0 K_2 + a_2 K_3)/a_0$.

The realization of the integer-order transfer function in Equation 1.19 could be easily performed by the FBD, which is depicted in Figure 1.20. This is a typical follow-the-leader-feedback (FLF) scheme, where the notation (xG_i) implies a scaled replica of the corresponding output current.

The transfer function is given by Equation 1.20 as

$$H(s) = \frac{(G_1/\tau_1)s^2 + (G_2/\tau_1\tau_2)s + (G_3/\tau_1\tau_2\tau_3)}{s^3 + (1/\tau_1)s^2 + (1/\tau_1\tau_2)s + (1/\tau_1\tau_2\tau_3)} \tag{1.20}$$

Comparing the coefficients of the corresponding terms in Equations 1.19 and 1.20, the values of time constants are easily obtained. An important thing is that the values of time constants and coefficients G_i are totally depended on the order (α). Taking also into account that the aforementioned variables are electronically controlled, it is concluded that the above topology is reconfigurable in the sense that it is capable of realizing any arbitrary order filter function through the adjustment of appropriate DC bias currents.

Using the concept of SD filtering, presentation of the frequency responses of a filter of order 1.3, 1.5, and 1.7 is depicted in Figure 1.21. The stopband attenuation was –7.6, –9.1, and –10.2 dB/octave, while the corresponding theoretical values are equal to –7.8, –9, and –10.2 dB/octave, respectively [42].

An important advantage of having the transfer function of the LP fractional-order filter is, an HP Butterworth filter of the order $1 + \alpha$ could be derived by employing the transformation $s \rightarrow 1/s$ into Equation 1.19. As a consequence, the cascade connection of LP and HP fractional-order filter results in a low-Q BP fractional-order filter with different slopes at each one of the stopband attenuation.

Also, the realization of a filter of order $n + a$ with Butterworth characteristics could be performed according to the formula given by

$$H_{n+a}^{LP}(s) = \frac{H_{1+a}^{LP}(s)}{B_{n-1}(s)} \tag{1.21}$$

where $B_{n-1}(s)$ is the corresponding Butterworth polynomial of order $n - 1$.

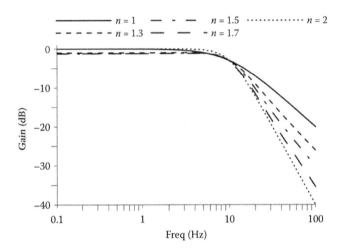

FIGURE 1.21
Simulated frequency responses of the filter in Figure 1.20, for $\alpha = 0.3$, 0.5, and 0.7.

Using Equation 1.21, the general form of a high-order fractional filter will be given by Equation 1.22

$$H_{n+a}^{LP}(s) \equiv \frac{K_1}{a_0} \frac{a_2 s^2 + a_1 s + a_0}{s^{n+2} + c_{n+1}s^{n+1} + \cdots + c_1 s + c_0} \qquad (1.22)$$

where the coefficients c_k ($k = 0, 1, \ldots, n + 1$) is a function of b_i ($i = 0, 1, 2$).
    The general expression of an nth order FLF filter is the following:

$$H_{n+a}^{LP}(s) = \frac{(G_1/\tau_1\tau_2\cdots\tau_n)s^2 + (G_2/\tau_1\tau_2\cdots\tau_{n+1})s + (G_3/\tau_1\tau_2\cdots\tau_{n+2})}{s^{n+2} + (1/\tau_1)s^{n+1} + (1/\tau_1\tau_2)s^n + \cdots + (1/\tau_1\tau_2\cdots\tau_{n+2})} \qquad (1.23)$$

and could be realized by the FBD given in Figure 1.22.
    Comparing the coefficients of the corresponding terms in Equations 1.22 and 1.23, the calculation of the required time constants could be readily performed. The simulated results of a fractional filter of order $3 + a$, are demonstrated in Figure 1.23. The slope of the stopband attenuation was –20.7, –21.6,

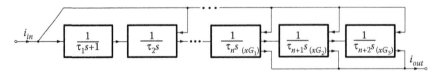

FIGURE 1.22
FBD of an n-th order FLF filter.

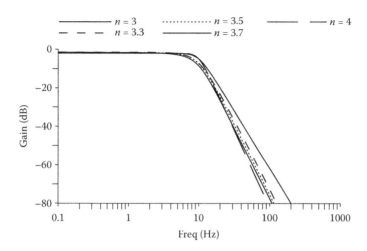

FIGURE 1.23
Simulated frequency responses of the fractional filter of order 3.3, 3.5, and 3.7.

and −21.9 dB/octave for orders 3.3, 3.5, and 3.7, respectively. The corresponding theoretical values were −19.8, −21, and −22.2 dB/octave, respectively.

The transfer function of a fractional-order differentiator is given by Equation 1.24 as

$$H(s) = (\tau < s)^a \tag{1.24}$$

where τ is the corresponding time constant and $0 < a < 1$ is the order of the differentiator.

Its magnitude response is given as $H(\omega) = (\omega/\omega_o)^a$, where the unity gain frequency has the same expression as in the case of its integer-order counterpart (i.e., $\omega_o = 1/\tau$). The phase response of the differentiator is also constant and equal to $a\pi/2$ and this implies a total reliance of phase from the fractional-order (α) against the constant value of phase of the conventional counterpart (90°). Comparing the expressions that give the magnitude responses of fractional- and integer-order differentiator, it is readily obtained that (at the same frequency) the fractional-order differentiator realizes a gain smaller than that achieved by its integer-order counterpart. Therefore, the high-frequency noise presented in a circuit will be relatively suppressed by a fractional-order differentiator and this is an important feature of this type of circuits especially in cases where differentiators are employed in a noisy environment. Also, its phase is constant and equal to $\alpha\pi/2$.

The corresponding expression for approximating variable s^a has been already given by Equation 1.18, which is a second-order approximation having attractive features with respect to circuit simplicity. Thus, the expression will be employed next for the fractional-order differentiator implementation.

Substituting Equation 1.18 into 1.24 the transfer function of a fractional-order differentiator becomes

$$H_{DIFF}(s) = \frac{(a^2 + 3a + 2/a^2 - 3a + 2)s^2 + (1/\tau)(8 - 2a^2/a^2 - 3a + 2)s + (1/\tau^2)}{s^2 + (1/\tau)(8 - 2a^2/a^2 - 3a + 2)s + (1/\tau^2)(a^2 + 3a + 2/a^2 - 3a + 2)}$$

(1.25)

The transfer function in Equation 1.25 could be performed by adopting the same procedure as that employed to LP fractional-order filter design. Therefore, the FBD of a fractional-order differentiator is given in Figure 1.24. The realized transfer function is given by Equation 1.26

$$H(s) = \frac{G_1 s^2 + (G_2/\tau_1)s + (G_3/\tau_1\tau_2)}{s^2 + (1/\tau_1)s + (1/\tau_1\tau_2)}$$

(1.26)

Comparing the coefficients of the corresponding terms in Equations 1.25 and 1.26, the values of time constants are easily obtained. An important thing is that the values of time constants and coefficients G_i are totally depended on the order (α) and, therefore, the above topology is totally electronically reconfigurable, providing the benefit for realizing any arbitrary order filter function through the tuning of the appropriate DC bias currents.

In a similar manner, fractional-order integrators and fractional-order filters with order equal to α, could be easily designed by substituting the same approximation in Equation 1.18 into the corresponding transfer functions, which are given by Equations 1.27 and 1.28, respectively.

$$H(s) = \frac{1}{(\tau \cdot s)^a}$$

(1.27)

$$H(s) = \frac{1}{(\tau \cdot s)^a + 1}$$

(1.28)

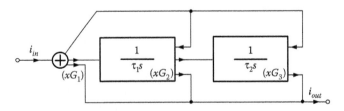

FIGURE 1.24
FBD for approximating a fractional-order differentiator.

The derived transfer functions are these in Equations 1.29 and 1.30, respectively

$$H_{INT}(s) = \frac{(a^2 - 3a + 2/a^2 + 3a + 2)s^2 + (1/\tau)(8 - 2a^2/a^2 + 3a + 2)s + (1/\tau^2)}{s^2 + (1/\tau)(8 - 2a^2/a^2 + 3a + 2)s + (1/\tau^2)(a^2 - 3a + 2/a^2 + 3a + 2)}$$

(1.29)

$$H_{LP}(s) = \frac{\begin{bmatrix} (a^2 - 3a + 2/2a^2 + 4)s^2 + (1/\tau)(8 - 2a^2/2a^2 + 4)s \\ + (1/\tau^2)(a^2 + 3a + 2/2a^2 + 4) \end{bmatrix}}{s^2 + (1/\tau)(8 - 2a^2/a^2 + 2)s + (1/\tau^2)}$$

(1.30)

Comparing the transfer functions in Equations 1.25, 1.29, and 1.30, it is obvious that they could be realized by the same core of the FLF topology in Figure 1.24. As a consequence, this topology is reconfigurable in the sense that the design of fractional-order differentiator, lossless, and lossy integrator of arbitrary order a could be performed through an appropriate adjustment of DC bias currents and scaling factors.

1.5.2 Biomedical System Design Example Using Fractional-Order Differentiators

The Pan–Tomkins algorithm is one of the most popular methods for detecting the QRS complexes of ECG signals. The QRS detection task is difficult due to the time-varying morphology of ECG, the physiological variability of the QRS complexes, and the noise presented in ECG signals. The main noise sources are muscular activity, movement artifacts, power line interference, and baseline wandering [43,44].

A typical preprocessing chain of the Pan–Tomkins algorithm is depicted in Figure 1.25a. It is constructed from a BP filter with 5–15 Hz bandwidth, a differentiator with unity gain frequency about 10 Hz, and a squarer. The

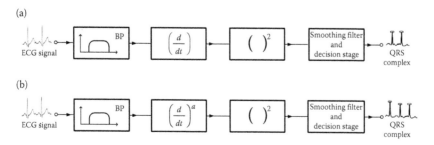

FIGURE 1.25
Preprocessing chain for realizing the Pan–Tomkins algorithm using (a) conventional integer-order differentiator and (b) fractional-order differentiator.

BP filter performs the selection of frequency band where most of the information of QRS complex is included and, simultaneously, attenuation of the low-frequency components of P, T waves and baseline wander, and the high-frequency components associated with electromyographic noise and power line interference. Thereafter, the signal is passed through a differentiator in order to detect the (positive and negative) high slopes of ECG, which is a unique characteristic of QRS complex in comparison to the other ECG waves. The squarer emphasizes the higher frequency content of the signal which is also a characteristic feature of QRS complex. The output of the squarer is fed into a digital smoothing (averaging) system which is the final stage before the decision.

Substituting the conventional differentiator with the fractional-order differentiator [45], the modified chain is given in Figure 1.25b. The second-order BP filter has been realized using the FBD in Figure 1.5. The implemented cutoff frequencies have been chosen to be equal to 5 and 15 Hz in order to select the frequency content of QRS complex. Therefore, considering that the order of the differentiator is equal to 0.5 and unity gain frequency 10 Hz, the output signal is passed through the squarer.

To demonstrate the operation of the proposed modification, let us consider a noisy ECG signal derived from [22] with the addition of Gaussian noise with SNR equal to 0 dB, which is depicted in Figure 1.26. The output waveforms derived after integer and noninteger differentiation and squaring are demonstrated in Figure 1.27.

Inspecting the plots provided in Figure 1.27, it is readily concluded that the fractional-order differentiator is more efficient for suppressing the noise components of ECG, in comparison with its integer-order counterpart.

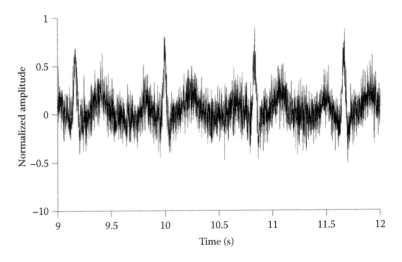

FIGURE 1.26
A noisy ECG input signal with SNR = 0 dB.

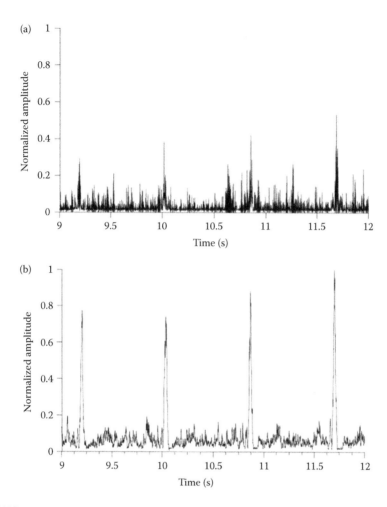

FIGURE 1.27
Output waveforms of the system in (a) Figure 1.25a and (b) Figure 1.25b.

1.6 Conclusions

The concept of SD filtering has been applied in this chapter for realizing several biomedical systems, including tinnitus detector, ECG signal acquisition chain, and RMS-to-DC converter for detecting arrhythmia. Moreover, fractional-order filters as well as integrators and differentiators have been introduced using the aforementioned concept. It was proved that the employment of fractional-order differentiators in the realization of the Pan–Tomkins algorithm is very efficient in terms of noise suppression. The employment of appropriate CDN allows the realization of extremely large time constants.

Taking also into account that the circuits are realized using MOS transistors operating in the subthreshold region, the capabilities of ultra-low (i.e., 0.5 V) voltage operation and power dissipation minimization are simultaneously offered. Therefore, all the presented circuits could be considered as attractive candidates for realizing fully integrable biomedical systems, which fulfill the nowadays industry trends.

References

1. Kardoulaki, E.M., Glaros, K.N., Katsiamis, A.G. et al. An 8Hz, 0.1 μW, 110 + dBs Sinh CMOS Bessel filter for ECG signals. *Int. Conf. Microelectron. (ICM)*, Marrakech, 2009, pp. 14–17.
2. Tsirimokou, G., Laoudias, C., Psychalinos, C. Tinnitus detector realization using Sinh-Domain circuits. *J. Low Power Electron.*, 9(4), 458–470, 2013.
3. Thanapitak, S., Toumazou, C. A bionics chemical synapse. *IEEE Trans. Biomed. Circuits. Syst.*, 7(3), 296–306, 2013.
4. Kafe, F., Psychalinos, C. Realization of companding filters with large time-constants for biomedical applications. *Analog Integr. Circuits. Signal Process.*, 78(1), 217–231, 2014.
5. Chow, H.C., Weng, P.N. A low voltage rail-to-rail OPAMP design for biomedical signal filtering applications. *Proc. 4th IEEE Int. Symp. Electron. Design, Test App.*, Hong Kong, January 23–25, 2008, pp. 232–235.
6. Setty S., Toumazou, C. N-Folded cascode technique for high frequency operation of low voltage opamps. *Electron. Lett.*, 32(11), 955–956, 1996.
7. Fried R., Enz, C. Nano-amp, active bulk, weak-inversion analog circuits. *Proc. IEEE Custom Integr. Circuits Conf.*, Santa Clara, California, USA, 1998, pp. 31–34.
8. Ker, M.D., Chen, S.L., Tsai, C.S. Design of charge pump circuit with consideration of gate-oxide reliability in low-voltage CMOS processes. *IEEE J. Solid-State Circuits*, 41(5),1100–1107, 2006.
9. Baswa, S., Lopez-Martin, A.J., Carvajal, R.G., Ramirez-Angulo, J. Low-voltage power-efficient adaptive biasing for CMOS amplifiers and buffers. *Electron. Lett.*, 40(4),217–219, 2004.
10. Vittoz, E., Fellroth, J. CMOS analog integrated circuits based on weak inversion operation. *IEEE J. Solid State Circuits*, 12(3),224–231, 1977.
11. Mulder, J., Serdijn, W.A., van der Woerd, A.C., van Roermund, A.H.M. A syllabic companding translinear Filter. *Proc. IEEE Int. Symp. Circuits Syst. (ISCAS'1997)*, Hong Kong, 1997, pp. 101–104.
12. Adams, R.W. Filtering in the log domain., Preprint #1470, *63rd AES Conference*, New York, NY, May 1979.
13. Katsiamis, A., Glaros K., Drakakis, E. Insights and advances on the design of CMOS sinh companding filters. *IEEE Trans. Circuits Syst.-I*, 55, 2539–2550, 2008.
14. Kasimis, C., Psychalinos, C. Design of Sinh-Domain filters using complementary operators. *Int. J. Circuit Theory Appl.*, 40(10),1019–1039, 2012.
15. Kasimis, C., Psychalinos, C. 1.2 V BiCMOS Sinh-Domain filters. *Circuits Syst. Signal Process.*, 31(4), 1257–1277, 2012.

16. Møller, A.R., Langguth, B. DeRidder, D. et al. *Textbook of Tinnitus.* Springer, New York, USA, 2011.
17. Snow, J. *Tinnitus: Theory and Management.* PMPH, BC Decker, Hamilton, Ontario, USA 2004.
18. Hiseni, S., Sawigun, C., Vanneste, S. et al. A nano power CMOS tinnitus detector for a fully implantable closed-loop neurodevice. *Proc. IEEE Biomed. Circuits Syst. Conf. (BioCAS)*, San Diego, California, USA, 2011, pp. 33–36.
19. Kasimis, C., Psychalinos, C. 0.65 V class-AB current-mode four-quadrant multiplier with reduced power dissipation. *Int. J. Electron. Commun.*, 65(7), 673–677, 2011.
20. Kafe F., Psychalinos, C. 0.5 V RMS-to-DC converter topologies suitable for implantable biomedical devices. *J. Low Power Electron*, 10(3), 373–382, 2014.
21. Freitas, D., Current, K. CMOS current comparator circuit. *Electron. Lett.*, 19(17),695–697, 1983.
22. PhysioNet web site (PhysioBank ATM). http://www.physionet.org/cgi-bin/atm/ATM.
23. Bailey, J., Berson, A., Garson, A. et al. Recommendations for standardization and specifications in automated electrocardiography: Bandwidth and digital signal processing. *J. Am. Heart Assoc., Cir.*, 81, 730–739, 1990.
24. Kligfield, P., Gettes, L., Bailey, J. et al. Recommendations for the standardization and interpretation of the electrocardiogram Part I: The electrocardiogram and its technology. *J. Am. Heart Assoc., Cir.*, 115, 1306–1324, 2007.
25. Kafe, F., Khanday, F.A., Psychalinos, C. A 50 mHz Sinh-Domain highpass filter for realizing an ECG signal acquisition system. *J. Circuits, Syst. Signal Process,* 33(12), 3673–3696, 2014.
26. Linares-Barranco, B., Serrano-Gotarredona, T. On the design and characterization of femtoampere current-mode circuits. *IEEE J. Solid-State Circuits,* 38(8),1353–1363, 2003.
27. Misal A., Sinha, G.R. Denoising of PCG signal by using wavelet transforms. *Adv. Comput. Res.*, 4, 46–49, 2012.
28. Zaman, T., Hossain, D., Arefin, T. et al. Comparative analysis of De-noising on ECG signal. *Int. J. Emerg. Technol. Adv. Eng.*, 2, 479–486, 2012.
29. Haddad, S., Serdijn, W. *Ultra Low-Power Biomedical Signal Processing: An Analog Wavelet Filter Approach for Pacemakers,* Springer, Dordrecht, The Netherlands, 2009, ISBN 1402090730.
30. Haddad, S., Houben, R., Serdijn, W. Analog wavelet transform employing dynamic translinear circuits for cardiac signal characterization. *IEEE Int. Conf. Circuits Syst. (ISCAS)*, Bangkok, Thailand, 2003, pp. 121–124.
31. Hongmin, L., Yigang, H., Sun, Y. Detection of cardiac signal characteristic point using log-domain wavelet transform circuits. *Circuits Syst. Sig. Process*, 27, 683–698, 2008.
32. Laoudias, C., Beis, C., Psychalinos, C. 0.5 V wavelet filters using current mirrors. *Proc. IEEE Int. Conf. Circuits Syst. (ISCAS)*, Rio De Janeiro, Brazil, 2011, pp. 1443–1446.
33. khanday, F.A., Pilavaki, E., Psychalinos, C. Ultra low-voltage ultra low-power sinh-domain wavelet filer for electrocardiogram signal analysis. *J. Low Power Electron.*, 9(3), 288–294, 2013.
34. Ortigueira, M. An introduction to the fractional continuous-time linear systems: The 21st century systems. *IEEE Circuits Syst. Mag.*, 8(3), 19–26, 2008.

35. Elwakil, A. Fractional-order circuits and systems: An emerging interdisciplinary research area. *IEEE Circuits Syst. Mag.*, 10(4), 40–50, 2010.
36. Freeborn, T. A survey of fractional-order circuit models for biology and biomedicine. *IEEE J. Emerg. Select. Top. Circuits Syst.*, 3(3), 416–424, 2013.
37. Biswas, K., Sen, S., Dutta, P. Realization of a constant phase element and its performance study in a differentiator circuit. *IEEE Trans. Circuits Syst.-II: Express Briefs*, 53(9), 802–806, 2006.
38. Radwan, A., Soliman, A., Elwakil, A. First-order filters generalized to the fractional domain. *J. Circuit. Syst. Comp.*, 17(1), 55–66, 2008.
39. Krishna, B., Reddy, K. Active and passive realization of fractance device of order 1/2. *Act. Passive Electron. Compon.*, 2008, DOI:10.1155/2008/369421.
40. Freeborn, T., Maundy, B., Elwakil, A. Field programmable analogue array implementation of fractional step filters. *IET Circuit. Devices Syst.*, 4(6), 514–524, 2010.
41. Maundy, B., Elwakil, A., Freeborn, T. On the practical realization of higher-order filters with fractional stepping. *Signal Process.*, 91(3), 484–491, 2011.
42. Tsirimokou, G., Laoudias, C., Psychalinos, C. 0.5-V fractional-order companding filters. *Int. J. Circuit Theory Appl.*, 2014, DOI: 10.1002/cta.1995.
43. Pan, J., Tompkins, W. A real-time QRS detection algorithm. *IEEE Trans. Biomed. Eng.*, 32(3), 230–236, 1985.
44. Hamilton, P., Tompkins, W. Quantitative investigation of QRS detection rules using the MIT/BIH arrhythmia database. *IEEE Trans. Biomed. Eng.*, 33(12), 1157–1165, 1986.
45. Tsirimokou, G., Psychalinos, C. Ultra-low voltage fractional-order differentiator and integrator topologies: An application for handling noisy ECGs. *Analog Integr. Circuit. Sig. Process. J.*, 81(2), 393–405, 2014.

2

Offset Reduction Techniques in Flash A/D Converters

Lampros Mountrichas and Stylianos Siskos

CONTENTS

2.1 Introduction

The increasing usage of digital signal processing in almost every aspect of the technology has inevitably increased the need for fast and accurate analog-to-digital converters (ADC). Moreover, the development of fast submicron complementary metal-oxide semiconductor (CMOS) technologies with constantly reducing headroom and increased device mismatch make the design of accurate ADCs a more difficult task than previously. The reduced headroom leads to decreased least significant bit (LSB) values. That results in the need for more accurate circuits in order to achieve the same resolution. At the same time, the increased mismatch makes these circuits less accurate. Now, the study of the offset cancellation techniques for high-speed ADCs is brought into focus more than ever.

2.2 Mismatch-Related Specifications

The architecture that achieves the highest sampling rate is the Flash ADC, presented in Figure 2.1. The resistive ladder generates the reference voltages required by the comparators to determine the input signal level. The output of a comparator is positive in the event that the input is larger than its reference voltage. Following the comparators, the digital encoder translates the output to binary format. Usually the comparator is comprised of a number of preamplifiers and a latched comparator, operated by clock. The latched comparator, by means of positive feedback, has a large gain and can provide a logic level output signal.

The main contributor in the resulting effective resolution of an ADC is the mismatch of the devices. When mismatch is present, the characteristics of the building blocks are altered. The resistor ladder generated values change, and offset is also introduced in the comparators. In return, the transfer function of the ADC can be degraded as shown in Figure 2.2. This causes integral nonlinearity (INL) and differential nonlinearity (DNL) errors. INL is the difference in LSB between the measured and ideal transition level. DNL is the difference between two consecutive code transition levels with respect to the ideal value. Usually, when designing an ADC, an INL and DNL value lower than 0.5 LSB is desired. The absolute maximum DNL that can be tolerated is 1 LSB. If the DNL is larger than 1 LSB, then the transfer function of the converter seizes to be monotonous.

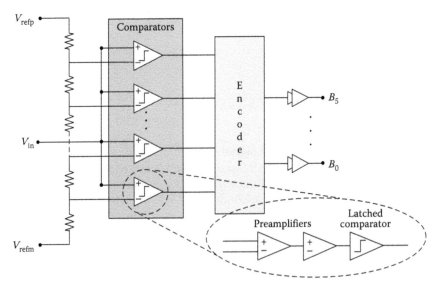

FIGURE 2.1
Flash ADC architecture.

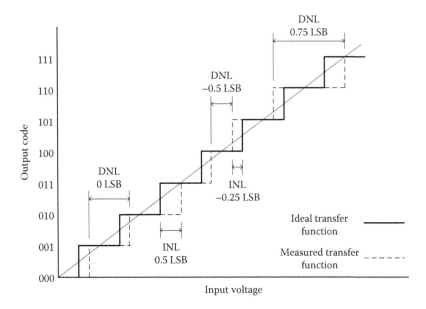

FIGURE 2.2
ADC transfer function.

To successfully design an ADC, it is crucial to calculate the maximum tolerable mismatch that still produces a converter that adheres to the specifications. Since in a typical flash ADC, the mismatch of the reference ladder is usually small enough not to affect the transfer function of the ADC, it can effectively be ignored in order to ease the calculations.

Without the effect of the resistor ladder mismatch, the maximum acceptable deviation (sigma) for the complete comparator chain can be easily calculated. Considering that offset voltage has a normal distribution and also that the offset voltage of the various amplifiers are independent variables, the probability of having an INL/DNL value lower than S is given by the following equations [1]:

$$P(\text{INL} \le S) = \left[\frac{1}{\sqrt{2\pi}\sigma(V_{\text{OS}})} \int_{-S \cdot V_{\text{LSB}}}^{S \cdot V_{\text{LSB}}} e^{-(u^2/2\sigma^2(V_{\text{OS}}))} du \right]^{2^{Nb}-1} \tag{2.1}$$

$$P(\text{DNL} \le S) = \left[\frac{1}{2\sqrt{\pi}\sigma(V_{\text{OS}})} \int_{-S \cdot V_{\text{LSB}}}^{S \cdot V_{\text{LSB}}} e^{-(u^2/4\sigma^2(V_{\text{OS}}))} du \right]^{2^{Nb}-2} \tag{2.2}$$

Figure 2.3 illustrates the yield of a 6-bit flash ADC for an INL and DNL of 0.5 LSB (S = 0.5).

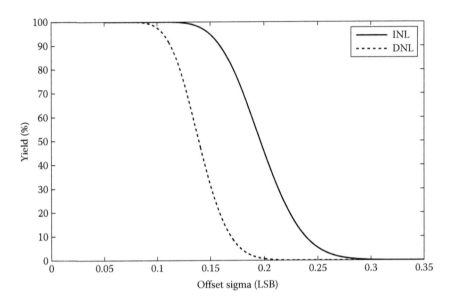

FIGURE 2.3
INL and DNL yield for a 6-bit flash A/D converter.

The DNL specification is more demanding. Specifically, for a 95% yield, about 0.1 LSB sigma offset variation is required. For the same sigma the INL yield is 99.9%. Using the following equations, the calculated sigma value can be translated to appropriate design parameters [2].

$$\sigma_X^2 = \frac{A_X^2}{W \cdot L} \tag{2.3}$$

The sigma of the resistors and capacitors can be estimated using the above equation, where to calculate the sigma of the transistors V_T and beta (β), Equations 2.4 and 2.5 can be used [2],

$$\sigma_{Vt}^2 = \frac{A_{Vt}^2}{W \cdot L} \tag{2.4}$$

$$\sigma_\beta^2 = \frac{A_\beta^2}{W \cdot L} \tag{2.5}$$

In Equations 2.3 through 2.5, σ_R and σ_C are the mismatch sigma of the resistor and the capacitor, σ_{Vt} and σ_β is the sigma of V_T and β for the transistors, and A_{Vt}, A_β, and A_R are the area proportionality constants for the specific technology. In addition, the sigma of a differential pair is given by [1]:

$$\sigma_{Vos}^2 = \sigma_{Vt}^2 + \frac{V_{OVD}^2}{4}[\sigma_\beta^2 + \sigma_R^2] \qquad (2.6)$$

Based on the equations above, the larger devices have less offset variation. For example, to decrease V_T sigma by 50%, an increase of 300% in area is required.

In retrospect, it can be concluded that the decision to disregard the resistor ladder induced offset was correct, since the usually large size of the resistors leads to small offset as shown in Equation 2.3. It can also be concluded that the main contributing factor in the offset voltage of a differential pair is the threshold voltage mismatch. For example, if $A_{Vt} = 5$ mV μm, $A_\beta = A_R = 2\% \cdot$ μm, $WL = 4$ μm², $W_R L_R = 2$ μm² and $V_{OVD} = 100$ mV using Equations 2.3 through 2.6, it can be calculated that $\sigma_{Vt} = 2.5$ mV and $\sigma_{Vos} = 2.65$ mV. Henceforth, the mismatch of the differential pairs will be calculated using only the V_T mismatch.

$$\sigma_{Vos}^2 \approx \sigma_{Vt}^2 \qquad (2.7)$$

Using Equations 2.2 and 2.7, it is calculated that in order to design a 6-bit flash ADC with 600 mV full input range, DNL ≤ 0.5 LSB and DNL$_{Yield} \geq 95\%$, the comparator devices should have an area of about 25 μm² ($\sigma_{Vos} \approx 1$ mV) leading to an ADC input capacitance of about 11 pF in a technology with $C_{ox} = 10$ fF/μm². To design a high-speed 7-bit converter with the above specification would require a considerable amount of current, since the input capacitance would be about 90 pF.

Moreover, an efficient high-speed comparator will have small size transistors, leading to high offset value. Typical values are ±40 mV. To ease the offset requirements for the comparator, a preamplifier can be introduced before the comparator. A preamplifier preceding the latched comparators reduces their input referred offset by the gain. Any number of preamplifiers can be introduced, leading to increased gain and as a result, a lower input referred offset. The total offset for a comparator chain with two preamplifiers is

$$\sigma_{Vos}^2 = \sigma_{Pre1\_Vt}^2 + \left(\frac{\sigma_{Pre2\_Vt}}{A_1}\right)^2 + \left(\frac{\sigma_{Vt}}{A_1 \times A_2}\right)^2 \qquad (2.8)$$

For $\sigma_{Pre1\_Vt} = \sigma_{Pre2\_Vt} = 0.5$ mV and $A_1 = A_2 = 4$, a latched comparator with $\sigma_{Vos} = 10$ mV can be used, leading to a total offset of $\sigma_{Vos} = 1$ mV. In that case, the comparator devices have 36 times smaller area, leading to lower current consumption and easier design. Of course, the same problems are now transferred in the design of the preamplifiers.

In conclusion, the performance of an ADC can be determined by speed, accuracy, and power consumption. The speed performance is mainly limited by the capacitive load of the building blocks. The accuracy is related to the offset that arise from device mismatches. To improve device matching, large transistors are needed. But large devices have large parasitic capacitances and thus degrade the speed of the building blocks. It is therefore obvious that if a low power, area efficient, high-speed, and high-resolution ADC is needed, some kind of offset reduction technique should be implemented to reduce feature sizes and at the same time maintain a relatively high accuracy. In the following section, such techniques will be described.

2.3 Offset Reduction in High-Speed Analog-to-Digital Converters

2.3.1 Averaging

It was shown that the offset voltage is related to the area of the components and in order to reduce it, larger devices must be used, which in turn increases parasitic capacitances, leading to larger power dissipation and to the reduction of the maximum operating frequency. A popular technique for reducing the offsets of the amplifiers is averaging [3–5].

Using this technique, the outputs of the preamplifiers are interconnected via an averaging network as illustrated in Figure 2.4. Usual implementations use either capacitors [6], resistors [7–13] or both [14] to form the network.

Unlike in a simple flash ADC, where the offset of each comparator chain is an uncorrelated random variable, the interconnection of the comparators via the averaging network forces each comparator output to be defined by a weighted sum, average of the neighboring unsaturated differential pair outputs. Thus, the offset and the zero crossings of the comparators become correlated variables.

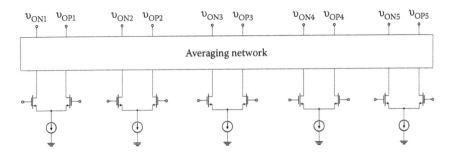

FIGURE 2.4
Averaging network. (From Springer Science+Business Media: *Offset Reduction Techniques in High Speed Analog-to-Digital Converters*, 2009, Figueiredo, P.M., Vital, J.C. With permission.)

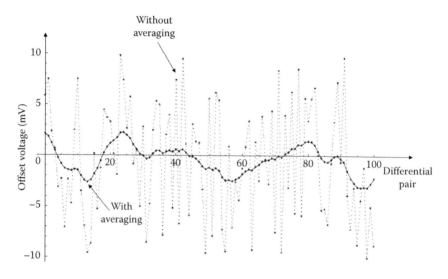

FIGURE 2.5
Averaging effect in comparator offset voltage. (From Springer Science+Business Media: *Offset Reduction Techniques in High Speed Analog-to-Digital Converters*, 2009, Figueiredo, P.M., Vital, J.C. With permission.)

The averaging effect can be appreciated as shown in Figure 2.5. The dashed line represents the offset voltage when averaging in not implemented. The solid line represents the offset voltage when averaging is used. The offset voltage of each differential pair of an ADC without averaging is random, but when averaging is used, the offset becomes correlated.

As mentioned above, the resistive networks are the most popular as shown in Figure 2.6. In resistive averaging, the offset reduction is a function of the value of the averaging resistors and the number of the neighboring unsaturated differential pairs.

More precisely, the offset reduction is a function of the ratio of the load resistor to the averaging resistor R_0/R_1. The larger the R_0/R_1 ratio, the smaller

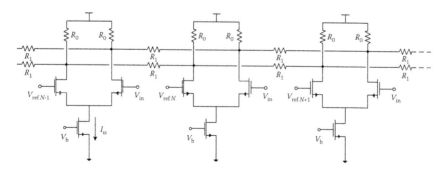

FIGURE 2.6
Resistive averaging network.

the effect of V_T, β, and R_0 mismatches becomes. This happens because the increase in R_0/R_1 allows a larger influence among the neighboring differential pairs. However, it leads to a slightly larger nonlinearity due to the current sources mismatch of the various differential pairs, I_{SS} and the increasing mismatch of R_1 as its value/area decreases. But at large, the nonlinearity decreases with increasing R_0/R_1 ratio.

Although a larger R_0/R_1 leads to a smaller preamplifier mismatch, it also decreases the preamplifier gain and thus leads to an increased input referred offset for the subsequent stages. This happens because in the case of resistive averaging, the equivalent load resistance, as seen by the differential pairs, is smaller ($R_{\text{load}} = R_0//R_1$) and thus, the gain $G_m \cdot R_{\text{load}}$ decreases.

The offset reduction is also a function of the number, N, of unsaturated neighboring differential pairs. The larger the number, the better the offset reduction, since the offset is averaged between a broader number of pairs.

The larger N also leads to larger gain. When averaging is employed, the transconductances of all nonsaturated differential pairs contribute to the overall gain. When N increases, the total transconductance of the circuit increases, leading to higher gain. This offers a positive side effect since the higher the preamplifier gain, the lower the offset contribution of the subsequent stages.

In conclusion, resistive averaging lowers the contribution of V_T, β, and R_0 mismatches, but also adds two new mismatch sources, I_{SS} and R_1. Usually this does not pose a problem. Also, it should be noted that a larger R_0/R_1, although it leads to a smaller mismatch, also decreases the preamplifier gain and thus leads to an increased input referred offset for the subsequent stages. This can be alleviated by increasing the number of unsaturated differential pairs by boosting the MOS overdrive voltage, in expense of increased current consumption.

Another advantage of the averaging technique is the effect upon DNL. As shown by Equation 2.2, the DNL specification is much harder to achieve when the mismatches are considered randomly uncorrelated variables. Since the averaging network introduces an interaction between neighboring differential pairs, the offset is not an uncorrelated variable any more. And since the DNL is all about neighboring pairs mismatch, the averaging technique greatly improves it.

In Figure 2.7, the INL and DNL of an ADC converter utilizing resistive averaging is illustrated. The comparator chain was designed with a final input referred offset V_{OS} of 1.367 mV. Figure 2.8 illustrates the INL and DNL of an ADC converter with the same final input referred offset but without averaging. It can be seen that although the INL values are about the same, the DNL of the averaging ADC is vastly superior.

2.3.1.1 Termination Strategies

The above analysis has merit for an infinite preamplifier array, meaning that all the preamplifiers have the same surrounding environment. Unfortunately,

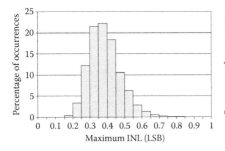
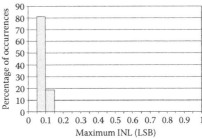

FIGURE 2.7
Averaging ADC, INL, and DNL performance $V_{OS} = 1.37$ mV. (From Springer Science+Business Media: *Offset Reduction Techniques in High Speed Analog-to-Digital Converters*, 2009, Figueiredo, P.M., Vital, J.C. With permission.)

in a real ADC implementation, the number of preamplifiers is finite and the preamplifiers at the edge of the array do not have the same environment as those in the middle. As a result, the comparison voltages (zero crossings) of the edge preamplifiers are shifted and a systematic mismatch is created. This leads to INL curvature.

To overcome this problem, dummy preamplifiers can be introduced at the edge of the functional array in order to extent the preamplifier chain and approximate an infinite array [15]. The minimum goal is every preamplifier to have the same number of neighboring unsaturated pairs. Thus, the minimum number of preamplifiers is equal to that number divided by two. In practice, to avoid the systematic INL curvature, a much larger number of preamplifiers must be introduced. Since these preamplifiers must be identical to the others, this termination technique increases the total power consumption and die area proportionally to the number of dummy preamplifiers. Moreover, since the extra preamplifiers require extra reference points, it can lead to difficulties in low voltage designs.

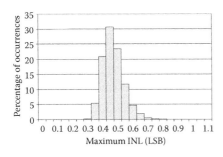
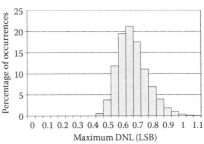

FIGURE 2.8
Simple flash ADC, INL, and DNL performance $V_{OS} = 1.37$ mV. (From Springer Science+Business Media: *Offset Reduction Techniques in High Speed Analog-to-Digital Converters*, 2009, Figueiredo, P.M., Vital, J.C. With permission.)

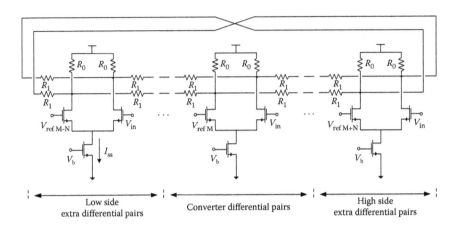

FIGURE 2.9
Mobius band averaging network.

To cut the number of amplifiers to the bare minimum of N, the mobius band technique is used [8,16–18]. In this technique, illustrated in Figure 2.9, the last of the extra preamplifiers from the low voltage reference side is cross connected to the last of the preamplifiers from the high voltage reference side. This connection ensures that the differential pairs after the extra preamplifiers, are in saturation. The cross connection is needed in order to achieve the right saturation level, either positive or negative. This technique closely approximates an infinite preamplifier array since following the extra N amplifiers, there are saturated differential pairs.

Other techniques approach the problem somewhat differently using a single extra differential pair and a different R_1 to connect it with the rest of the chain [10,13]. The linearity of those methods can be improved by adding extra differential pairs before the final preamplifiers.

The termination technique proposed in Reference 15, uses an equivalent resistor along with an appropriate current source to approximate the array (Figure 2.10). In this termination scheme, a number of N extra differential pairs are added to ensure that all the functional differential pairs have the same number of unsaturated neighbors. Furthermore, to simulate the completely unbalanced pairs that would exist in an infinite averaging network, an equivalent circuit is added. This is comprised of two resistors and a current source in each end of the preamplifier array. The resistor and the current of the current source can be calculated using the following equations.

$$R_{eq} = \frac{R_1 + \sqrt{R_1(R_1 + 4R_0)}}{2} \tag{2.9}$$

$$I_{TERM} = I_{SS}\frac{R_0}{R_{eq}} \tag{2.10}$$

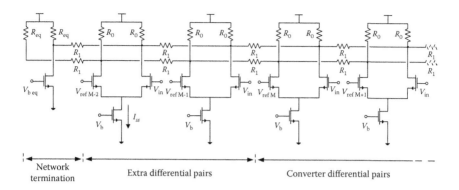

FIGURE 2.10
Termination via equivalent circuits.

This termination technique provides superior results compared to those of the previously presented solutions.

2.3.2 Offset Reduction by Sampling

Another way of minimizing the offset of preamplifiers incorporates capacitors to deduct the offset after sampling it. Input offset sampling (IOS) and output offset sampling (OOS) cancellation techniques, both utilize capacitors to sample and then subtract the offset of the preamplifiers preceding the comparators [1,17].

2.3.2.1 Input Offset Sampling

The input offset storage technique incorporates the use of a sampling capacitor in order to store the offset at the input of the preamplifier and then subtract it from the input signal. Figure 2.11 illustrates the use of IOS to minimize the offset of a preamplifier driving a latched comparator. C_s is the offset storage capacitors and C_p is the parasitic capacitance of each node.

While in offset storage phase, signal S1 is high and a unity gain loop is closed around the amplifier. At the same time, the input terminals are shorted to a common voltage. In this case, the differential input/output voltage of the amplifier is equal to

$$\frac{A \cdot V_{OS}}{A + 1} \tag{2.11}$$

In Equation 2.11, A is the gain of the amplifier and V_{OS} its offset voltage. The higher the gain, the closer the input is to the offset voltage, while the residual offset, equal to $V_{OS}/(A + 1)$, decreases. In the second phase, signal S2 is high. The capacitors conserve their charge and the input voltage appears shifted

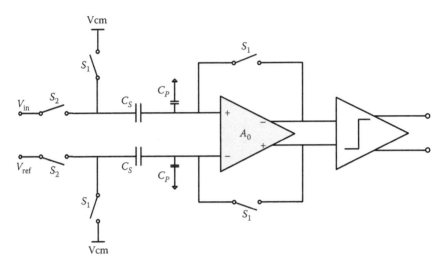

FIGURE 2.11
Input offset storage technique.

by the offset voltage at the amplifier input. The total input referred offset for the comparator and latch is given by

$$V_{OS} = \left(1 + \frac{C_P}{C_S}\right)\left[\frac{V_{OSa}}{1 + A} + \frac{V_{OSl}}{A}\right] + \frac{\Delta q}{C_S} \tag{2.12}$$

where Δq is the charge transferred to the capacitors by the switches, and V_{OSl} is the offset voltage of the latched comparator.

Given that V_{OS} in Equation 2.12 is inversely proportional to the amplifier gain, the IOS technique is better suitable for offset cancellation of high gain amplifiers. Moreover, using relatively high value sampling capacitors further minimize the residual offset. For that reason, IOS is mainly used in medium speed ADCs [2,19–21].

2.3.2.2 Output Offset Sampling

The output offset storage technique also uses sampling capacitors, but unlike IOS, the offset is stored after the preamplifier. A typical application of OOS technique is illustrated in Figure 2.12. A sampling capacitor is placed at the output terminal of the amplifier and switches are present both before and after the device. While signal S_1 is high, the amplifier inputs are shorted and the device amplifies its offset voltage, $V_{out} = A \cdot V_{OS}$. At the same time, the other terminal of the capacitors is placed at V_{cm} and in return, the amplified offset is stored in the capacitors. At the next phase, signal S_2 is high. The input voltage is applied at the preamplifier terminals and the output is equal

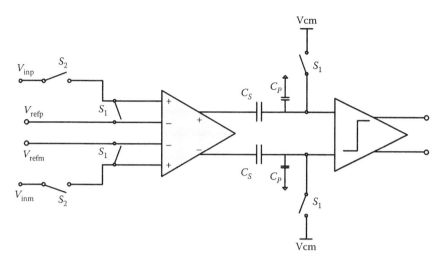

FIGURE 2.12
Output offset sampling technique.

to the amplified input plus the amplified offset, $V_{out} = A \cdot (V_{in} + V_{OS})$. After the capacitor, the offset voltage is subtracted and the resulting output is equal to the amplified input and a small residual offset. The operation of the OOS is illustrated in Figure 2.13.

The total input referred offset voltage is

$$V_{OS} = \left(1 + \frac{C_P}{C_S}\right)\frac{V_{OSl}}{A} + \frac{\Delta q}{A \cdot C_S} \tag{2.13}$$

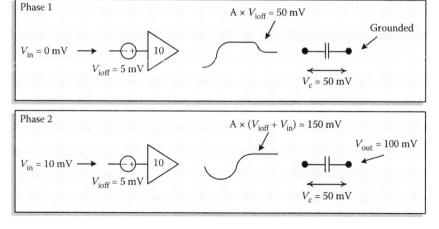

FIGURE 2.13
Output offset sampling operation.

where Δq is the charge transferred to the capacitors by the switches, and V_{OSL} is the offset voltage of the latched comparator. Naturally, the higher the gain, the lower the input referred offset of the latched comparator. An additional advantage is the division of the switch charge induced offset by the gain factor. The total offset is lower than that of the IOS technique, but care must be given so that the gain and the offset of the amplifier are not high enough to saturate the output. In such a case, the offset will not be stored correctly and there will be considerable residual offset remaining.

In conclusion, in high speed analog-to-digital converter design, the OOS technique is advantageous since the devices that are incorporated in such converters are usually low gain. Usually, OOS and IOS are used in combination [22,23].

2.3.3 Feedback Assisted Offset Cancellation Techniques

In the techniques mentioned in the previous sections, the offset of the latched comparator remains unaffected. The goal was to sufficiently amplify the signal before it reaches the comparator, while at the same time introduce minimal offset.

Utilizing a feedback loop around the preamplifier/comparator, it is possible to suppress the combined offset of both the preamplifier and the latched comparator by properly adjusting the preamplifier offset so the total offset is zero. In such a case, the offset voltage is not a design parameter and thus, minimum size devices can be used, maximizing in return, the operating frequency of the converter, while minimizing the current consumption at the same time.

The usual way to achieve this is by adjusting the current of each output branch of the preamplifier, thus changing the offset value, utilizing an auxiliary differential pair, or a digital-to-analog converter (DAC) controllable by calibration logic [1,24]. The use of an auxiliary pair is illustrated in Figure 2.14. This secondary g_m stage has reduced tail current and depending on the control voltage, it injects a different amount of current to the positive

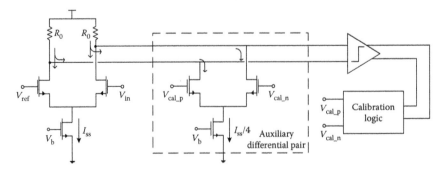

FIGURE 2.14
Auxiliary differential pair.

or negative output branch of the preamplifier, effectively changing the offset voltage. This is done in such a way that the offset of the preamplifier counteracts the offset of the latched comparator.

Other ways to trim the offset of the preamplifier are to

1. Vary the value of the load resistors [25,26] or if an active load is used, vary the bias voltage of the MOS device [27].
2. Vary the threshold voltage of the differential pair devices by changing their substrate voltage [28–30].
3. Vary the threshold voltage of the differential pair devices, by using and programming floating gate transistors (FGMOS) [31].

2.4 Design Examples

2.4.1 Averaging

A 6-bit 3.5-GS/s Flash ADC utilizing the averaging technique is presented in Reference 9. The architecture of the converter is illustrated in Figure 2.15. A track-and-hold (T/H) is used to aid the dynamic performance of the converter. By holding the input voltage, errors associated with clock and signal

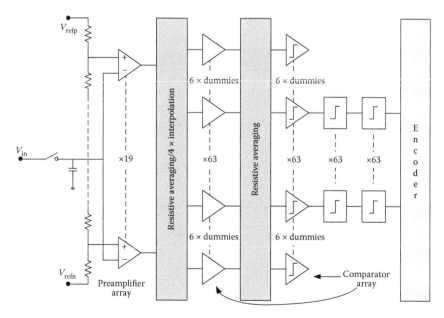

FIGURE 2.15
Analog-to-digital converter architecture.

skew are minimized. Following the T/H, 4-input amplifiers amplify the difference between the differential input signal and the reference voltage acting as the preamplifying stage. Averaging is used to minimize the offset of the amplifiers. A stage of averaged comparators follows. To decrease the bit error rate, two more regenerative stages are added. For simplicity, single-ended devices are depicted.

The first averaging stage also acts as an interpolation stage. Interpolation consists of the splitting of the averaging resistors, forming resistive ladders between the outputs of neighboring preamplifiers as shown in Figure 2.16. By interpolation, extra intermediate crossing points are created between the consecutive preamplifiers, permitting the use of a lower number of input preamplifiers, reducing the input capacitance, die area, and power dissipation.

Nevertheless, some tradeoffs must be taken into account in order to choose the appropriate interpolation factor (IF) along with the averaging/interpolation resistor value. As stated in Section 2.3.1, a large R_0/R_1 ratio minimizes the effect of the differential pair mismatch, but allows the current source mismatch to affect the total offset. The appropriate R_0/R_1 ratio to minimize the sum of these mismatches must be found, preferably by simulations. In Reference 9, it is shown that the mismatch sum has a minimum value independent of the IF as shown in Figure 2.17. In return, the IF is optimized by minimizing die area and interpolation induced errors, analyzed below.

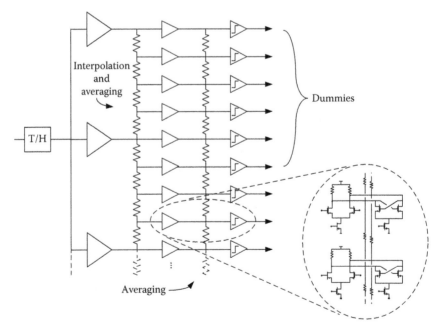

FIGURE 2.16
Averaging and interpolation.

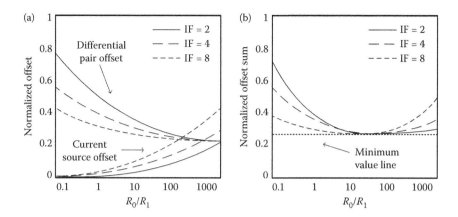

FIGURE 2.17
(a) Differential pair and current source normalized offset and (b) Normalized offset sum.

The term "dead area" is used for the area occupied by interconnection, isolation, and so on. The larger the IF, the smaller this area is. In Reference 9, this area is estimated at 10.8% for an IF of 4 and 5.4% for an IF of 8. Another consideration is the systematic offset introduced by interpolation [17]. This error is caused by the nonlinearity of the input–output characteristics of the differential pairs. An IF factor of 2 does not introduce any amount of error. For an IF factor over 2, the error becomes progressively larger. In this design, although an IF of 8 results in smaller area, an IF of 4 is chosen because it complies with the error budget of this A/D converter.

To further reduce the input referred offset, a second averaging network is used with the comparator array. As was the case with the interpolation network, the appropriate R_0/R_1 ratio must be chosen in order to minimize the offset of the latched comparator. To terminate the averaging network 12 dummy devices are used.

The converter was designed in a 90 nm CMOS technology. A DNL of +0.50/ − 0.48LS and an INL of +0.96/ − 0.39 LSB is achieved. The SNDR at 3.5 GS/s with the nyquist rate input frequency is 31.18 dB, whereas the SFDR reaches 38.67 dB. The power consumption is 98 mW from a 0.9 V supply. A comparison with other A/D converters using similar architecture is presented in Table 2.1. Excluding the design presented in Reference 8 which has a much lower sampling frequency and is designed in a more advance technology node, the converter in Reference 9 offers better figure-of-merit (FOM).

2.4.2 Offset Sampling

A 6-bit 2.2-GS/s Flash A/D converter in Reference 32 and 33 uses a special offset sampling technique that combines the OOS with a distributed track-and-hold architecture, alleviating the need of a track-and-hold circuit and a

TABLE 2.1

Averaging A/D Converter Comparison

	[11]	[34]	[8]	[9]
Sampling rate	4 GS/s	4.2 GS/s	1.2 GS/s	3.5 GS/s
Supply	1.8 V	1.2 V	1.2 V	0.9 V
ENOB at DC	3.7	4.2	5.7	4.89
Power	43 mW	180 mW	28.5 mW	98 mW
FOM [pj/conv.-step]	2.14	2.8	0.46	0.95
Offset reduction technique	Res. averaging	Res. interpolation	Res. averaging/ interpolation	Res. averaging/ interpolation
Technology	180 nm	130 nm	45 nm	90 nm

separate offset reduction technique. The architecture is illustrated in Figure 2.18. Sampling capacitors are situated after the first preamplifier along with a common mode voltage sampling switch. The circuit operates in two phases. At the input sampling phase, the switch is closed, and the capacitors sample the amplified input voltage. In this phase, the operation resembles a distributed track-and-hold architecture. At the hold phase, the input is shorted to a DC reference voltage, while the sampling switch opens. This operation

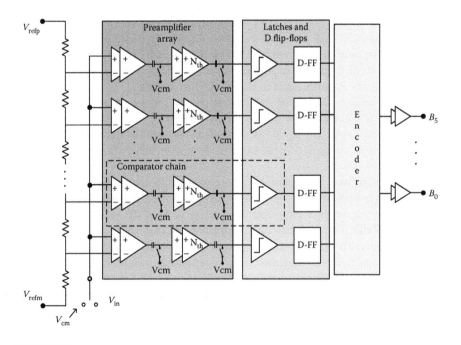

FIGURE 2.18

Analog-to-digital converter architecture. (Reprinted from *Microelectron. J.*, 44(12), Mountrichas, L., Siskos, S. A high-speed offset cancelling distributed sampleand-hold architecture for flash A/D converters, 1123–1131, Copyright 2013, with permission from Elsevier.)

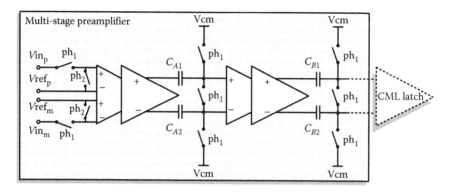

FIGURE 2.19
Multistage preamplifier. (Reprinted from *Microelectron. J.*, 44(12), Mountrichas, L., Siskos, S. A high-speed offset cancelling distributed sampleand-hold architecture for flash A/D converters, 1123–1131, Copyright 2013, with permission from Elsevier.)

cancels the amplifier offset and at the same time samples the input voltage. For simplicity, single-ended devices are depicted.

To further attenuate the latched comparator input referred offset, a multistage preamplifier is employed as shown in Figure 2.19. The first stage uses the modified technique while the second stage is based on the OOS. In both cases, a double amplifier is used in order to minimize the usage of sampling capacitors along with the signal attenuation associated with them.

The operation of the multistage preamplifier is divided into two phases as shown in Figure 2.20. At phase 1, the amplified input voltage is sampled by the first stage capacitors, while at the same time, the capacitors of the second stage are storing the amplified offset voltage associated with this stage. At phase 2, all the capacitors conserve their charge. The rightmost terminals of the first set of capacitors are shifted according to the offset voltage. That way the sampled and offset free input voltage enters the second stage which further amplifies the signal. Since the offset of this stage is already stored in the associated capacitors, the output remains offset free. As in the simple OOS technique, it is essential that the offset is not large enough to saturate the output of the amplifiers.

To avoid multiple design iteration, some effects must be considered while calculating design parameters. In real implementations, finite device bandwidth and switch mismatches result in a small amount of residue offset voltage as shown in Equation 2.13. Furthermore, capacitive attenuation due to parasitic capacitance at the output node of the devices leads to attenuation of the total gain. To avoid meticulous calculations, some preliminary simulations can be used to estimate the effect of the above mentioned effects. Simulations show that the preamplifier gain is reduced by 30% after each offset cancellation stage, when 50 fF (such as in this case) sampling capacitors are used. Moreover, Monte Carlo simulations reveal that about 10%

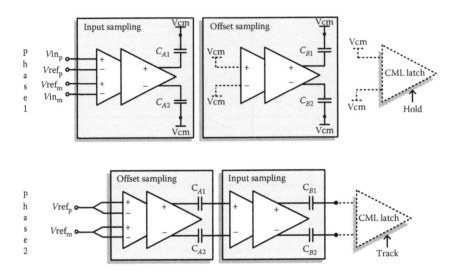

FIGURE 2.20
Multistage preamplifier phases. (Reprinted from *Microelectron. J.*, 44(12), Mountrichas, L., Siskos, S. A high-speed offset cancelling distributed sampleand-hold architecture for flash A/D converters, 1123–1131, Copyright 2013, with permission from Elsevier.)

of the residue offset remains in every offset cancelation stage. Therefore, Equation 2.8 for this particular architecture becomes

$$V_{\text{ioff}} \approx 0.1 \times \left(V_{\text{off}_1} + \frac{V_{\text{off}_2}}{A_1} \right) + \frac{0.14 \times (V_{\text{off}_3} + (V_{\text{off}_4}/A_3))}{A_1 \times A_2} + \frac{2 \times V_{\text{iofflatch}}}{A_1 \times A_2 \times A_3 \times A_4}$$

$$(2.14)$$

In reference to device sizing, we can no longer use the long channel equations to calculate transistor sizes. In submicron technologies, a simulation-based technology aware methodology is more appropriate. By first setting the system-wide parameters, such as sampling frequency and maximum allowed offset, we can easily obtain the design parameters for the preamplifiers and the latched comparators.

The following parameters should be considered when sizing the preamplifiers.

The desired gain

$$A = g_m \times R_{\text{tot}} \qquad (2.15)$$

The desired bandwidth

$$f_t = \frac{1}{2\pi \times R_{\text{tot}}(C_{\text{out}} + C_L)} \qquad (2.16)$$

The maximum allowed amount of offset

$$V_{\text{ioff max}} = \pm 3\sigma = \pm 3 \cdot \frac{A_{Vt}}{\sqrt{W \times L}} \qquad (2.17)$$

Using Equation 2.14 and allocating the gain and offset budget of the various preamplifiers and the latched comparator, the device area can be calculated using Equation 2.17. By knowing the device area, the load of each preamplifier can be calculated. From Equation 2.16 and knowing the desired bandwidth, the load resistance can be calculated. By means of Equation 2.15, the g_m and thus the current of the transistors can be calculated.

The design of the reference ladder is based upon two constraints. The random variation of the resistor value affecting the reference voltage and the reference voltage disturbance by noise. The random variation is chosen much smaller than the converters LSB value. Based on technology information and using Equation 2.3, it is possible to calculate the minimum resistor area. To complete the design of the reference ladder, the current and the unit resistor value must be set. Noise coupled to the reference voltage causes voltage disturbances along the reference ladder. This noise is minimized by lowering the reference ladder resistance, trading current consumption for a more stable reference voltage. Setting the reference voltage noise to a small value compared to the LSB sets the unit resistor value and the reference ladder current.

The converter was designed in a 90 nm CMOS technology, achieving +0.69/ − 0.48 LSB DNL and +0.32/ − 0.48 LSB INL. The SFDR at 2.2 GS/s with the Nyquist rate input frequency is 41.4 dB, where the SNDR is 33.4 dB. The power consumption is 135 mW from a 1.2 V supply. Excluding the converter presented in Reference 22 which utilizes a subranging architecture, this ADC provides better FOM compared to converters with similar offset cancellation technique and architecture as shown in Table 2.2.

TABLE 2.2

Offset Sampling A/D Converters Performance Comparison

	[22]	[6]	[19]	[32]
Sampling rate	125 MS/s	500 MS/s	1 GS/s	2.2 GS/s
Supply	1.2/2.5 V	1.8 V	1.2 V	1.2 V
ENOB at DC	7.6	6.35	5.16	5.64
Power	21 mW	160 mW	112 mW	135 mW
FOM [pj/conv.-step]	0.54	4.17	3.13	1.29
Offset reduction technique	IOS/OOS Res. averaging	IOS/Cap. averaging	IOS	Improved OOS
Technology	130 nm	180 nm	130 nm	90 nm

2.4.3 Feedback Assisted: Bulk Trimming

In Reference 28, a 6-bit A/D converter using bulk voltage trimming is presented. By trimming the bulk voltage of a preamplifier preceding the latched comparator, the offset voltage of the whole comparator chain is compensated. The offset calibration is applied to the preamplifier as shown in Figure 2.21.

Since the calibration feedback loop checks the output of the latched comparator and acts accordingly, the offset of the entire comparator chain is corrected. This has the added advantage that the preamplifier does not need to have high gain in order to minimize the input referred offset of the latched comparator, leading to decreased device area and current consumption. Since the design uses a triple-well technology, the preamplifier uses triple-well N-MOS input transistors, instead of P-MOS devices [29,30]. The use of N-MOS transistors allows the preamplifier to operate at a higher frequency.

The offset calibration is achieved by a digital feedback loop. During the calibration phase, all inputs are shorted at a common voltage. Due to mismatches in the preamplifier and the latched comparator, the output would be either positive or negative. By adjusting the bulk voltage of the input transistors, it is possible to change the threshold voltage and thus control the offset voltage of the preamplifier. If the output is positive, this means that the combined offset is positive. To fix this, the differential output of the preamplifier must decrease and that can be accomplished by increasing the current on the rightmost output branch of the preamplifier. Lowering the threshold voltage of the right branch transistors leads to increased current conduction and as a result, a lower output voltage. The bulk voltage is changed until the comparator barely changes state. In that moment, minimum offset has been achieved.

The offset calibration is implemented digitally with the configuration as shown in Figure 2.22. The calibration circuit consists of a digital counter connected to the input of a DAC, an edge detection circuit, and two multiplexers. When the procedure is initiated, the inputs of the differential

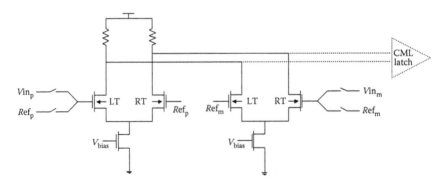

FIGURE 2.21
Calibrated preamplifier.

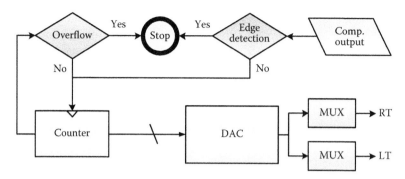

FIGURE 2.22
Offset calibration feedback loop.

pairs are shorted. Whether the comparator output is positive or negative controls which multiplexer will be active determining which transistors will be calibrated. The counter starts, changing the input word of the DAC and thus effectively changing the bulk voltage of the appropriate input devices. The counter uses a slower clock compared to the analog clock so that the output of the DAC and the comparator settles before a new value is generated. When the comparator changes sign, triggering the detection circuit or when the counter has overflowed, the calibration procedure is halted. The calibration procedure is illustrated in Figure 2.23.

For proper operation, it must be ensured that the bulk source voltage of the triple well transistors does not become forward biased. In this technology, the maximum bulk source voltage allowed is 400 mV. To ensure acceptable margin, the maximum bulk voltage is kept under 400 mV while the drain voltage of the current source is 300 mV. Thus the maximum bulk source voltage for the input transistor is 100 mV.

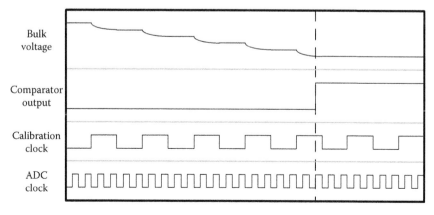

FIGURE 2.23
Calibration procedure.

To calculate the minimum preamplifier and latch comparator device area so that the A/D converter still adheres to the specifications, the maximum calibration range must be known. The bulk voltage range is associated with the maximum allowed input referred offset and thus the minimum transistor area. In this design, the 400 mV bulk source voltage range allows for the calibration of ±35 mV of input referred offset. Using this value with the modified version of Equation 2.8 for one preamplifier stage, Equation 2.18 becomes

$$\sigma^2_{Vos} = \sigma^2_{Pre1\_Vt} + \left(\frac{\sigma_{Vt}}{A_1}\right)^2 \tag{2.18}$$

The allowed amount of offset can be calculated as well as the corresponding transistor area. Monte Carlo simulations verify the maximum offset before calibration at ±35 mV. The remaining offset after the calibration procedure ranges from −0.5 to 3 mV with a deviation of about 800 µV.

It is worth mentioning, that the offset voltage of the input stage illustrated in Figure 2.21, is not given by Equation 2.6, or by simply using Equation 2.4. The offset of such stage is increased by $\sqrt{2}$ [17] and as such, the offset can be estimated using Equation 2.19

$$\sigma^2_{Vt} = \sqrt{2} \times \frac{A^2_{Vt}}{W \cdot L} \tag{2.19}$$

The converter was designed in a 90 nm triple-well CMOS technology. It achieves ±0.16 LSB DNL and ±0.13 LSB INL with a yield of 93% and 99%, respectively. The SFDR at 5 GS/s with a 500 MHz input frequency is 38.56 dB, where the SNDR is 30.13 dB. The power consumption is 108 mW from a 1.2 V supply. Compared to other converters utilizing similar calibration techniques, this design provides the lower FOM. Detailed comparison is made in Table 2.3.

TABLE 2.3

Trimming A/D Converters Performance Comparison

	[30]	[27]	[28]
Sampling rate	5 GS/s	5 GS/s	5 GS/s
Supply	1.2 V	1.2 V	1.2 V
ENOB at DC	3.71 dB	3.93 dB	5.786 dB
Power	86 mW	34.3 mW	108 mW
FOM [pj/conv.-step]	1.32 pj/conv.-step	0.45 pj/conv.-step	0.39 pj/conv.-step
Offset reduction technique	P-MOS bulk trimming	Load transistor trimming	N-MOS triple well bulk trimming
Technology	90 nm	65 nm	90 nm

2.5 Conclusions

The development of submicron CMOS technologies with constantly reducing headroom and increasing device mismatch makes the design of accurate ADCs a more difficult task than before. There are various ways to overcome this issue, but while the classic techniques of averaging and offset storage work well, they do so by increasing design complexity and current consumption. Furthermore, the increasing speed requirements of the analog–digital interfaces, will force the use of minimum transistor size, inevitably leading to extensive utilization of digital calibration techniques, and exploiting the superior speed and linearity associated with such converters.

References

1. Figueiredo, P.M., Vital, J.C. *Offset Reduction Techniques in High Speed Analog-to-Digital Converters.* Netherlands: Springer, 2009.
2. Pelgrom, M. J. M., Duinmaijer, Aad C.J., Welbers, A.P.G. Matching properties of MOS transistors. *IEEE J. Solid-State Circuit*, 24(5), 1433–1439, 1989.
3. Tang, H., Peng, Y., Lu, X. et al. Quantitative analysis for high speed interpolated/averaging ADC. Paper presented at *2013 IEEE 10th International Conferences. on ASIC (ASICON)*, Shenzhen, 2013.
4. Figueiredo, P.M., Vital, J.C. Averaging technique in flash analog-to-digital converters. *IEEE Trans. Circuits Syst. I: Regular Papers*, 51(2), 233–253, 2004.
5. Kattmann, K., Barrow, J.A technique for reducing differential non-linearity errors in flash A/D converters. Paper presented at *1991 IEEE International Solid-State Circuits Conference, Digest of Technical Papers*, 38th ISSCC, San Francisco, CA, 1991.
6. Lee, C.-C., Yang, C.-M., Kuo, T.-H. A compact low-power flash ADC using auto-zeroing with capacitor averaging. Paper presented at *2013 IEEE International Conference on Electron Devices Solid-State Circuits (EDSSC)*, Hong Kong, 2013.
7. Chen, C.-C., Chung, Y.-L., Chiu, C.-I. 6-b 1.6-GS/s flash ADC with distributed track-and-hold pre-comparators in a 0.18 μm CMOS. Paper presented at *ISSCS 2009 on International Symposium, Signals, Circuits Syst.*, Iasi, 2009.
8. Veldhorst, P., Guksun, G., Annema, A.-J. et al. The impact of CMOS scaling projected on a 6b full-nyquist non-calibrated flash ADC. Paper presented at *ProRISC 2009, 20th Annual Workshop on Circuits Systems Signal Processing*, Veldhoven, Netherlands, 2009.
9. Deguchi, K., Suwa, N., Ito, M. et al. A 6-bit 3.5-GS/s 0.9-V 98-mW flash ADC in 90-nm CMOS. *IEEE J. Solid-State Circuits*, 43(10), 2303–2310, 2008.
10. Ismail, A., Elmasry, M. A 6-bit 1.6-GS/s low-power wideband flash ADC converter in 0.13- m CMOS technology. *IEEE J. Solid-State Circuits*, 43(9), 1982–1990, 2008.

11. Sheikhaei, S., Mirabbasi, S., Ivanov, A. A 43 mW single-channel 4GS/s 4-bit flash ADC in 0.18 μm CMOS. Paper presented at *IEEE Custom Integrated Circuits Conference (CICC)*, San Jose, CA, 2007.

12. Lin, Y.-Z., Liu, Y.-T., Chang, S.-J. A 6-bit 2-GS/s flash analog-to-digital converter in 0.18-μm CMOS process. Paper presented at *IEEE Asian Solid-State Circuits Conference (ASSCC)*, Hangzhou, 2006.

13. Scholtens, P.C.S., Vertregt, M. A 6-b 1.6-Gsample/s flash ADC in 0.18 μm CMOS using averaging termination. *IEEE J. Solid-State Circuits*, 37(12), 1599–609, 2002.

14. Wang, Z., Chang, M-C.F. A 600-MSPS 8-bit CMOS ADC using distributed track-and-hold with complementary Resistor/Capacitor averaging. *IEEE Trans. Circuits Syst. I: Regular Papers*, 55(11), 3621–3627, 2008.

15. Figueiredo, P.M., Vital, J.C. Termination of averaging networks in flash ADCs. Paper presented at *Proceedings of the 2004 International Symposium. on Circuits Systems (ISCAS)*, Vancouver, 2004.

16. Ismail, A., Elmasry, M. Analysis of the flash ADC Bandwidth–Accuracy tradeoff in deep-submicron CMOS technologies. *IEEE Trans. Circuits Syst. II: Express Briefs*, 55(10), 1001–1005, 2008.

17. Van de Plassche, R. High-speed A/D converters. In *CMOS Integrated Analog-to-Digital and Digital-to-Analog Converters*. US: Springer, 2003, pp. 121–122 and 131–132.

18. Choi, M., Abidi, A.A. A 6-b 1.3-Gsample/s A/D converter in 0.35-um CMOS. *IEEE J. Solid-State Circuits*, 36(12), 1847–1858, 2001.

19. Lien, Y.-C., Lin, Y.-Z., Chang, S.-J. A 6-bit 1GS/s low-power flash ADC. Paper presented at *International Symposium on VLSI Design, Automation and Test (VLSI-DAT)*, Hsinchu, 2009.

20. Nagaraj, K., Martin, D. A., Wolfe, M. et al. A dual-mode 700-Msamples/s 6-bit 200-Msamples/s 7-bit A/D converter in a 0.25 μm digital CMOS process. *IEEE J. Solid-State Circuits*, 35(12), 1760–1768, 2000.

21. Mehr, I., Dalton, D. A 500-MSample/s, 6-bit nyquist-rate ADC for disk-drive read-channel applications. *IEEE J. Solid-State Circuits*, 34(7), 912–920, 1999.

22. Mulder, J., Ward, C.M., Lin, C.-H. et al. A 21-mW 8-b 125-MSample/s ADC in 0.09-mm² 0.13 μm CMOS. *IEEE J. Solid-State Circuits*, 39(12), 2116–2125, 2004.

23. Nagaraj, K., Chen, F., Viswanathan, T.R. Efficient 6-bit A/D converter using a 1-bit folding front end. *IEEE J. Solid-State Circuits*, 34(8), 1056–1062, 1999.

24. Shiwen, L., Hua, D., Peng, G. et al. Design and implementation of a CMOS 1Gsps 5bit flash ADC with offset calibration. Paper presented at *IEEE International Conference on Green Computers Communication, (GreenCom)*, Beijing, 2013.

25. Lin, Y.-Z., Lin, C.-W., Chang, S.-J. A 5-bit 3.2-GS/s flash ADC with a digital offset calibration scheme. *IEEE Trans. Very Large Scale Integration (VLSI) Syst.*, 18(3), 509–513, 2010.

26. Lin, Y.-Z., Lin, C.-W., Chang, S.-J. A 2-GS/s 6-bit flash ADC with offset calibration. Paper presented at *IEEE Asian Solid-State Circuits Conference (ASSCC)*, Fukuoka, 2008.

27. Yao, J., Liu, J. A 5-GS/s 4-bit flash ADC with triode-load bias voltage trimming offset calibration in 65-nm CMOS. Paper presented at *2011 IEEE Custom Integrated Circuits Conference (CICC)*, San Jose, CA, 2011.

28. Vassou, C., Mountrichas, L., Siskos, S. A NMOS bulk voltage trimming offset calibration technique for a 6-bit 5GS/s flash ADC. Paper presented at *2012 IEEE International Instrument Measurement Technology Conference (I2MTC)*, Graz, 2012.

29. Xu, Y., Belostotski, L., Haslett, J.W. Offset-corrected 5GHz CMOS dynamic comparator using bulk voltage trimming: Design and analysis. Paper presented at *IEEE 9th International New Circuits Systems Conference (NEWCAS)*, Bordeaux, 2011.

30. Yao, J., Liu, J., Lee, H. Bulk voltage trimming offset calibration for high-speed flash ADCs. *IEEE Trans. Circuits Syst. II: Express Briefs*, 57(2), 110–114, 2010.

31. Brady, P., Hasler, P. Offset compensation in flash ADCs using floating-gate circuits. Paper presented at *IEEE International Symposium on Circuits Syst. (ISCAS)*, Kobe, 2005.

32. Mountrichas, L., Siskos, S. A high-speed offset cancelling distributed sample-and-hold architecture for flash A/D converters. *Microelectron. J.*, 44(12), 1123–1131, 2013.

33. Simitsakis, P., Liolis, S., Psyllos, D. et al. Design of a 1.2-V 60 GHz transceiver in a 90 nm CMOS RF technology. Paper presented at *18th IEEE International Conference on Electron. Circuits Systems (ICECS)*, Beirut, 2011.

34. Lin, Y.-Z., Liu, Y.-T., Chang, S.-J. A 5-bit 4.2-GS/s flash ADC in 0.13-µm CMOS. Paper presented at *IEEE Custom Integrated Circuits Conference (CICC)*, San Jose, CA, 2007.

3

Protecting Mixed-Signal Technologies against Electrostatic Discharges: Challenges and Protection Strategies from Component to System

Marise Bafleur, Fabrice Caignet, Nicolas Nolhier, Jean-Phillppe Laine, and Patrice Besse

CONTENTS

3.1 Introduction

In the automotive industry, complex electronic systems are developed to enable advanced remote applications, such as airbag control, antilock braking systems, relay drivers, transmission, cruise controls, etc. that require mixed-signal technologies and more particularly, smart power technologies. In these applications, environment is quite severe in terms of temperature and types of stress (electromagnetic compatibility (EMC), electrostatic discharges (ESD), load dump, etc.), whereas reliability is critical to ensure the safety of the vehicle with a drastic objective of zero ppm failure. Regarding ESD protection, requirements are much more demanding than in consumer applications with at least a 2-kV human body model (HBM) and specific pins required to withstand up to 15-kV contact discharge and 25-kV air discharge. Designing ESD protections in smart power technologies is thus very challenging. In this chapter, we will review these challenges and give some important design guidelines and methodologies for efficient ESD protection both at chip and system levels.

3.2 ESD Qualification Techniques

Automotive qualification stress tests for integrated circuits (ICs) are defined in the AEC-Q100 (Automotive Electronics Council) documents [1]. They include ESD test requirements performed at the IC level. HBM [2] and charged device model (CDM) [3] are now the standards currently used for ESD qualification of ICs. A machine model (MM) standard [4] that was originally created for automotive applications is now abandoned. Each standard defines a specific dynamic current waveform to be injected to the IC pins. For each type of waveform, a dedicated model specifies a capacitor that is preliminarily charged and a "discharge" network (generally a series resistor) [5]. In the case of the HBM and MM stress, the equivalent capacitor that is charged is a human or a machine, respectively. In the case of the CDM stress, it is the IC itself that is initially charged and then discharged by a contact with a conductive surface. The CDM current depends on the equivalent capacitance of the whole die that increases with the size of the package. To this simple model, parasitic elements (RLC) need to be added to correctly represent the actual dynamic behavior of the stress.

Figure 3.1a presents the equivalent electrical circuit that can represent the various waveforms of HBM, MM, or CDM stress using the related values of the parasitic components [6]. These values are summarized in Table 3.1. Figure 3.1b shows the comparison of the various waveforms of HBM, MM, and CDM stress for the respective standard level required for consumer products.

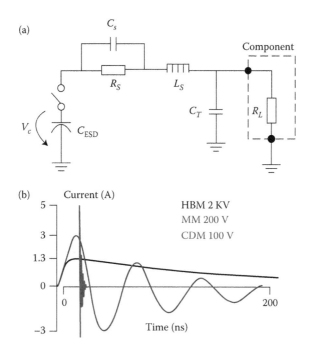

FIGURE 3.1
Equivalent RLC electrical schematic (a) and resulting waveforms (b) for HBM, MM, and CDM models.

Pins of the device under test (DUT) are classified into two categories: supply pins and no supply pins. An ESD zap between pins and groups of pins is applied. Combinations depend on the pin category. Typical maximum required HBM stress levels are 1–2 kV. Common requirements for CDM are 250 V and it can be up to 500 V for all pins and 750 V for corner pins.

Other standards are defined at a system level (vehicle), such as IEC 61000-4-2 [7] and the ISO 10605 [8], that is more dedicated to automotive systems. In this case, an ESD zap is applied at the system level, for example, at a connector, with an ESD generator (Gun), and the current is diverted through the different elements of the printed circuit board (PCB). A residual current of this stress can reach the ICs and generate permanent damage. There are multiple test setups, powered, unpowered conditions, contact discharge, air

TABLE 3.1

Characteristics of HBM, MM, and CDM Discharge Models

Model	C_{ESD} (pF)	R_S (Ω)	L_S (μH)	C_S (pF)	C_T (pF)
HBM	100	1500	5	<5	<30
MM	200	0	0.5/2.5	0	<30
CDM	10	10	0.01	0	0

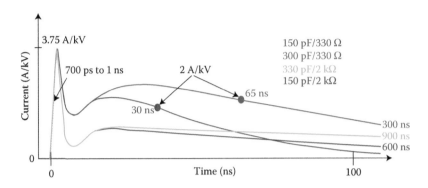

FIGURE 3.2
Comparison of various waveforms required by the ISO 10605 standard.

discharge, with several types of ground connections. For these standards, the same equivalent electrical circuit as shown in Figure 3.1 applies. For the IEC 61000-4-2 standard, the discharge is realized with a gun RC module of 150 pF and 330 Ω. However, for the automotive ISO 10605, all combinations of 150 and 330 pF capacitors with 330-Ω and 2-kΩ resistors are possible, respectively. This greatly impacts the dynamic behavior of the stress waveforms as reported in Figure 3.2.

For a system, different levels of functionality are required after or during the ESD. For the most severe requirements all functions should operate normally during and after the ESD stress.

3.3 Investigation Methodologies

3.3.1 Electrical Characterization

Standard ESD tests at the IC level such as HBM, MM, or CDM are not suitable to help IC designer to optimize its ESD protection strategy. Indeed these tests characterize the ESD robustness level of a given protection but do not provide crucial data such as threshold voltage V_{T1}, holding voltage V_H if a snapback occurs, or ON-resistance during ESD stress. The maximum current capability I_{T2} of the protection can be extracted from its HBM robustness but the relationship between these two values is not always verified.

All these parameters are necessary to check if the protection is compliant with the ESD design window. Today, the integration of these parameters is not standardized yet. But an approach is ongoing to incorporate ESD parameters in standard models like input/output buffer information specification (IBIS) models already used in the EMC field.

Specific characterization tools have been developed at a wafer level to extract these intrinsic parameters, preventing parasitic effects from packaging and are presented hereafter.

3.3.1.1 Transmission Line Pulse Testing: TLP and VF-TLP

Transmission line pulse (TLP) testing [9] that is based on time-domain reflectometry (TDR) method is a commonly used characterization tool and its principle is described in Figure 3.3. The key idea is firstly to charge a 50-Ω transmission line to a given high voltage (up to 1 kV or more), and then discharge it into the DUT. The line length defines the pulse duration (typically 100 ns) and a passive filter controls its rise time. Voltage and current sensors are located on a coaxial connection between the pulse generator and the DUT. The length of this connection is smaller regarding the transmission line in the generator. So, there is an overlap of the incident pulse from the generator with the pulse reflected from the device. Sensors directly read the current and voltage at the device level. A current/voltage point is extracted for a given time from the beginning of the pulse, when the device is generally in steady state. By increasing the generator high voltage, a full quasi-static I–V curve is obtained up to the device destruction (I_{T2}, V_{T2}). A leakage current is also monitored between TLP pulses and acts as a failure criterion. Even if their waveforms are strictly different, 100 ns TLP stresses are similar to HBM ones in terms of energy and timing.

To characterize the behavior of the device regarding faster ESD event like CDM, an extension of the TLP setup has been proposed [10]. Very-fast TLP (VF-TLP) is based on the same architecture as TLP, but the generated pulse is shorter (1–5 ns) using a smaller transmission line. The current and voltage measurements are more difficult because the incident and reflected pulses are totally separated in the time domain. Postprocessing is needed to recalculate the current/voltage value. Also, some hardware solutions exist to merge the pulses using delay line and a combiner. VF-TLP characterizations at the wafer level require high-bandwidth probes with special care for the ground

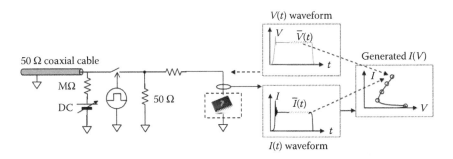

FIGURE 3.3
TLP testing principle.

path. Ground/signal 50-Ω probes can be used for the best results but involve a special design of the pad access. Generally VF-TLP allows characterizing ESD protections at a higher current level since the pulse is shorter, the power dissipated by the device is smaller. But the state of the device could be different, that is, in TLP the device is in a quasi-static mode while during VF-TLP it could be in a transient situation.

3.3.1.2 Overshoot Issues: Gate Monitoring and Transient TLP

Even if the ESD protection exhibits low impedance when turned on, during its triggering an overvoltage can occur due to capacitive effects. Typically this critical phase is a few hundred picoseconds after the beginning of the ESD event. This overvoltage can induce the destruction of the IC. As a result, ESD protection designers should be aware of this phenomenon and take it into account.

One method to check the protection efficiency is to implement a gate monitor [11]. The gate and the source of an NMOS transistor are coupled in parallel with the ESD protection. The drain of the MOS transistor is open but connected to an external pad for a functional test. The intrinsic TLP curves of each component are already known. Then a TLP characterization of this association is performed until failure. If the ESD protection is firstly destroyed, there is no overvoltage issue. If the MOS transistor fails, for a quasi-static voltage higher than its V_{T2} value, there is no overvoltage issue but this means that the ON-resistance of the ESD protection is too high. If the failure quasi-static voltage is lower than the MOS V_{T2}, we are facing a destructive overvoltage but without knowing its value.

A transient TLP method has been developed to evaluate this overvoltage value [12]. It uses a simplified VF-TLP setup combined to a postprocessing technique in order to increase time accuracy. Firstly, RF scattering parameters (DC to 4 GHz) of each part of the VF-TLP setup are characterized using a vector network analyzer; this is the calibration step. During transient characterization, the waveform of the incident and reflected pulses are digitized with a high-bandwidth oscilloscope. A fast Fourier transform (FFT) is then applied in order to convert signals from the time domain to the frequency domain. Data coming from the vector network analyzer for characterization are then used to correct the signals with a complex algorithm. And finally an inverted FFT is processed in order to come back in the time domain. With this technique, measurement limitation is coming from the oscilloscope bandwidth (6 GHz) and the time resolution is about 60 ps, which is enough to estimate the transient overvoltage [13].

3.3.1.3 Near-Field Scan

One of the main issues when designing ESD protection strategies at the system level is to identify the propagation path of an ESD stress within the system and to extract the precise resulting waveform and its impact on the embedded ICs. Near-field scanning is a noninvasive measurement technique based on the use of a magnetic field probe [14]. It allows current monitoring at a PCB level or even at an IC level if the probe is miniaturized.

The same probe as the one used in the EMC standard [15] allows capturing the magnetic field emission while the current propagates. It is made up of a semirigid copper coaxial cable (Figure 3.4). The inner conductor of the coaxial cable is twisted and shorted to the outer shield, thus creating a current loop. Such a probe can be modeled using an inductor and a capacitor to represent the loop and a transmission line to take into account the RG402 high-frequency coaxial cable. In terms of transmitted power, it exhibits a pure magnetic coupling effect up to 1 GHz that is mainly inductive. This injection model was already validated in previous works [16].

Based on the purely inductive behavior of the probe up to 1 GHz, we directly reconstruct the current waveform flowing through the PCB line by integrating the measured magnetic field.

With regard to system level ESD stresses, it turns out that a 1 GHz probe bandwidth is sufficient for low discharge levels (2 kV) and integral reconstruction waveform can be used. For high levels (8 kV), it is more appropriate to work in the frequency domain by performing a FFT of the probe dynamic measurement, correcting it with the probe frequency response and finally getting the current waveform using an FFT^{-1}.

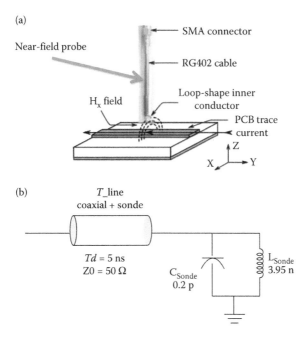

FIGURE 3.4
(a) Homemade magnetic field probe. (b) Equivalent model of the probe.

As a validation of this method, we present a measurement comparison with the 1-Ω-resistor method on a simple commercial inverter submitted to the injection of a square current pulse (1 A, 100 ns) using a VF-TLP tester. The injection is applied to one output pin and current is measured at the ground pin. As shown in Figure 3.5, there is a good agreement between the current measurement obtained by the field probe and the 1-Ω resistor. It has to be noticed that the external decoupling capacitance on the V_{DD} pin dramatically changes the waveform of the injected square pulse. Such a waveform could not be predicted by the ESD protection strategy of the chip.

To understand how the current of an ESD stress propagates within a system at the board level, we developed a near-field scanning system where the field probe is mounted on a three-axis table with a maximum spatial

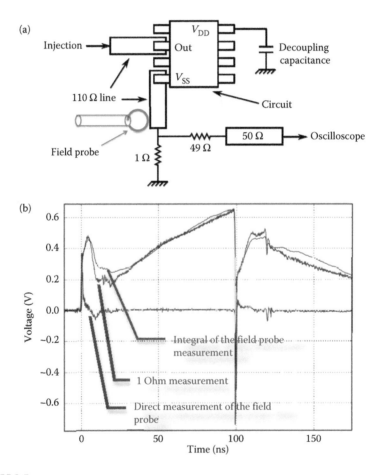

FIGURE 3.5
(a) Schematic diagram of the test setup for a commercial inverter using a 1-Ω resistor and a magnetic field probe. (b) Current measurement comparison for a VF-TLP (1 A, 100 ns) injected into out pin.

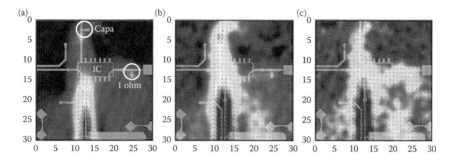

FIGURE 3.6
Pictures from the movie of the current propagation into the board: current injection (a), decoupling capacitor absorption (b), and current split between the capacitor and 1-Ω resistor (c).

resolution of 100 µm. The TLP generator synchronizes the measurement test bench. Each time a stress is injected into the connector of the PCB, H_X and H_Y magnetic fields are captured using a 12-GHz oscilloscope. By repeating this operation step by step, the scan of the PCB is performed. From the registered data, a movie can be built showing the evolution of the emitted field during an electrical fast transient.

Figure 3.6 reports three pictures showing the current propagation into the system of Figure 3.5 at different times and in particular, the impact of the decoupling capacitor. The near-field scanning system is a powerful tool for system level ESD investigation and for understanding the system behavior under different types of stresses.

3.3.2 Debug and Investigation Methods

The simulation of functional failures during ESD events remains very challenging for ICs due to the frequency domain and the high current injection mechanisms. Hence, the root cause of a functionality loss after an ESD event can be very difficult to understand.

Combining TLP and emission microscopy (EMMI) can provide good results (Figure 3.7). EMMI is an optical analysis technique used to detect and localize IC defects. It is noninvasive and this phenomenon can be observed in real time, integrating the photoemission generated by the IC. Photons are generated during gate oxide defects, polysilicon filaments, substrate damage, junction avalanche, active bipolar transistors, saturated MOS device, etc.

Combined TLP + EMMI technique consists in taking a reference picture of the photon emission while the IC is submitted to a TLP zap at a level just before a failure occurs. Then, at a higher TLP zap level another picture is taken once a malfunction is detected. Both pictures are compared to identify where the photon activity has changed at blocks or elements levels. Figure 3.7 shows an example with the emission activity of an internal 5 V regulator when a TLP pulse is applied on a global pin. Figure 3.7a shows a normal

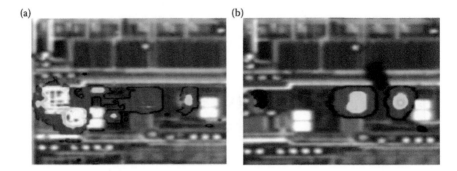

FIGURE 3.7
Coupling TLP and EMMI for debugging. In GaAs sensor 20×, 30 s integration: (a) TLP injection level below fault generation threshold and (b) TLP injection level increased until the soft failure occurs.

activity of the block, whereas Figure 3.7b exhibits the emission change after a "reset" generation.

The analysis of these pictures allowed understanding that the 5-V regulated voltage was decreasing until the undervoltage detection and then creating a reset. After a thorough investigation into schematics and layout, a parasitic coupling on the current reference circuit was identified as the root cause of the failure. This study is detailed in Reference 17.

3.4 ESD Design in Smart Power Technologies

3.4.1 ESD Design Window

The role of ESD devices is the protection of ICs not only against any ESD events during manufacturing and assembly but also during the IC operation. An ESD design window defining a safe operating area of the protection device in the IC was proposed to properly optimize its voltage/current characteristics. Figure 3.8 describes the ESD design window with the lower and upper voltage limits. The lower voltage limit is the maximum pin power supply voltage or the maximum pin voltage rating. In contrast, the upper voltage limit is the lowest breakdown voltage of the IC (oxide or junction breakdown of IC devices). With these elements, any designer can develop an ESD structure with specific voltage characteristics between these voltage limits. By this way, the ESD protection scheme provides protection of the IC while being compatible with the electrical pin constraints.

The ESD design window is generally defined at an ambient temperature. However, ICs are also required to handle ESD stresses during their operation and temperature in some applications such the automotive can be as

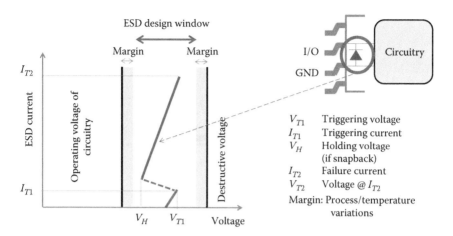

FIGURE 3.8
ESD design window.

high as 125°C or even more. An illustration of the impact of temperature on the efficiency of a protection is the case of a MOS power clamp (PC). MOS devices are very sensitive to temperature. Indeed, a MOS PC exhibiting 1.74 A failure current and 4.4-Ω ON-resistance at 25°C sees its robustness and ON-resistance degraded down to 1.34 A and 7.1 Ω at 200°C, respectively [18,19]. If this temperature behavior is not taken into account, it could induce detrimental effects such as a lower failure current and even could lead to not providing the expected protection due to the noncompliance with the ESD design window. To compensate this effect, the size of the PC protection would have to be increased.

ESD protection strategy placement consists in implementing the appropriate ESD protection structures for a safe ESD discharge current path while providing proper voltage clamping at the I/O pins to be protected. A good protection strategy should provide an efficient current path during any ESD events between I/O and supply pins whereas optimizing ESD protection structures placement to minimize die size.

There are two different ESD protection strategies at any pin: local and distributed. The local ESD protection strategy is the basic technique consisting in placing the ESD protection structure at the pin to be protected. It requires for the protection structure to be bidirectional. This technique presents the advantage to simplify the definition of the ESD architecture placement using the common metal ground bus around the chip (Figure 3.9a). It also limits any coupling between the different pins. However, it implies significant design efforts to develop different types of ESD protection structures with multiple I/O or power pins. As a consequence, die size may not be optimized.

Conversely, the distributed ESD protection strategy uses diodes to divert any ESD current from any I/O pin to a single or distributed clamp generally placed at the power pins (Figure 3.9b). Diodes only work in the forward mode

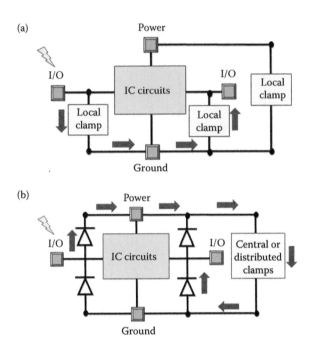

FIGURE 3.9
ESD protection strategies: local (a) versus distributed (b). In both cases, the arrows indicate the ESD current path for an ESD zap between the two I/Os.

then conferring a high robustness and requiring a small silicon footprint. Design efforts are focused on the clamps at power pins and on the common metal routing for power and ground bus. This strategy has the advantage to reduce the ESD protection development time since design efforts are focused on the PC. In addition, since the protection strategy uses active devices from the technology library, it can be easily simulated.

However, the use of this technique is limited with the number of voltage levels applied at different power pins (more than 3). It increases the complexity of bus architecture by increasing additional common metal bus ring connected at different power-pin voltage levels. In addition, this technique cannot be applied for I/O pins isolated from power pins in some cases.

3.4.2 ESD Integrated Protection

Although in some cases active devices can be self-protected by an appropriate design, this is not generally the case and specific ESD protections have to be implemented even for power devices. A major difficulty of ESD protection design is that it should be done with a small footprint and free device for a given technology, that is, it should not involve any extra mask or technology process step. As local protections are placed under the pads, the maximum-targeted surface is equal to pad one, generally closed to $100 \times 100\ \mu m^2$. ESD design specifications involve both high power dissipation and high current density. As a result, snapback ESD clamps are popular bipolar-based protection devices. In addition, since multiple voltage domains could be integrated on the same chip, ESD protections with tunable operating voltage are highly desirable. We will discuss hereafter the main issues for proper ESD device design, their advantages and drawbacks, and main use.

3.4.2.1 Power MOS

The typical ESD protection clamp used from several products is the power MOS transistor. This kind of protection is often self-protected during any transient ESD current event.

This technique uses a high gate resistance between the gate terminal and the source terminal of the N-channel MOS transistor (Figure 3.10). The size of a MOS transistor is optimized to get not only a low ON-resistance when activated but also to have high capacitive coupling between the drain terminal and the gate terminal for fast activation.

The MOS transistor performance protection depends on its area form factor. It was demonstrated that using a specific area configuration of the grounded-gate NMOS transistor provides the best ESD performance [20,21].

Designing a grounded-gate NMOS transistor with $F = 6$ improves by a factor of 3, the HBM ESD performance in comparison with the one designed with $F = 0.5$ (Figure 3.11).

3.4.2.2 NPN-Based ESD Protections

The main advantage of snapback bipolar-based devices is related to the conductivity modulation mechanisms [22] that allow drastically reducing their ON-resistance and then their silicon footprint. In CMOS technologies, parasitic bipolar transistors intrinsic to MOS structures are used for

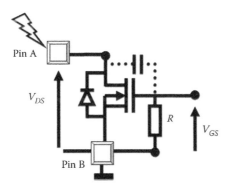

FIGURE 3.10
ESD clamp—self-protected NMOS transistor.

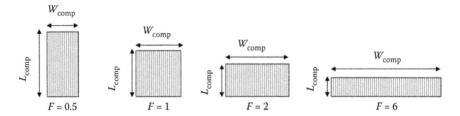

FIGURE 3.11
Grounded-gate NMOS transistor.

ESD protection. In the case of smart power technologies, bipolar transistors are part of the library of active devices. They just need to be adapted for pulsed operation to provide high power dissipation and high current density (10^5–10^6 A · cm^{-2}). To this aim, they are generally self-triggered by operating them in conditions of avalanche injection. The resulting high power generation in local regions of the device requires specific design solutions such as proper energy balance inside the device structure and current ballasting implemented via a proper layout using contact diffusion regions and backend metallization.

Figure 3.12 shows a typical NPN transistor of a smart power technology that is currently used as a self-biased ESD protection. When the voltage V

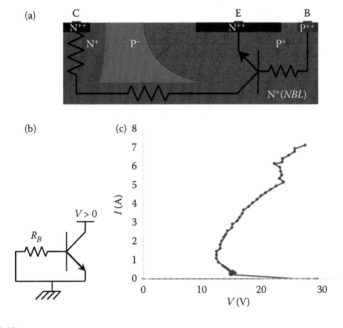

FIGURE 3.12
Grounded-base NPN transistor used as for ESD protection: cross section (a), electrical schematic (b), and typical TLP characteristics (c).

on the collector pin increases and reaches the breakdown voltage of the collector–base junction, an avalanche current is generated and flows through the intrinsic base resistance (R_B on schematic). When its value is such that the voltage drop on this resistance is higher than 0.6 V, the bipolar transistor turns on and as a result its *I–V* characteristics undergoes a strong snapback as shown by the TLP curve shown in the same figure. In the case of the protection of high voltage pins, this strong snapback is an important drawback since it is generally not compliant with the requirement of the ESD design window for a snapback voltage higher than the operation voltage. To cope with this issue, a classical solution is to stack as many NPN structures as necessary in series. This allows increasing the snapback voltage but concurrently the ON-resistance and the silicon footprint. An alternative solution consists in using devices exhibiting a weak snapback or even a no-snapback device.

An avalanche diode could have been an interesting component as a no-snapback protection. A high robustness should be attained given its excellent current uniformity due to a constant breakdown voltage increase with current and temperature. However, the current-induced field modification leads to low R_{ON} only for structures closed to the p⁺-i-n⁺ configuration. Otherwise, the R_{ON} could be extremely high [23,24].

In bipolar devices, the simplified implicit equation that governs the snapback behavior is the following:

$$\alpha M = 1 + \frac{I_B}{I_E}$$ (3.1)

where α is the common base gain (in a self-biased configuration), M is the multiplication factor, I_B is the current at the base contact, and I_E is the current at the emitter contact.

According to this equation, two methods can be contemplated to increase the holding voltage of a bipolar structure.

- First, lower the gain α (in a self-biased configuration), to increase the corresponding M value and then the voltage.
- Second, use a device with a lower avalanche multiplication factor M.

3.4.2.3 PNP-Based ESD Protections

Using a PNP bipolar transistor instead of a NPN is an attractive solution. Its gain is generally lower than the NPN, given the lower holes diffusion coefficient. The multiplication factor is also lower in a PNP than in a NPN, since holes' impact ionization coefficient is smaller than the electrons. The main issue to develop PNP-based ESD protections concerns the reduction of the ON-resistance (R_{ON}) value, which is intrinsically higher than the NPN. Two major reasons explain this difference: firstly, the only available PNP transistor is generally a lateral one then resulting in higher access resistances and secondly, the conductivity modulation mechanisms (beneficial to the R_{ON}) are less pronounced in PNP devices. To counterbalance these drawbacks, the following design guidelines can be implemented:

- Use of multifingers interdigitated structures with minimum dimensions for emitter and collector widths without any ballasting.
- Suppression of base contact since PNP devices exhibit a weak snapback or no snapback at all. This allows saving up to 25% silicon area.
- Abrupt collector profile and low base doping to favor high injection effects.

These design guidelines have been implemented on PNP devices issued from SmartMOS® 8-MV smart power technology [25]. Figure 3.13 presents the cross section of a PNP bipolar transistor from the library without a base

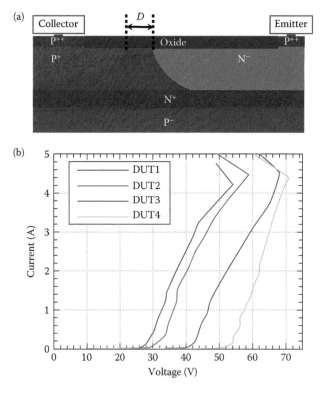

FIGURE 3.13
Optimized PNP-based ESD protection in 0.8 μm smart power technology: cross section (a) and measured TLP characteristics. (b) Distance D allows tuning V_{t1} triggering voltage.

contact and its measured TLP characteristics. For this transistor, distance D allows tuning the triggering voltage of the structure. In the four different cases corresponding to different distances D, the structures do not show any snapback behavior and the triggering voltage value has almost no impact on the failure current, which corresponds to an HBM robustness of about 7 kV. Table 3.2 summarizes the main electrical parameters of the DUT4 structure. Finally, the leakage current of the PNP without a base contact was measured to be lower than 1 pA (limit of the measuring setup) that is compatible with standard leakage specifications.

To further improve the performance of such PNP-based ESD protection, it is worthy to efficiently couple the intrinsic vertical diode with the lateral PNP bipolar transistor [26]. Figure 3.14a presents the basic principle of the structure. Its optimization first requires adjusting the voltage triggering of both devices: making them equal provides the best trade-off. As the breakdown voltage of the vertical diode is defined by the doping profile of the different layers, the only way to tune this parameter is to vary the distance D that defines the collector–emitter PNP breakdown voltage. The second parameter to optimize is the current density flowing in each device: it controls the ON-resistance of the global structure. This is achieved by adjusting the width of the anode contact L. Increasing L allows significantly decreasing the ON-resistance, the optimum being defined by the following factor of merit:

$$F_{R_{ON}} = \frac{R_{ON} \cdot S}{V^2} \tag{3.2}$$

where R_{ON} is the ON-resistance of the structure, S is its silicon area, and V is the clamping voltage. The simulation results of Figure 3.14b show the homogeneous current distribution over both devices in such optimized structure.

Figure 3.15a illustrates the significant improvement obtained with such optimization. The measured PNP/diode having an optimum L of 10 µm and 10,000 µm² silicon footprint results in a very low ON-resistance of 1-Ω and 8-kV HBM robustness. It was successfully implemented in a commercial circuit to protect an 80-V I/O pin using two stacked PNP/diode devices, one

TABLE 3.2

Measured Electrical Parameters for the PNP-Based ESD Protection of Figure 3.13

PNP Bipolar Transistor (DUT4)					
V_{T1} (V)	V_{T2} (V)	I_{T2} (A)	R_{ON} (Ω)	I_{T2}/S (mA µm⁻²)	$R_{ON} \cdot S^a$ (m$\Omega \cdot$ cm²)
49	70.7	4.42	4.9	0.28	0.77

[a] The PNP bipolar transistor is made up of 17 100-µm-long fingers resulting in a 15,700 µm² surface.

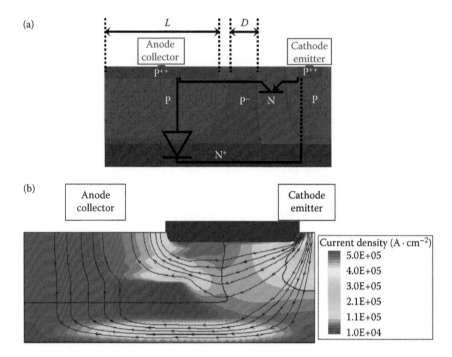

FIGURE 3.14
Proposed PNP/diode ESD protection: schematic cross section (a) and TCAD simulation show-ing the current distribution over the two elements, diode and PNP (b).

with a V_{T1} of 32 V and the second with a V_{T1} of 54 V. As shown in Figure 3.15b, it provided the required robustness of 2-kV HBM (~1.3 A) for a silicon footprint of 21,000 μm². This has to be compared with the original imple-mented protection (NPN stacked with an LDMOS) that required a silicon footprint of 40,000 μm² for the same HBM robustness.

3.4.2.4 Silicon-Controlled Rectifier

Silicon-controlled rectifiers (SCRs) are very attractive devices. Their excellent clamping capabilities and high ESD robustness allow designing efficient and compact protections [27–29]. However, their main drawback is that most of the SCR-based protections exhibit a low holding voltage V_H (~1.2 V), which makes them prone to latch-up and not compliant with the ESD design win-dow. In deep-submicron CMOS technologies, the problem is solved since the power supply is below 1 V and SCR-based protections are now widely used [30]. In high voltage technologies, many research studies have been carried out to increase either the holding voltage [29,31,32] or the holding current of SCR devices [15,33]. The holding current of the parasitic SCR is defined by the following parameters and equation [19]:

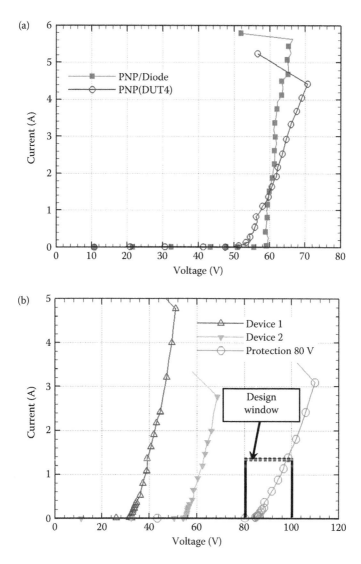

FIGURE 3.15
PNP/diode experimental results: TLP characteristic comparison with a single PNP for a five-finger structure with 10,000 µm² footprint (a) and practical implementation for the protection of an 80-V I/O pin of a power-over-Ethernet commercial circuit (b).

$$I_H = \frac{\beta_p(\beta_n + 1) \cdot I_{NW} + \beta_n(\beta_P + 1) \cdot I_{PW}}{\beta_n\beta_p - 1} \tag{3.3}$$

where β_n and β_p are the current gains of the parasitic NPN and PNP bipolar transistors and I_{NW} and I_{PW} are the currents flowing into the N-well

(PNP base) and P-well (NPN base), respectively. In the same way, the holding voltage V_H is linearly proportional to I_H.

As a result, to increase I_H or V_H, following two options are available:

- Reduce the current gain of the respective bipolar transistors or
- Reduce as much as possible the resistance of the NPN and PNP bases

The specific design of the bidirectional SCR presented in Figure 3.16 allows achieving a high V_H: it is made up of two P-wells isolated by a deep N+ diffusion and an N+-buried layer [34,35]. It is composed of three bipolar transistors: a lateral NPN Q1 (P-well as a base, N++ as an emitter, and central deep N+ as a collector), a lateral PNP Q2 whose base is the central deep N+ and a vertical PNP Q3 (P-well as a collector, N+-buried layer (NBL) as a base, and P-substrate as an emitter). This protection structure allows the protection against both positive and negative ESD stresses.

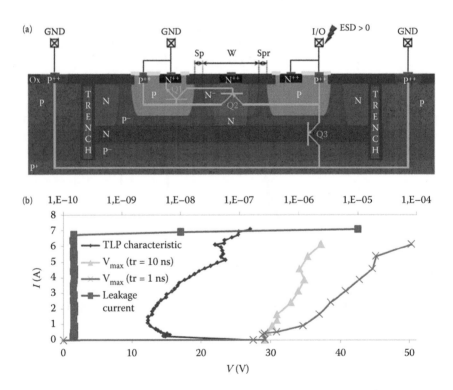

FIGURE 3.16
High holding voltage bidirectional SCR-based ESD protection showing active bipolar transistors under a positive ESD zap (a). Measured TLP characteristics and associated overshoot, V_{max}, according to TLP duration (b).

Upon a positive ESD stress, this protection is triggered via the self-biasing of the lateral NPN Q1 when the ESD voltage reaches its base–collector breakdown voltage that is defined by the distance S_p. Once turned on, both lateral and vertical NPN transistors inject current into the PNP base. As this latter is highly doped, the PNP current gain is very low and as a result induces a high holding voltage (12.5 V). In addition, the vertical PNP transistor Q3 provides a secondary current path. Figure 3.16 also presents the TLP characteristics of this device realized in 0.25-μm SmartMOS® technology for a distance S_p that results in a triggering voltage of 28.9 V and a holding voltage of 12.5 V. These structures can be stacked to meet different voltage targets.

The main issue of a SCR device is the overshoot that can be induced before its triggering. In the graph of Figure 3.16, is also plotted the overshoot under two different TLP pulses with a 10 ns and a 1 ns rise time, respectively. This overshoot has to be carefully characterized and taken into account in the ESD design window or reduced by implementing efficient SCR triggering techniques [36]. Specific characterization methods such as the transient TLP described in Section 3.2 are essential to assess this overshoot.

3.5 System Level ESD Protection Strategies

With the wide dissemination of electronic products and the stringent failure requirements of original equipment manufacturers (OEMs), it is important to estimate what would be the impact of an ESD stress on ICs depending on their implementation in the system. Predicting IC's ESD robustness has become a challenge for both IC manufacturers and OEMs. ESD requirements are even more severe on ICs with "global pins" tied to connectors and directly exposed to external aggressions of a system.

As a result, the demand for robustness against the system level, defined by IEC 61000-4-2 standard (some other standards like ISO 10605 are also used for automotive applications), is increasingly shifted to the IC component itself [37,38].

Performing ESD system-level reliability prediction remains a very challenging topic. Being able to do so could bring further improvements to existing protection approaches [39], while reducing costs. The ESD stress level required during system qualification is increasing over the years. Therefore, taking into account these disturbances during the system design phase becomes mandatory to improve the system immunity, which tends to decrease with technology nodes, as observed by Camp [40].

Up to now, there is no methodology to anticipate system ESD failures, which results in redesign if the system does not fulfill the standard. To implement such a predictive approach, both modeling and characterization methodologies should be developed.

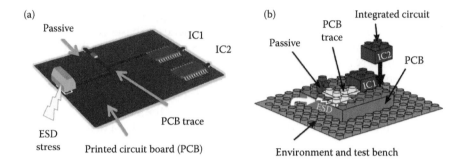

FIGURE 3.17
ESD system level modeling methodology: view of a printed circuit board (a) and its hierarchical representation (b).

To this aim, a codesign approach mixing IC design and PCB design is necessary to predict ESD current paths and failures at the system level.

To ensure that chips can handle system level ESD events, OEMs face the problem of the lack of information concerning the internal ESD strategy of the ICs. The same problem arose years ago concerning signal integrity (SI) issues and gave birth to the IBIS standard [41]. The IBIS file contains behavioral data of parasitic elements (of package, inputs and outputs) and description of buffers that are needed for SI simulation. Keeping the concept of the IBIS description, we proposed to develop IC models to perform system efficiency ESD design (SEED) [42] according to the methodology presented in Figure 3.17. This methodology is intended to model a system composed of ICs, active components, and external protection elements mounted on a PCB. The main principle consists in modeling each system part separately and assembling all parts hierarchically by following the system topology [43]. Each part of the system can be identified as an independent block as reported in Figure 3.17 ("Lego block" system approach).

The main element of the system is the IC, and a dedicated attention has to be paid to be sure that the current flow through this element is correctly reproduced. Depending on the internal protections, the current can use different paths. The proposed IC model has to maintain intellectual property rights of several manufacturers. For this reason, it is based on the existing methodology provided by the IBIS model. Some of the information provided by the IBIS can be kept, but to perform SEED additional information has to be added.

3.5.1 Advantages and Limitations of IBIS Models

An RLC network (given by the IBIS) models the package of the device. The series inductance is the only element that has the most significant impact on the waveforms. Typical values are close to a few nH [44], which is enough to induce a significant voltage spike across its terminals during transient current injection (i.e., TLP) [43].

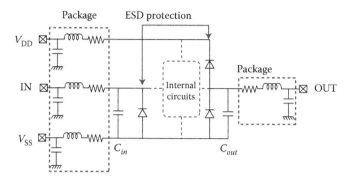

FIGURE 3.18
Equivalent IC model obtained from IBIS.

ESD signals exhibit high dI/dt and as a result, parasitic elements from the IBIS are important to estimate the dynamic of the current that flows inside the chip and through the IC protections. Let us show in a simple example the importance of package parasitic elements. The typical inductance value of a surface-mount technology (SMT) package is around 3 nH. For 8-kV gun stress, the current rises up to 25 A in less than 1 ns. This means that the overshoot voltage across the inductance is instantaneously around 75 V.

The IBIS includes I–V data of the diodes between input/output and V_{DD} (called power clamp in the IBIS), and to ground. These diodes are static ones and are only modeled from $-V_{DD}$ to $2 \times V_{DD}$, which is a too small range for ESD events. Moreover, IBIS suffers from a lack of information concerning the central PC protection between V_{DD} and V_{SS}, which is crucial for the ESD strategy. An equivalent model of the commercial inverter IC used in Reference 43, extracted from the IBIS file is reported in the schematic shown in Figure 3.18.

IBIS suffers from a lack of information concerning input and output ESD protections. The philosophy of previous work [45] was to extract a behavioral description of ESD protections between two pins from TLP measurements. The new I–V curves will replace the I–V characteristics from the IBIS and protections that are not defined will be added and in particular, the PC between V_{DD} and V_{SS}. Then, a full behavioral description of the ESD strategy is built. To keep the IBIS concept, no prior knowledge of IC internal structures is needed, which is often the case for system designers at the OEM level. Specific measurements are needed to extract information about ESD protections of the chip. TLP testing can be used to extract quasi-static and static characteristics between each pair of IC pins.

3.5.2 Behavioral Description of the Protections from TLP Measurement

Various types of components can be extracted from TLP measurement, like simple diodes, or more complex structures with a snapback, like a SCR, or structures that clamp with dynamic conditions (triggered MOSFET – PC).

Piecewise linear curves provide sufficient information while preserving the intellectual property of manufacturers. For example, only two I–V points could describe a simple diode. For improved accuracy and more complex devices, as many points as necessary can be added.

From all these structures, we proposed to use the simplest I–V characteristics as much as possible. Given the high current level of ESD stresses, small discrepancies are not important regarding the accuracy. The parasitic elements of the PCB (line and passives) play a much more important role than a small variation on the ESD protection ON-resistance as discussed later on. Moreover, behavioral models do not reveal any proprietary knowledge on the ESD protections and are exchangeable between IC manufacturers and OEMs.

3.5.2.1 Diodes

Diodes are good examples to expose the philosophy. Only two parameters can be given like the triggering voltage, V_{th}, and the ON-state resistance, R_{ON}. This defines a very simple two states machine diagram (Figure 3.19). When the voltage across the diode is below V_{th}, the diode is off (State 0), no current flows into the protection. Otherwise, we are in state 1 with the equation $V = R_{ON} \cdot I$.

3.5.2.2 Snapback Devices

For snapback protection devices, the same philosophy is used and defines four states, reported on the measured I–V curves (Figure 3.20a) and following the state diagram reported in Figure 3.20b. Six couples of voltages and currents are defined as parameters. These (V_x, I_x) couples define the inflection points of the SCR and the equivalent equations for states 0, 1, 2, and 3. The model has been used in previous works for simulation of ESD stress injection (TLP and IEC 61000-4-2) [46]. Such model could suffer from convergence issues that are solved by adding in parallel to the SCR, its equivalent parasitic capacitance (often given into IBIS). It prevents the internal voltage of the SCR from a strong drop.

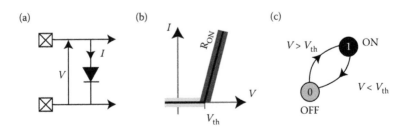

FIGURE 3.19
Schematic diagram of a diode (a), ideal I–V curve (b), and state diagram (c).

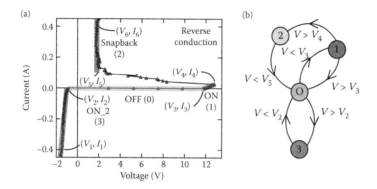

FIGURE 3.20
TLP measurement (a) and extracted state machine diagram (b).

3.5.2.3 Power Clamp

PCs are based on MOS transistors with complex triggering conditions. The proposed behavioral model is a two-state machine, defined like diodes, but the triggering condition is set on a dV/dt. A similar structure has been implemented into the study of paper [42] where the susceptibility of failure is compared with measurement data into the direct power injection (DPI) configuration [48]. The PC drives part of the current during the stress, reducing the failure level that could be predicted using only IBIS information.

3.5.3 Synthesis of IC Model

The combination of the different blocks forms the complete IC model introduced previously. While keeping the philosophy of IBIS files, this approach proposes improved IC model with full behavioral ESD protections description that preserves the intellectual property of semiconductor's manufacturers. Based on the state-machine diagram, with a minimum set of parameters like resistances and inflection points, these models allow predicting with good accuracy the ESD current path and its impact on the IC robustness. Figure 3.21 summarizes the proposed methodology and the input needed to build up the IC model. Both system designers and IC suppliers can use such models. This work is under development at the ESDA working group WG14 – "System Level ESD." This approach can be used to simulate the injection of an electrical fast transient (TLP, IEC 61000-4-2, etc.) directly on the bare die.

3.5.3.1 Passive Elements and PCB Traces

For ESD pulses with rise time close to the nanosecond, high-frequency (HF) effects of passive elements are not negligible. The use of perfect models for passive components (capacitors and inductors) results in unrealistic simulations. Indeed, the impedance measurement of a 47-nF decoupling capacitor

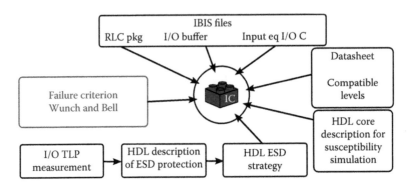

FIGURE 3.21
IC model elements.

versus frequency is moving from a capacitive behavior to an inductive one at 110 MHz. In this case, only a HF model allows visualizing the actual transient response [48]. In the same way, for the package inductance, the high-frequency inductance is nonnegligible for the ESD stress. Nevertheless, it is not necessary to use an HF model for every single passive element of the circuit. Only decoupling capacitors need to be modeled this way, apparently being the main passive element with a significant impact.

Similarly, it is really important to take into account PCB traces, even short ones (< cm), since they can introduce nonlinear overvoltages when a measurement is performed. Depending on their dimensions, PCB traces have different equivalent impedances and their equivalent model can be complex [49]. In a high current regime, using VF-TLP testing and TDR method, we showed that PCB lines can be modeled using basic LC lumped elements, calculated using analytical formulations of [46], to take into account propagation and coupling effects.

PCB traces induce a delay proportional to their length (~33 ps mm^{-1} for 110-Ω PCB lines). Moreover, on the propagation path of an ESD stress, when a PCB line separates a load from a measurement point, a residual voltage peak can be generated on the measurement.

3.5.3.2 Validation on a Simple Circuit

This modeling methodology was applied to the commercial inverter IC whose IBIS model is given in Figure 3.18. Based on static and TLP measurement on each pair of combination of pins, we were able to identify the related ESD protections. Three types were identified: diodes, an RC-triggered PC, and a SCR. Each "device-type" model is assembled in a classical protection strategy as shown in Figure 3.22. The conductions of SCR and PC are modeled by a diode for their forward-biased operation.

To validate our simulation methodology, this commercial inverter was implemented on a PCB including a 1-Ω resistor between the ground plane

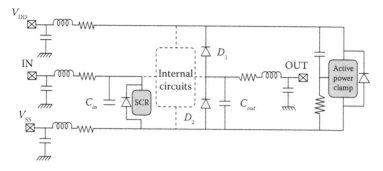

FIGURE 3.22
Full model of the commercial inverter after including the IBIS model of Figure 3.18, the ESD protections extracted from TLP measurement.

and V_{SS} for a current measurement comparison. Using the previously described methodology, we modeled the whole system by assembling the different blocks: parasitic elements of the package, ESD protection strategy and IC core, PCB lines, cables, measurement test bench, and ESD generator [43]. With this behavioral model, we studied the impact of the value of the decoupling capacitor (50 nF, 14 nF, and no capacitance) on the ESD current propagation path. Figure 3.23 presents the comparison between measurement and simulation of the resulting current in the 1-Ω resistor when a 1A TLP pulse is applied between the output of the inverter and ground. We can

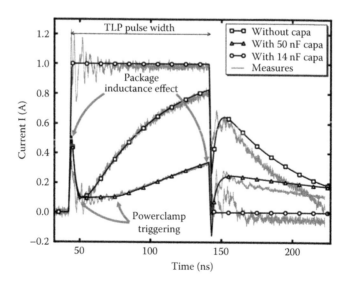

FIGURE 3.23
Measurement versus simulation comparison of the current in the 1-Ω resistor for 1 A TLP pulse injected from output to ground, with 14 nF, 50 nF, and without a decoupling capacitor.

note the good agreement between measurement and simulation, the error being smaller than 20%. Without the external decoupling capacitance, all the current flows through the ESD protection strategy network as expected. By adding an external decoupling capacitance between V_{DD} and ground of the IC, the current waveform strongly differs depending on the capacitance value, and on the parasitic elements of the system. When the TLP pulse ends, a continuous current is observed over twice longer time. Even if the energy is the same, the resulting thermal effect of this second current pulse could lead to the destruction of the chip since the time to failure is temperature dependent as described by Wunch and Bell [50].

3.5.4 Discrete Protection: Compatibility with the Device

To cope with the high current of gun stress applied to ICs, some system engineers propose to add external discrete protections or transient voltage suppressors (TVS) on the PCB. Such discrete protection is designed to clamp a high ESD current. The choice of the proper discrete protection is not trivial and depends on the knowledge of the on-chip ESD protection characteristics.

A system-efficient ESD design (SEED) approach proposes the TLP information exchange of the on-chip ESD protection characteristics between the system engineer and on-chip design engineer.

Regarding the optimization between discrete protection and on-chip ESD protection structure, there are three possible scenario cases that are described hereafter.

First, or bad scenario case means that the IC ESD protection will absorb the gun stress current instead of the discrete protection (Figure 3.24a). Low resistive voltage clamping characteristic of the IC ESD protection prevents the high voltage activation of the discrete protection. Failure will be expected at the on-chip ESD protection structure.

With resistive on-chip ESD protection structure characteristics as shown in Figure 3.24b, the discrete protection can be activated at a high current. By this way, the on-chip ESD protection is only activated at a low current level. Experiences from paper [51] show that this kind of protection works for the short-time high current of the 150 pF/330-Ω gun model and not for

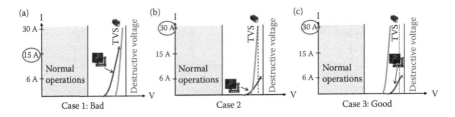

FIGURE 3.24
Various scenarios of ESD discrete protection: bad case (a), mitigated case (b), and good case (c).

the long-time discharge current of the 2 kΩ/300 pF gun model. It was demonstrated that the long-time current discharge due to this 2-kΩ resistor gun model (>1 μs) at a low current level is applied at the on-chip ESD protection structure instead of the discrete one. As a result, the on-chip ESD protection has to absorb the energy provided by the end of this 2 kΩ/330 pF gun current pulse. TLP measurement with different pulse widths from 100 ns up to 1 μs allows extracting the time to failure curves as mentioned by Reference 50 and can help the system engineer to check the robustness of the on-chip ESD protection over longer pulse durations.

Last scenario case (case 3, Figure 3.24c) implies that low resistive voltage characteristics of discrete protection provide good protection of the IC protection thanks to high triggering voltage of the on-chip ESD protection structure. This case presents the drawback for the system engineer to have additional cost due to the expensive low resistive voltage clamping of discrete protection.

3.6 Conclusion and Perspectives

ESD protection strategies should be elaborated with a global approach taking into account the final application of the IC. Indeed, this would save on the one hand, silicon footprint dedicated to integrated protections since the ESD robustness of the system is not directly related to the one of the IC and on the other hand, it would avoid costly design iterations at the system level. Such a global approach is already used for electromagnetic compatibility (EMC) qualification and we have shown that the same modeling approach should be implemented for ESD qualification. Regarding ESD, the main focus is on the development of protection strategies against various types of stresses. The drastic growing of embedded electronic systems in a large variety of applications requires a more global approach for their ESD qualification that is basically part of the EMC one. In the EMC field, propagation phenomena are prevailing and a large number of standards already exist, whereas this is not the case for the system level ESD (except IEC 61000-2-4 or ISO 10605). The idea is to directly reuse the experience and methodologies developed in the EMC field to achieve the ESD system level objectives.

References

1. AEC-Q100-REV-D. Stress Test Qualification for Integrated Circuits. Automotive Electronic Council, Component Technical Committee, August 25, 2000.
2. EIA JEDEC. JESD22-A114D: Electrostatic Discharge (ESD) Sensitivity Testing Human Body Model (HBM), 2006.

3. ANSI/ESDA/JEDEC JS-002-2014, ESDA/JEDEC Joint Standard for Electrostatic Discharge Sensitivity Testing—Charged Device Model (CDM)—Device Level, ISBN: 1-58537-276-5.

4. ANSI/ESD STM5.2-2012, ESD Association Standard Test Method for Electrostatic Discharge Sensitivity Testing—Machine Model (MM)—Component Level, ISBN: 1-58537-239-0.

5. Hyatt, H. ESD: Standards, threats and system hardness fallacies. *Electrical Overstress/Electrostatic Discharge (EOS/ESD) Symposium*, Charlotte, USA, 2002, pp. 175–182.

6. Barth, J., Richner, L., Henry, G., Kelly, M. Real HBM and MM waveform parameters. *J. Electrostat.*, 62, 195–209, 2004.

7. IEC 61000-4-2. *Electromagnetic Compatibility (EMC) Part 4—2: Testing and Measurement Techniques—Electrostatic Discharge Immunity Test.* 2.0 Edition, 2008–2012. ISBN 978-2-88910-372-0.

8. ISO 10605 Road vehicles—Test Methods for Electrical Disturbances from Electrostatic Discharge, 2.0 Edition, 2008-07-15, Reference number ISO 10605:2008(E), Published in Switzerland. www.iso.org

9. Maloney, T. J., Khurana, N. Transmission line pulsing techniques for circuit modeling of ESD phenomena. *Proceedings of the Electrical Overstress/Electrostatic Discharge (EOS/ESD) Symposium*, Minneapolis, USA, 1985, pp. 49–54.

10. Gieser, H., Haunschild, M. Very-fast transmission line pulsing of integrated structures and the charged device model. *Proceedings of the Electrical Overstress/Electrostatic Discharge Symposium*, Orlando, USA, September 10–12, 1996, pp. 85–94.

11. Zhou, Y., Hajjar, J. -J., Ellis, D. F., Olney, A. H., Liou, J. J. A new method to evaluate effectiveness of CDM ESD protection. *Proceedings of the 32nd Electrical Overstress/Electrostatic Discharge (EOS/ESD) Symposium*, Reno, USA, October 3–8, 2010, pp. 1–82010.

12. Delmas, A., Tremouilles, D., Nolhier, N., Bafleur, M., Mauran, N., Gendron, A. Accurate transient behavior measurement of high-voltage ESD protections based on a very fast transmission-line pulse system. *Proceedings of the 31st Electrical Overstress/Electrostatic Discharge (EOS/ESD) Symposium*, Anaheim, USA, August 30–September 4, 2009, pp. 1–8.

13. Delmas, A., Gendron, A., Bafleur, M., Nolhier, N., Gill, C. Transient voltage overshoots of high voltage ESD protections based on bipolar transistors in smart power technology, *Proceedings of Bipolar/BiCMOS Circuits and Technology Meeting (BCTM)*, Austin, USA, October 2010, pp. 253–25614.

14. Caignet, F., Monnereau, N., Nolhier, N. Non-invasive system level ESD current measurement using magnetic field probe. *International Electrostatic Discharge Workshop (IEW)*, Tutzing (Allemagne), May 10–13, 2010.

15. IEC TS 61967-3:2014, Integrated circuits—Measurement of electromagnetic emissions—Part 3: Measurement of radiated emissions—Surface scan method, Edition 2.0, 2014-08, International Electrotechnical Commission. ISBN 978-2-8322-1809-9.

16. Lacrampe, N., Boyer, A., Vrignon, B., Nolhier, N., Caignet, F., Bafleur, M. Investigation of the indirect effects of a VF-TLP ESD pulse injected into a printed circuit board. *Workshops of International Symposium on Electromagnetic Compatibility, Immunity at the IC level, EMC Europe*, Barcelona, Spain, September 2006, pp. 538–545.

17. Besse, P., Abouda, K., Abouda, C. Identifying electrical mechanisms respon-sible for functional failures during harsh external ESD and EMC aggressions. *Microelectron. Rel.*, 51(9–11), pp. 1597–1601, 2011.
18. Houssam, A., Bafleur, M., Tremouilles, D. Zerarka, M. Combined MOS-IGBT-SCR structure for a compact high-robustness ESD power clamp in smart power SOI technology. *IEEE Trans. Device Mat. Rel.*, 14(1), pp. 432–440, 2014.
19. Troutman, R. R. *Latchup in CMOS Technology: The Problem and Its Cure*, Vol. 13. Springer, USA, 1986.
20. Besse, P. Tenue en énergie de structures LDMOS avancées de puissance inté-gréedans les domainestemporels de la nanoseconde à la milliseconde. Doctorat, Université Paul Sabatier, Toulouse, Janvier 28, 2004, 156p.
21. Besse, P., Patrice, N., Nolhier, M., Bafleur, M., Zecri, M., Chung, Y. Investigation for a smart power and self-protected device under ESD stress through geom-etry and design considerations for automotive applications. *Proceedings of the Electrical Overstress/Electrostatic Discharge (EOS/ESD) Symposium*, Charlotte, USA, 2002, pp. 351–356.
22. Vashchenko, V. A., Sinkevitch, V. F. *Physical Limitations of Semiconductors Devices*. ISBN 978-0-387-74513-8, Springer Science + Business Media, 2008.
23. Bowers, H. C. Space-charged-induced negative resistance in avalanche diodes. *IEEE Trans. Electron Devices*, ED-15(6), 343–350, 1968.
24. Gendron, A., Salamero, C., Renaud, P., Besse, P., Bafleur, M., Nolhier, N. Area-efficient, reduced and no-snapback PNP-based ESD protection in advanced smart power technology. *Proceedings of the Electrical Overstress/Electrostatic Discharge (EOS/ESD) Symposium*, Tucson, USA, September 10–15, 2006, pp. 69–76.
25. Gendron, A. Structures de protection innovantescontre les décharges électro-statiquesdédiées aux entrées/sorties hautes tensions de technologies Smart Power, PhD thesis, University Paul Sabatier, Electrical Engineering Department, March 29, 2007.
26. Gendron, A., Renaud, P., Besse, P. Semiconductor device structure and inte-grated circuit therefor. U.S. Patent No. 2007104342, September 20, 2007.
27. Jensen, N., Groos, G., Denison, M., Kuzmik, J., Pogany, D., Gornik, E., Stecher, M. Coupled bipolar transistors as very Robust ESD protection devices for auto-motive applications. *Proceedings of the Electrical Overstress/Electrostatic Discharge (EOS/ESD) Symposium*, Las Vegas, USA, 2003, pp. 54–63.
28. Vashchenko, V., Kuznetsov, V., Hopper, P. Implementation of dual-direction SCR devices in analog CMOS process. *Proceedings of the Electrical Overstress/Electrostatic Discharge (EOS/ESD) Symposium*, Anaheim, USA, 2007, pp. 75–79.
29. Liu, Z., Liou, J., Vinson, J. Novel silicon-controlled rectifier (SCR) for high-volt-age electrostatic discharge (ESD) applications, *IEEE Electron Device Lett.*, 29(7), 2008, pp. 753–755.
30. Bourgeat, J., Jimenez, J., Dudit, S., Galy, P. New beta-matrix topology in CMOS32nm and beyond for ESD/LU improvement. *International Semiconductor Conference (CAS)*, Sinaia, Romania, October 14–16, 2013, Vol. 2, pp. 159–162.
31. Meneghesso, G., Tazzoli, A., Marino, F. A., Cordoni, M., Colombo, P. Development of a new high holding voltage SCR-based ESD protection struc-ture. *Proceedings of the International Reliability Physics Symposium*, Phoenix, USA, 2008, pp. 3–8.

32. Ko, J.-H., Kim, H.-G., and Jeon, J.-S. Gate bounded diode triggered high holding voltage SCR clamp for on-chip ESD protection in HV ICs. *Proceedings of the 35th Electrical Overstress/Electrostatic Discharge (EOS/ESD) Symposium*, Las Vegas, USA, 2013, pp. 1–8.

33. Mergens, M. P. J., Russ, C. C., Verhaege, K. G., Armer, J., Jozwiak, P. C., Mohn, R. High holding current SCRs (HHI-SCR) forESD protection and latch-up immune IC operation. *Microelectron. Rel.*, 43(7), pp. 993–1000, 2003.

34. Gendron, A., Gill, C., Zhan, C., Kaneshiro, M., Cowden, B., Hong, C., Ida, R., Nguyen, D. New high voltage ESD protection devices based on bipolar transistors for automotive applications. *Proceedings of the 33rd Electrical Overstress/ Electrostatic Discharge (EOS/ESD) Symposium*, Anaheim, USA, 2011, pp. 1–10.

35. Laine, J.-P., Salles, A., Besse, P., Delmas, A. Impact of snapback behavior on system level ESD performance with single and double stack of bipolar ESD structures. *Proceedings of the 34th Electrical Overstress/Electrostatic Discharge Symposium (EOS/ESD)*, Tucson, USA, 2012, pp. 1–5.

36. Bourgeat, J., Entringer, C., Galy, P., Fonteneau, P. Bafleur, M. Local ESD protection structure based on silicon-controlled rectifier achieving very low overshoot voltage. *Proceedings of the 31st Electrical Overstress/Electrostatic Discharge (EOS/ESD) Symposium*, Anaheim, USA, 2009, pp. 314–321.

37. Mergens, M. P. J., Mayerhofer, M. T., Willemen, J. A., Stecher, M. ESD protection considerations in advanced high-voltage technologies for automotive. *Proceedings of the Electrical Overstress/Electrostatic Discharge (EOS/ESD) Symposium*, Tucson, USA, 2006, pp. 54–63.

38. Smedes, T., Van Zwol, J., De Raad, G., Brodbeck, T., Wolf, H. Relations between system level ESD and (vf-) TLP. *Proceedings of the Electrical Overstress/Electrostatic Discharge (EOS/ESD) Symposium*, Tucson, USA, 2006, pp. 136–143.

39. Industry Council on ESD Target Levels. White Paper 3: System Level ESD. Part I: Common Misconceptions and Recommended Basic Approaches, December 06, 2010.

40. Camp, M. Garbe, H., Nitsch, D. Influence of the technology on the destruction effects of semiconductors by impact of EMP and UWB pulses. *IEEE Symposium on Electromagnetic Compatibility*, August 19–23, 2002, Vol. 1, pp. 87–92.

41. IBIS (Input Output Buffer Information Specification). ANSI/EIA-656B, www.eigroup.org/IBIS.

42. Industry Council on ESD Target Levels. White Paper 3: System Level ESD. Part II: Implementation of Effective ESD Robust Designs., October 08, 2012.

43. Monnereau, N., Caignet, F., Trémouilles, D., Nolhier, N., Bafleur, M. Building-up of system level ESD modeling: Impact of a decoupling capacitance on ESD propagation. *Microelectron. Rel.*, 53(2), 2013, pp. 221–228.

44. Pavier, M., Woodworth, A., Sawle, A., Monteiro, R., Blake, C., Chiu, J. Understanding the effect of power MOSFET package parasitics on VRM circuit efficiency at frequencies above 1 Mhz. *Proceedings of the PCIM Europe Conference*, Nuremberg, Germany, 2003, pp. 279–284.

45. Monnereau, N., Caignet, F., Nolhier, N., Tremouilles, D. Behavioral modeling methodology to predict ESD susceptibility failures at system level: An IBIS improvement. *IEEE Symposium on Electromagnetic Compatibility (EMC), Europe*, York, UK, 2011, pp. 457–463.

46. Besse, P., Lafon, F., Monnereau, N., Caignet, F., Laine, J.P., Salles, A., Rigour, S., Bafleur, M., Nolhier, N., Trémouilles, D. ESD system level characterization and modeling methods applied to a LIN transceiver. *Proceedings of the Electrical Overstress/Electrostatic Discharge (EOS/ESD) Symposium*, Anaheim, USA, 2011, pp. 329–337.

47. IEC 62132-4:2006. *Integrated Circuits—Measurement of Electromagnetic Immunity 150 kHz to 1 GHz—Part 4: Direct RF Power Injection Method*, 1st edition, 2006, Reference number CEI/IEC 62132-4:2006, www.iec.ch.

48. Bèges, R., Caignet, F., Bafleur, M., Nolhier, N., Durier, A., Marot, C. Practical transient system-level ESD modeling—Environment contribution. *Proceedings of the Electrical Overstress/Electrostatic Discharge (EOS/ESD) Symposium*, Tucson, USA, September 7–12, pp. 1–10, 2014.

49. Bakoglu, H. B. *Circuits, Interconnections and Packaging for VLSI*. Addison-Wesley, Reading, MA, 1990, ISBN-10: 0201060086.

50. Wunsch, D. C., Bell, R. R. Determination of threshold voltage levels of semiconductor diodes and transistors due to pulsed voltages. *IEEE Trans. Nucl. Sci.*, NS-15(6), pp. 244–259, 1968.

51. Laine, J. -P., Besse, P., Salles, A. System efficient ESD design (SEED) including 2 kΩ/330 pF RC Gun Module. International ESD Workshop (IEW), Villard de Lans (France), May 19–22, 2014.

4

Variability and Reliability Issues in Mixed-Signal Circuits

Georgios D. Panagopoulos

CONTENTS

4.1 Introduction

In the last half-century, we have become increasingly dependent on the plethora of electronic applications to improve the quality of life. Examples of such electronic devices include computers, laptops, digital cameras, and mobile phones. When we are interested in purchasing such a device, we are usually interested in the performance and lifetime of the product. Thus, as an end user, we ask questions such as, how much it cost, what resolution does the display provide, how long does the battery last between recharges, and for how long our device would perform functionally without observing any degradation or malfunction. The last two questions are tightly related with the device and circuit degradation which lead to product degradation and malfunction. On the other hand, for manufacturers it is essential to know the useful life of their products since this will determine the warranty. Apart from that the manufacturers also need to know the yield of their product, namely how many parts/dies of their production line/wafer fail to operate reliably. Answers to these questions are essential since they are related to the cost of a product. Hence, it is obvious that variability and reliability issues and their analysis from device up to system level are crucial during design phase of a commercial product.

Another factor that influences the yield of a product arrives from the continuous demand for high-performance, low-power, and low-cost electronic applications, which drives the transistor scaling to sub-50 nm regime [1–4]. In the past few decades, aggressive device scaling has achieved these goals providing higher integration density and improved performance of integrated circuits (ICs). However, these advantages have been achieved at the cost of severe variability and reliability issues [4–7]. These issues in scaled technologies are attributed to (a) the high electric fields causing hot carrier injection (HCI), (b) the small number of dopants in the channel region leading to a phenomenon known as random dopant fluctuations (RDF), (c) the countable number of traps at the silicon/oxide interface, resulting in bias temperature instability (BTI), (d) the oxide thickness scaling, causing a phenomenon called time-dependent dielectric breakdown (TDDB), and (e) the thin metal lines that cause electromigration (EM). A categorization of these variations as spatial and temporal is presented in Figure 4.1. Unfortunately, all these variations at technology and device levels affect the nominal operation of digital, analog, mixed signal, and RF circuits. This is due to the wide distributions in delay, "on" and "off" (leakage) current, noise immunity, transconductance, gain, and slew rate [4]. Hence, as we approach the physical limits of device scaling, it is essential and imperative to develop accurate

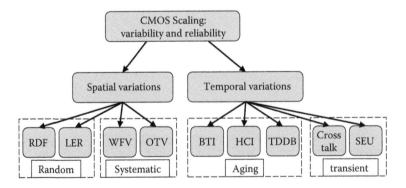

FIGURE 4.1
CMOS device scaling leads to spatial and temporal variations. Spatial variations are categorized as random and systematic while the temporal as aging and temporal.

models for the design and simulation of nanoscaled circuits which are able to account for the above variability and reliability effects. In this chapter, we concentrate on modeling the variability and aging effects of CMOS technologies for circuit level simulations.

To cover these topics both from device and circuit abstraction level, we have split this chapter into two parts. In the first part of this chapter, we have concentrated on the development of a framework/subcircuit which models the most known variability and reliability issues of CMOS transistors. Using this subcircuit, we perform simulations on traditional digital and mixed signal circuits showing how variability and reliability affect their circuit performance figure of merits. These results are presented and discussed in detail in the second part of this chapter. This framework/subcircuit is the "bridge" between the device and circuit levels. The framework refers to the device level while the subcircuit refers to the circuit level.

4.2 Technology and Device Level

Transistor scaling is the most efficient way to continuously follow Moore's law. Unfortunately, device scaling leads to limitations that can be overcame by the introduction of new device structures and materials. However, these new techniques increase spatial and temporal variations which are shown in Figure 4.1. Examples of spatial variations include RDF, LER (line edge roughness), WFV (work-function fluctuations), and OTV (oxide thickness fluctuations), and examples of temporal variations are BTI, HCI, TDDB, and SEU (single event upset).

In this part, we present the framework which captures the most important variability and reliability issues that appear at the technology and device level and are encapsulated in the proposed subcircuit, shown in Figure 4.2, to simulate their influence at the circuit level. This framework has been constructed by mainly physics-based models and can accurately predict the parametric variations due to RDF, BTI, and TDDB (Figure 4.3). The device structure and the geometry of each device, such as the width (W), length (L), and the oxide thickness (T_{ox}), constitute the main size-related input parameters for the proposed tool. The bias voltage and temperature conditions are operation-related input parameters and the channel doping profile is an input parameter as well. Using the developed tool, we are able to estimate the effects of variations and aging on circuits. Specifically, the proposed framework provides the threshold voltage (V_{th}) and gate leakage (I_G) distributions which can be used directly at the circuit level. Note that V_{th} and I_G variations at the device level strongly affects the nominal operation at the circuit level [8,9]. The correlation between RDF and BTI as well as BTI and TDDB can

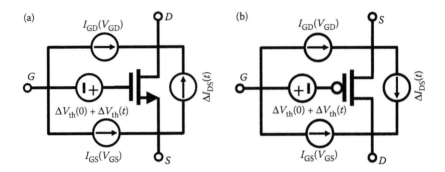

FIGURE 4.2
(a) NMOS and (b) PMOS nominal transistors wrapped up with voltage and current sources which are able to emulate the spatial and temporal effects. Spatial effects are modeled by $\Delta V_{th}(0)$ parameters and temporal by $\Delta V_{th}(t)$, I_{GS}, I_{GD}, and $\Delta V_{DS}(t)$. Statistical data for these sources are generated by the framework shown in Figure 4.3.

also be incorporated in this framework. However, in this chapter we will not describe these phenomena but a detailed description can be found in [11].

4.2.1 Random Dopant Fluctuations (RDF)

The increasingly transistor miniaturization has led to random variations in the transistor threshold voltage [9–20]. This is mainly due to random

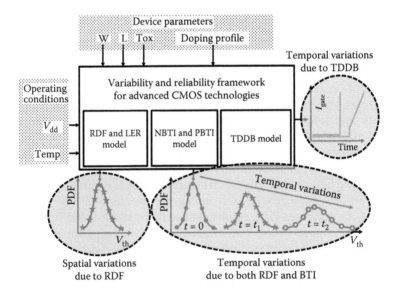

FIGURE 4.3
Flowchart of the proposed framework. Circuit level operating conditions and device level parameters are inputs. Variability and reliability data is generated to be used by the subcircuits shown in Figure 4.2.

phenomena such as RDFs in the device channel region. In scaled technology nodes fluctuations in the source and drain regions also contribute to V_{th} variations. Note that this variability is caused by both the number and location of dopant atoms [5,11–13,21–37]. Hence, to capture the effects of RDF we need to develop 3D models that take into account the location of the individual dopant atoms. The random location of dopant atoms (random profile) affects both the electrostatic potential and the electric field at the Si–oxide interface of the transistor and these in turn affect the threshold voltage (V_{th}). Hence, the proposed framework provides statistics for V_{th} due to RDF. In addition, this framework offers information and statistics for quantities such as the total number of dopants, the electrostatic potential, and the electric field which might not needed immediately in the circuit level but gives information for the statistics of the simulated device.

In the next paragraphs, we present key derived parameters such as depletion width, number of dopants, electrostatic potential, and electric field as a function of the input parameters, that is, the width W and the length L, the channel doping profile and oxide thickness T_{ox}.

Depletion Width. The maximum depletion width W_{dm} of a transistor is expected to vary spatially due to nonuniform doping profile. However, we can approximate W_{dm}, by the following expression that is used for 1D uniform channel region [38–40]:

$$W_{dm} = \sqrt{4\varepsilon_{Si}kT \ln \frac{N_A}{n_i} q^2 N_A}$$

(4.1)

where q is the electronic charge, ε_{Si} is the permittivity of silicon, k is the Boltzmann constant, T is the temperature, N_A is the doping concentration, and n_i is the intrinsic carrier density. This approximation is valid because the contribution of the ionized dopants away from the surface and close to the edge of the depletion region, which is not as significant in comparison to those which are closer to the oxide interface. Also, on average, the dopants that are ionized and outside the depletion region are canceled out by those that are neutral and inside the depletion region. This approximation is essential to eliminate the otherwise needed iterations for the exact calculation of the spatially varying depletion width speeding up the computational time of the model. Note that it is essential to reduce the computational time at the device level to get fast statistical data.

Channel region. The channel region is discretized (meshed), and at each vertex of the generated mesh, we assign an identification number i and a binary random variable X_i to denote the existence or absence of a dopant, with i running from 1 to N, where N is the total number of possible positions that a dopant can be placed in. Thus we assign the value "1"/"0" to the random variable X_i to denote the existence/absence of a dopant at the corresponding vertex. All the random variables X_i follow the Bernoulli distribution, and they are defined as [43]

$$X_i = \begin{cases} 0, & \text{dopant existence} \\ 1, & \text{dopant extinction} \end{cases} \qquad (4.2)$$

For example, if the channel is designed to follow a uniform doping profile then the associated probabilities are defined as $P[X_i = 1] = s, P[X_i = 1] = 1 - s = t$, and their corresponding first and second statistical moments can be obtained as $E[X_i] = s$ and $\text{Var}[X_i] = s(1 - s) = s \cdot t$, respectively. If the channel follows a nonuniform doping profile (this is the case for scaled devices to control the average V_{th} and reduce the short-channel effects) then the probabilities s_i for each random variable X_i have to be assigned properly. Such a doping profile can be step, retrograde, Gaussian, and Halo [40]. For instance, if a Gaussian doping profile is assumed, then the concentration and the corresponding probabilities s_i are given by the following expression:

$$s_i = \frac{a^3 N_0}{\sqrt{2\pi}\sigma} e^{-(z_i - z_c^2)/(2\sigma^2)} \qquad (4.3)$$

where z_c is the center of the Gaussian doping profile, σ is its deviation from the center, and a is the semiconductor lattice constant.

To understand the concept of our framework, let us consider an example of two different transistor instances (Figure 4.4a and d) at the same technology node and minimum size ($W = L = 22$ nm), and let us assume uniform doping concentration ($N_A = 10^{19}$ cm^{-3}). These channel region instances are generated by performing two iterations, one for each instance. In addition, we define a vector $\{X\}$ of the aforementioned random variables X_i: $\{X\} = \{X_1...X_N\}^T$. We will use this vector in subsequent sections to formulate the problem. Having discussed the tools that are required to build a 3D model for RDF, various derived random variables are defined in the next sections, which are necessary to build our model. These variables are categorized as quantitative variables (e.g., number of dopants and effective doping concentration) and electrical variables (e.g., surface potential, electric field at the surface, and threshold voltage).

Number of Dopants D_p. The total number of ionized dopants in the channel region is obtained by the sum of X_i over all the N vertices of the generated mesh.

$$D_p = \sum_{i=1}^{N} X_i \qquad (4.4)$$

Doping concentration C_p. The "effective" doping concentration can be defined as the ratio of the total number of dopants in the depleted channel region to the volume of this region. C_p is also a random variable related to the

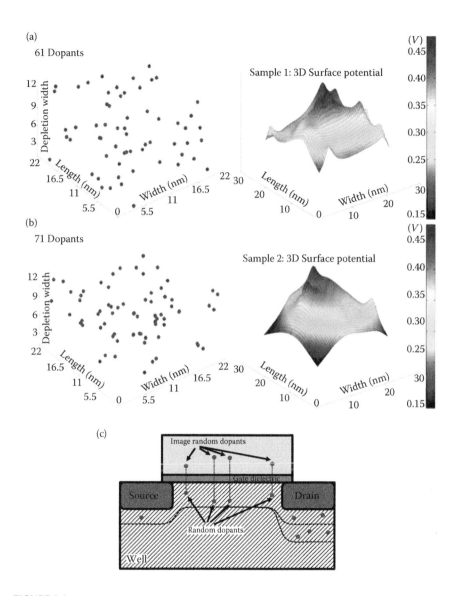

FIGURE 4.4
(a,b) Two random patterns from the channel region of transistors of the same technology node (22 nm) showing the random number and position of dopants. The corresponding 3D potential at the surface are also illustrated. (c) Vertical cross section illustrating real and image random dopants used for modeling RDF. The depletion width is modulated by V_{ds}.

random variable D_p by the following expression:

$$C_p = \frac{D_p}{WLW_{dm}} = \frac{1}{N\alpha^3}\sum_{i=1}^{N}X_i \qquad (4.5)$$

Surface potential Φ. The surface potential as well as the electric field are determined by Coulomb's law [17,18,41,42]:

$$E = Q_i/(4\pi\varepsilon_{Si}R_i^2) \text{ and } \Phi = Q_i/(4\pi\varepsilon_{Si}R_i)$$

where Q_i is the charge of the ith dopant and R_i is the distance of the ith dopant to the Si–oxide interface. By applying the preceding two fundamental relations, analytical expressions for the electrical properties of the transistor are derived in (4.6) through (4.8). In addition, for the derivation of analytical relations we use the "delta depletion approximation" [38–40] (no mobile carrier in the channel region for gate voltages is less than V_{th}). Both the electric field and surface potential are essential for the definition of threshold voltage. Note that due to the random position of the dopants, both the electric field and the electrostatic potential at the Si–oxide interface are spatially non-uniform. These quantities are also expected to be spatially correlated, that is, they are dependent on their corresponding values at the neighboring points (Figure 4.4c). Let us now define the electrical variables beginning with the surface potential. Using the superposition principle the surface potential is given by [16,40]

$$\Psi(x,y) = \Phi(x,y) + U_D + U_S + U_B \tag{4.6}$$

where Φ is the contribution to the total potential due to random dopants, and U_D, U_S, U_B are solutions that satisfy the boundary conditions at the drain, source, and body electrodes, respectively, and they are bias dependent. The first term Φ at a point (x,y) at the Si–oxide interface is calculated by the sum of the potential contribution of each dopant in the depleted channel region

$$\Phi(x,y) = \sum_{i=1}^{N} [\Phi_i(x,y) + \Phi'_i(x,y)] \tag{4.7}$$

where $\Phi_i(x,y)$ is the contribution of the ith dopant to the surface potential at position (x,y), and $\Phi'_i(x,y)$ is the contribution of the corresponding image charge, which resulted from the method of images [41,42]

$$\Phi_i(x,y) = \frac{-qX_i}{4\pi\varepsilon_{Si}R_i(x,y)} \qquad \Phi'_i(x,y) = \frac{-qX_i}{4\pi\varepsilon_{Si}\Phi'_i(x,y)} \tag{4.8}$$

where R_i is the distance from a dopant i to a point (x,y) on the surface, and R'_i is the corresponding distance of the image charge (Figure 4.4c). The rest of the terms of Equation 4.6 are given by solving the Poisson equation with applied voltages at the contacts as boundary conditions

$$U_D = f_1(\Phi(x,y))\sin h\left(\frac{\pi(L-x)}{W_{dm}+3T_{ox}}\right)\Bigg/\sin h\left(\frac{\pi L}{W_{dm}+3T_{ox}}\right)$$

$$U_S = f_2(\Phi(x,y))\sin h\left(\frac{\pi(L-x)}{W_{dm}+3T_{ox}}\right)\Bigg/\sin h\left(\frac{\pi L}{W_{dm}+3T_{ox}}\right)$$
(4.9)

U_B is usually an order of magnitude smaller than U_S and U_D and has been ignored in this analysis (Figure 4.5) [40].

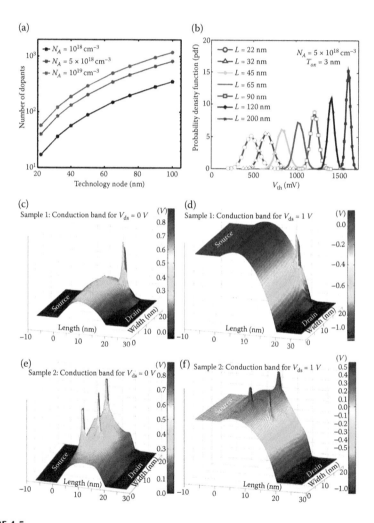

FIGURE 4.5
(a) Statistics of the random variables "number of dopants" in the channel region. (b) Threshold voltage distributions obtained changing the model parameters length and width ($W = L = [22 - 200 \text{ nm}]$). (c)–(f) Three-dimensional band diagram for low and high V_{th} of two samples of 20-nm transistors.

Electric field EZ. The electric field is a vector and consists of three components (x, y, and z directions). Even though all these components can be calculated from the proposed model, the most important for the determination of the threshold voltage is the component vertical to the Si–oxide interface (z-component). Hence, the electric field is given by the following equation:

$$E(x,y) = \sum_{i=1}^{N}\left[E_i(x,y) + E_i'(x,y)\right] = \sum_{i=1}^{N}\left[\frac{qX_i}{4\pi\varepsilon_{Si}R_i^2} + \frac{qX_i}{4\pi\varepsilon_{Si}R_i'^2}\right] \qquad (4.10)$$

Threshold voltage V_{th}. In the final step of the proposed framework, we define the threshold voltage V_{th}. The definition of the threshold voltage is not unique; it can be extracted from current–voltage characteristics [2,22], or it can be computed using percolation paths, which connect the source and the drain terminals [2,15].

For the determination of V_{th}, we need to define the random variable $V_t(x,y)$, which represents the applied gate voltage that inverts the channel locally. This local V_t can be defined as the voltage drop across the bulk together with the boundary conditions plus the voltage drop across the oxide:

$$V_t = \Phi(x,y) + U + \frac{\varepsilon_{Si}E_z(x,y)}{C_{ox}} \qquad (4.11)$$

where $C_{ox} = \varepsilon_{ox}/T_{ox}$. Using now the definition of V_t the threshold voltage can be calculated either from the maximum of the V_t for each dy segment or from the average of V_t for all the dy segments. Such formulation can be described mathematically as

$$V_{th} = \frac{1}{W}\int_0^W \max_{x=(0,\ldots,L)}\left(\Psi + \frac{\varepsilon_{Si}E_z(x,y)}{C_{ox}}\right)dy \qquad (4.12)$$

and

$$V_{th,lin} = \frac{1}{WL}\int_0^W\int_0^L\left(\Psi + \frac{\varepsilon_{Si}E_z(x,y)}{C_{ox}}\right)dxdy \qquad (4.13)$$

The later definition is more suitable when V_{ds} is small, that is, when linear threshold voltage is of interest. To conclude this subsection, we should mention that the proposed framework is not limited by the definition of V_{th}. If an analytical solution is used then the simple well-known expression for σ_{Vth} can be used by the proposed framework,

$$\sigma_{Vth} = \frac{qT_{ox}}{\varepsilon_{ox}} \sqrt{\frac{N_A W_{dm}}{4L_{eff} W_{eff}}} \tag{4.14}$$

4.2.2 Bias Temperature Instability (BTI)

Bias temperature instability (BTI) is one of the most severe reliability issues of modern digital, analog, and RF circuits. In modern CMOS technologies it appears in both PMOS and NMOS transistors and it is called NBTI (negative bias temperature instability) and PBTI (positive bias temperature instability), respectively.

Several models exist that describe and predict the physical mechanism behind BTI. Among them, reaction–diffusion (RD) model [47–49] as well as many other RD alternates [50–53], dispersive and hole trapping models [54–57], are the most widely accepted ones. In this book, we describe the RD model as the key model without limiting the applicability of other physical or empirical models. Note that for circuit level simulations we use the exponential with the time nature of BTI and usually the other parameters of the model are extracted from the experimental data. On the other hand, proper physical modelling is mandatory if we obtain measurement data from stressed devices and we need to extrapolate them at nominal conditions.

According to RD model, BTI is a chemical reaction that takes place near or at the Si–oxide interface and generates H or H_2 atoms when vertical electric field is applied. The generated hydrogen atoms diffuse into the oxide, and their production is terminated when all the Si–H bonds at the interface have been broken. When no vertical electrical field is applied then the hydrogen atoms diffuse back toward the Si–oxide interface and recombine with the broken Si bonds. The generation of charges at the Si–oxide interface during circuit operation results in V_{th}-shift. Figure 4.8a–d shows NMOS and PMOS transistor cross sections and the BTI mechanisms that take place. The differential equations that describe these two phenomena (reaction and diffusion) are expressed by

$$\partial N_{IT}/\partial t = k_f (N_0 - N_{IT}) - k_r Y_{IT} N_{IT} (\partial H(\mathbf{r},t)/\partial t) + \delta \mathbf{j}(\mathbf{r},t) = 0$$

These equations can be extended to the stochastic differential equations (SDE) and their solutions will provide more statistical information due to random motion of few H atoms in the oxide [58]. Finally, the self-consistent solution of these two differential equations leads to an exponential time dependence of V_{th} known as V_{th} shift

$$\Delta V_{th} = \alpha e^{\Delta H/kT} E_G^\beta t^n \tag{4.15}$$

where ΔV_{th} is the threshold voltage shift, n is the time exponent, E_G is the vertical applied field, ΔH is Arrhenius activation energy, k is the

Boltzmann constant, and T the temperature. α and β are empirical fitting parameters. From the above equation it is obvious that the vertical electric field (E_G) enables BTI, and as such this kind of degradation occurs even when a circuit operates in the static mode (e.g., SRAM operating in the hold mode). As we approach the physical limits of device scaling due to higher electric fields, BTI becomes more severe and thus it is essential and imperative to simulate sensitive circuits to ensure that they will operate not only when the transistors are fresh ($t = 0$) but also during a specified lifetime. It is important to note here that BTI degradation exhibits logarithmic dependence on time. In Figure 4.6, we observe how the average V_{th} of NMOS and PMOS devices changes with time for different applied voltages and temperature conditions. The BTI model provides the V_{th} shift depending on the operating voltage of each transistor and the temperature of the circuit/chip.

4.2.3 Hot Carrier Injection (HCI)

Hot carrier injection (HCI) degradation appears when the MOSFET is biased in strong inversion with large lateral (drain–source) electric field. At these conditions, some carriers achieve high velocity and kinetic energy (hot carriers) which accelerate themselves at the region near the drain junction. Hence, under the influence of high lateral electric fields these carriers collide with other atoms and carriers, and lead to additional carrier generation due to impact ionization (Figure 4.8e) [59]. If there are carriers that have kinetic energy larger than the silicon-insulator barrier height, then these carries are injected into the gate dielectric creating traps at the gate dielectric interface and the dielectric bulk. Some other carriers can flow through the oxide contributing on the gate leakage or flow to the substrate contact contributing on the substrate current. Phonons can also be generated which propagate into the device creating electron–hole pairs. As a result, each of these phenomena affects key transistor parameters. The accumulation of traps in the oxide and at the interface results in V_{th} increase, mobility degradation, and subthreshold slope reduction. In addition, carrier flow to the substrate results in substrate voltage drop which might be significant depending on the substrate doping level. This voltage drop might turn-on the source–substrate diode, cause additional impact ionization, and lead to snapback breakdown.

Since HCI is an electric field driven phenomenon, devices with the minimum length face more prominent degradation. The worst conditions occur when the MOSFET is on with $V_G < V_D$ and V_{DS} is high. According to the description given above, the HCI degrades both the drain current and the threshold voltage. The percentage change of the drain current is given by

$$\Delta I_{DS} = \alpha e^{V_{DS}/v0} t^n \tag{4.16}$$

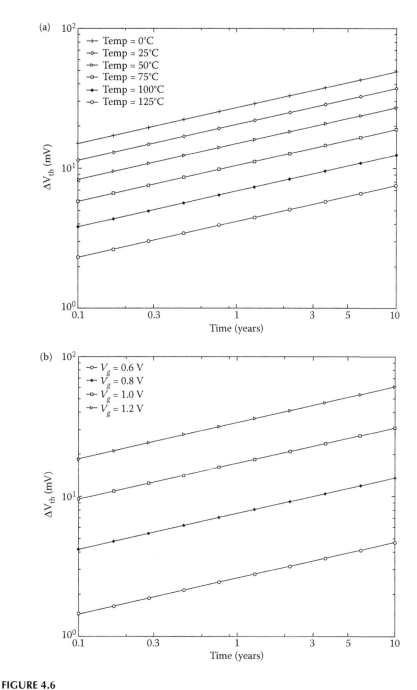

FIGURE 4.6
(a), (b) PMOS threshold voltage shift due to NBTI versus time as a function of temperature and gate voltage. (c), (d) NMOS threshold voltage shift due to PBTI versus time as a function of temperature and gate voltage. *(Continued)*

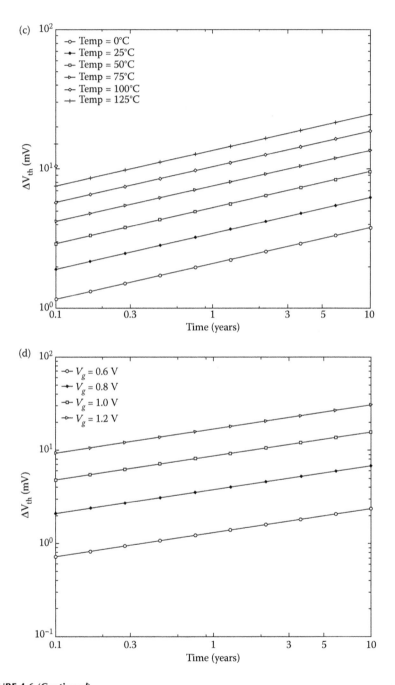

FIGURE 4.6 (*Continued*)
(a), (b) PMOS threshold voltage shift due to NBTI versus time as a function of temperature and gate voltage. (c), (d) NMOS threshold voltage shift due to PBTI versus time as a function of temperature and gate voltage.

where V_{DS} is the applied drain–source voltage and α, $v0$, and n are fitting parameters. Furthermore, ΔI_{DS} follows exponential law with time (BTI follows also exponential law). Note that the exponential parameter n for NBTI, PBTI, and HCI has different values that depend on the particular technology process. Therefore, all degradation phenomena might occur at the same time, but the dominant one depends on the quality of the process which is captured from the aforementioned model parameters.

To capture this effect in the circuit level, we add a dependent current source in parallel to the intrinsic MOSFET device and in opposite direction of the drain–source current. The V_{th}-shift is embodied as one more voltage source at the gate terminal. This voltage source will be in series with the voltage source which appears due to BTI (Figure 4.2).

4.2.4 Time Dependent Dielectric Breakdown (TDDB)

As we already have discussed in the previous sections, with technology, scaling the gate-oxide (T_{ox}) becomes thinner, and thus TDDB is one of the key reliability issues in CMOS technologies. TDDB generates traps within the oxide, resulting in breakdown (BD) gate current ($I_{G,BD}$) through the oxide film and generation of a percolation path. TDDB in ultrathin oxides ($T_{ox} < 2$ nm) consists of soft BD (SBD) and hard BD (HBD). Since SBD is characterized by a much smaller change of $I_{G,BD}$ compared to HBD, SBD may not result in circuit failure [60–62]. However, at the circuit level, the increase in $I_{G,BD}$ due to TDDB may result in degradation of the circuit parameters such as power consumption, increase in quiescent current (I_{DDQ}), changes in the oscillation frequency of ring oscillators, and degradation in read/write FOMs of SRAMs. It is worth pointing out that proper analysis of TDDB for ultrathin oxides would require not only the information about the increase in $I_{G,BD}$ due to SBD or HBD [63–67], but also the time required to generate a percolation path (t_{BD}, usually called as time to breakdown) statistically and the location of the path.

In the previous two subsections, we discussed how aggressive transistor scaling has led to random variations in the transistor threshold voltage (RDF effect) and bias temperature instabilities (BTI effect). In this section, we concentrate on the effect of oxide scaling on transistor reliability. This is mainly due to generation of percolative conduction paths between the gate electrode and the channel region of a transistor [67]. It has been shown that TDDB (together with BTI) are the most severe reliability-related phenomena leading to increased gate leakage current [68–73]. The time needed (t_{FBD}) to generate a path and observe a larger gate current (I_G) depends on (a) applied gate voltage, (b) temperature, (c) oxide thickness and gate oxide area (A_{ox}). In this section, we present the modeling of TDDB incorporating not only the time to failure of a transistor but also the increase gate current during stress conditions and the location of the path in the channel under a common framework. Finally, the TDDB model along with the RDF and NBTI model is embodied together into a common framework (Figure 4.3). This framework forms the basis of the proposed tool which can be used to perform variability and reliability simulations at the circuit level.

The proposed framework determines time to breakdown (t_{BD}) by applying a percolation model [21,70]. As shown in Figure 4.3, simulation starts giving at the input the geometrical parameters and stress conditions which are necessary for determining the statistics of t_{BD} and the postbreakdown gate currents ($I_{G,BD}$). In the next step, a defect is generated at a specific location in the oxide using the anode hole injection (AHI) model [71]. The increase in the number of defects follows a time-dependent power law [72]

$$D(t) = \frac{T_{BD}^{-a}}{\Delta t}\left(\frac{WL}{a_0^2}\right)^{a_0/T_{ox}} t^{\alpha} \qquad (4.17)$$

where a is the trap generation power-law exponent, a_0 is the diameter of the defects, WL is the oxide area, and T_{BD} is given by

$$T_{BD} = \frac{qT_{ox}N_{BD}(W,L,T_{ox})}{J_e(V_G,T_{ox})\gamma(V_G)} \qquad (4.18)$$

Here N_{BD} is the critical breakdown defect density (depends on W, L, and T_{ox}), J_e is the current density (depends on V_G and T_{ox}), and γ is the defect generation efficiency and in general is measured experimentally (depends only on V_G). The only unknown from the above equations is N_{BD} that is obtained by

$$N_{BD} = \frac{T_{ox}}{a_0^3}\exp\left[-\frac{a_0}{T_{ox}}\log\left(\frac{WL}{a_0^2}\right)\right] \qquad (4.19)$$

Note that we have connected the bias conditions with defect generation probability given by Equations 4.17 through 4.19. The generated defects would either overlap with existing defects (existing cluster), overlap with no other defects (new cluster), or overlap with more than one defect (merge cluster). If any of the existing clusters connects the two sides of oxide (a BD path is generated), we determine t_{BD}, the location of BD (x_{BD}), and the envelope of this path in the channel. Having the envelope of the BD path, we are able to determine the potential barrier height. The potential barrier height can be used to determine $I_{G,BD}$ using the quantum point contact (QPC) model [73]. In the proposed framework, $I_{G,BD}$, t_{BD}, and t_{BD} are final outputs, which can be fed to the subcircuit shown in Figure 4.3.

To model the post-BD gate current (i.e., $I_{G,BD}$), we need to know the local change in the band diagram of the insulator due to BD. Figure 4.7 shows the

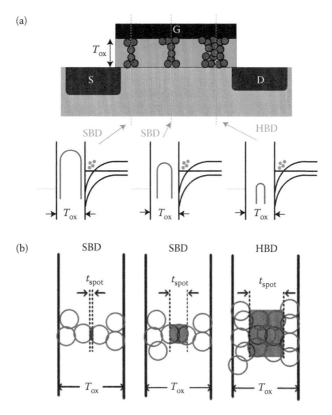

FIGURE 4.7
(a) The shape of BD determines the shape of the potential barrier which is responsible for the I–V curve. (b) The shape of the BD path given by the percolation model. The corresponding rectangular area is used to determine the potential barrier.

thickness and the height of the potential barrier for various post-BD shapes. Depending on the height and the width of the potential barrier, the BD can be classified as soft or hard. With the increase in the width of percolation path, the height and the width of the potential barrier which can become lower and smaller, resulting in larger $I_{G,BD}$. Once percolation path is formed, we can first calculate the minimum constriction thickness of the BD path (t_{spot}). As shown in Figure 4.7, the dark rectangular area marks the constriction using the BD path from our percolation model,

$$t_{spot} = \min W_{BD}(x) \qquad (4.20)$$

$W_{BD}(x)$ is the thickness of BD across the oxide. Also the minimum thickness of the BD path

$$\Phi_{spot} = \frac{\pi^2 \hbar^2}{2qm^* t_{spot}^2} \tag{4.21}$$

where q is electron's charge, m^* is the effective mass of electron in the oxide, and $\hbar$ is Planck's constant. Note that the expression for Φ_{spot} in (4.21) is similar to the expression describing particles confined in a potential well. We have assumed that the potential barrier has a quadratic form. Therefore, using the quadratic form, we can readily derive analytical solutions for $I_{G,BD}$ [67] as

$$I_{G,BD} = \frac{2q}{h} \left\{ q(V_{GS} - V_0) + \frac{1}{a} \ln\left[\frac{1 + \exp\{a[\Phi_{spot} - \beta q(V_{GS} - V_0)]\}}{1 + \exp\{a[\Phi_{spot} + (1 - \beta)q(V_{GS} - V_0)]\}} \right] \right\} \tag{4.22}$$

where $a = \xi \pi t_{spot} / \hbar \sqrt{m_z / 2\Phi_{spot}}$, V_0 is the voltage at which we have calibrated our model, and β is the fraction of V_{ox} (the voltage drop cross the oxide) that drops on the source side of the constriction, ξ is the fitting parameter, and m_z is the mass of the particles in the oxide. The typical values of such parameters can be obtained from experimental results. Figure 4.8f shows device-level cross section with a conductive path. The $I_{G, BD}$ from (4.22) is split into two bi-directional current sources – gate to source current (I_{GS}) and gate to drain current (I_{GD}). Both these currents depend on the BD location that is given by

$$I_{GS}(V_{GS}) = \frac{L - x_{BD}}{L} I_{G,BD}$$
$$\tag{4.23}$$
$$I_{GS}(V_{GS}) = \frac{x_{BD}}{L} I_{G,BD}$$

Note that x_{BD} is a random variable since the location of the BD spot depends on the random generation of defects in the oxide. It is important to note here that the proposed framework is able to handle both TDDB and BTI effects simultaneously at the circuit level.

4.2.5 Electromigration

Electromigration (EM) is the gradual displacement of metal atoms in semiconductor devices. It occurs when the current density is high enough to cause the drift of metal ions in the direction of the electron flow, and is characterized by the ion flux density. This current density depends on the magnitude of forces that tend to hold the ions in place, that is, the nature of the conductor, crystal size, interface and grain-boundary chemistry, and the magnitude of forces that tend to dislodge them, including the current density, temperature, and mechanical stresses.

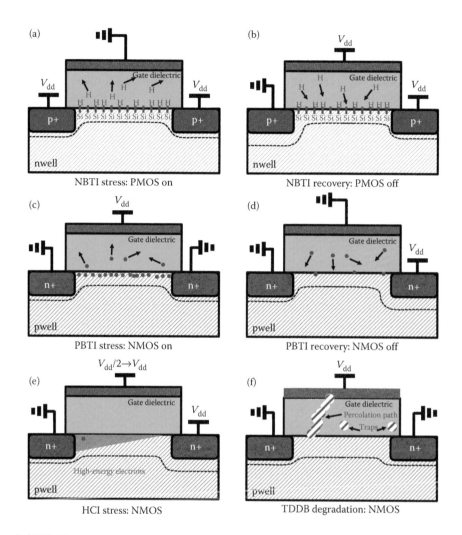

FIGURE 4.8
Transistor cross sections showing the physical mechanisms, the carries that take place and the bias conditions that enable CMOS transistor degradation. (a) PMOS in "on" state results in NBTI stress mode and (b) in "off" state results in NBTI recovery, (c) and (d) show the stress and recovery mode of a NMOS transistor, (e) HCI, and (f) TDDB stress mode.

4.2.6 Overview of Aging Mechanisms

To summarize the first part of this chapter we present an overview with all the physical mechanisms that cause device aging depicted in Figure 4.8. In these cross sections, we present both the bias conditions that trigger each mechanism and the internal device/material degradation. Note that most of the effects occur near the oxide–channel interface and inside the oxide.

4.3 Circuit Level

In this section, we present the effect of device variability and reliability presented in B on key parameters of circuits found every day. Hence, we deal with the most often used circuit components in mixed-signal and RF circuits such as inverters, nands, OPAMPs, LNAs, current mirrors, and ring oscillator. All these circuits are simulated using the proposed framework to evaluate their robustness and reliability. For all the simulations we use 22 nm PTM models [46]. Due to the generality of the proposed framework (Figure 4.3), it can be used for any circuit that could be found in a real product.

4.3.1 Basic Combinational Logic

PBTI and NBTI affect both the speed and leakage power of any combinational circuit that consists of digital gates such as inverters, NANDs, NORs, and XORs. In this subsection, we analyze and discuss the degradation of an inverter and a two-input NAND gate. Similar analysis can be performed for any other digital gate, combinational or sequential digital circuits.

4.3.1.1 Inverter

When digital "0" is applied at the input terminal of an inverter then the PMOS transistors (M_2) is on and stressed due to NBTI. NBTI on M_2 results in V_{th} shift ($?V_t$, NBTI), weaker PMOS transistor and slower transition from "0" to "1." Similarly, when digital "1" is applied then the NMOS transistor (M_1) on and it is degraded due to PBTI. PBTI on M_1 results also in V_{th} shift ($?V_t$, PBTI) that makes the transition from "1" to "0" slower. It is obvious that both input patterns degrade either the NMOS or the PMOS transistor depending on the input making the inverter slower.

There are in addition two more very important factors that have to be taken into account during the degradation analysis. The first one is the switching factor. Switching factor designate how much time a signal is "1" and "0." Note that, due to the recovery mechanism of BTI, the switching factor of the input signals affect the degradation of the gate. The second one is the frequency of the input signals. Figures 4.9 and 4.10 depict the simulation testbench of the inverter and the delay increase due to BTI over time. Note that for the testbench there are three inverters but only the one in the middle is degraded.

4.3.1.2 Two-Input NAND

The analysis of the 2-input NAND gate is a generalization to inverter's analysis. However, we have decided to present it here because it has more that one

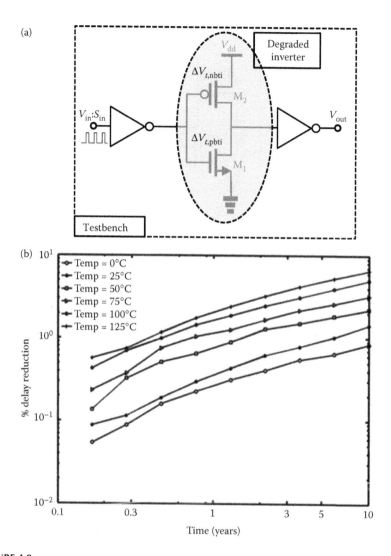

FIGURE 4.9
(a) Testbench used to simulate performance and leakage degradation of an inverter. Input and output inverters are used to generate the proper input signal slope and output capacitive load and (b) the percentage change of delay and leakage due to BTI degradation of M_1 and M_2.

input. Note that for each input we need to apply two signals with different switching factors and different signal frequencies. This information can be readily obtained from a logic simulator. Figure 4.10 shows the delay degradation of a 2-input NAND gate.

If now a chain of digital gates switch on and off then this path become slower over time. Hence, if this path is the critical path that is, the slower path of a digital circuit, then the whole circuit is slower.

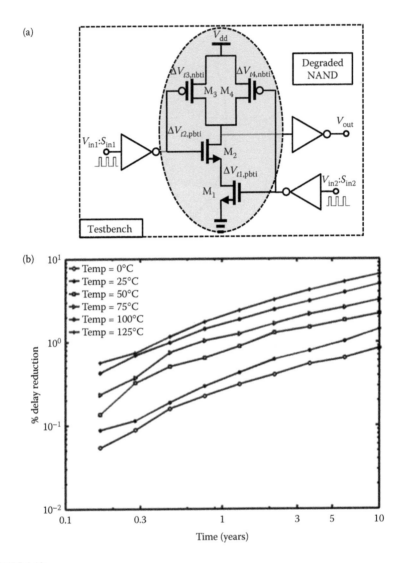

FIGURE 4.10
(a) Testbench used to simulate performance and leakage degradation of a two-input NAND. Input and output inverters are used to generate the proper input signal slope and output capacitive load and (b) the percentage change of delay and leakage due to PBTI degradation of M_1 and M_2 and NBTI degradation of M_3 and M_4, for 01, 10, and 11 input vectors.

4.3.2 Low-Noise Amplifier Circuit (LNA)

Low-noise amplifier circuit (LNA) is usually appears at the first stage of the receive path coming after the antenna of a wireless system and hence it plays a critical role in the overall performance of the RF transceiver. Noise figure, voltage gain, input return loss, stability, and bandwidth are the most

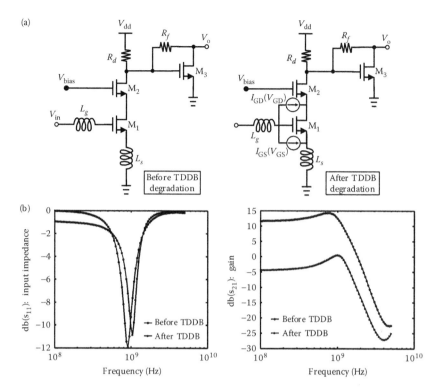

FIGURE 4.11
(a) LNA circuit schematic to study the effect of TDDB on the s-parameters at the in and out ports and (b) Frequency dependence of S11 and S21 showing the performance before and after dielectric breakdown.

important figure of merits that determine the performance of LNAs. In this subsection, we will show how the degradation of a device might affect the reliability of LNA. The circuit diagram of an LNA is shown in Figure 4.11a. The sizes of active and passive elements have been chosen to design an LNA for GSM applications operating in the 900 MHz band. Performing s-parameter analysis for a fresh circuit we get the frequency dependence of S_{11} and S_{11} depicted in Figure 4.11b and c.

When TDDB creates conductive paths from the gate to the channel of M_1, then both the input impedance and the voltage gain degrade as shown in Figure 4.11. With the proposed framework the degradation of M_2 and M_3 can also be modeled.

4.3.3 Operational Amplifier (OPAMP)

In this subsection, we study the effect of RDF and BTI on OPAMP the most important component to build high-precision analog and mixed signal blocks. Both RDF and BTI lead to device mismatch and threshold voltage shift that result in DC offsets and DC offset drift.

Consider the OPAMP shown in Figure 4.12a. If the circuit is simulated without any device mismatch then when $V_{in} = V_{in,p} - V_{in,m} = 0$ results in $V_{out} = 0$. However, RDF introduces imperfections in the MOS devices and hence $V_{out} \neq 0$, or to set $V_{out} = 0$ we need to apply an input voltage equal to the DC offset $V_{os,in} = V_{os,out}/A_v \cdot V_{os,in}$ which is called input-referred offset voltage and $V_{os,out}$ output-referred offset voltage. In the circuit level, V_{os} reduces the accuracy and limits the precision in which signals can be sampled.

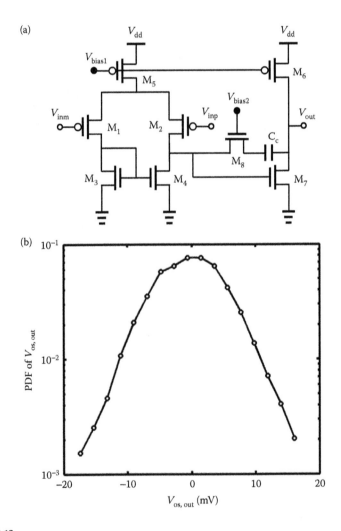

FIGURE 4.12
(a) The OPAMP circuit used for the simulation of V_{os} and (b) the V_{os} distribution due to RDF of all the transistors. Note that RDF depends on the transistor size.

Assuming that all the transistors operate in the saturation region and focusing only on the mismatch introduced by the first stage of the OPAMP we can readily conclude that [42–45]

$$V_{os,in} = \Delta V_{th,1-2} + \Delta V_{th,3-4} \frac{g_{m3}}{g_{m1}} + \frac{V_{ov,1-2}}{2} \left[\frac{\Delta S_{th,3-4}}{S_{th,3-4}} - \frac{\Delta S_{th,1-2}}{S_{th,1-2}} \right] \quad (4.24)$$

The first term is introduced due to RDF and BTI in the input transistors. BTI can occur when asymmetric signals are applied in the input pair, for example, in a comparator. The second term is introduced due to LER in the current-mirror-load transistors and the third term due to LER in the input differential pair. The DC offset can be minimized by operating the input transistors in lower overdrive voltages (reducing the tail current) or increasing the size of the transistors trading gate capacitance, speed and power. Additional mismatch is introduced by the current sources of OPAMP which is studied in the next subsection. To overcome this issue, there are offset cancelation techniques which are not studied in this book. Figure 4.12 shows the OPAMP circuit used for the simulation of V_{os} and the corresponding V_{os}-distribution due to RDF. Note that all the transistors have been affected by RDF but the RDF effect depends on their size.

4.3.4 Current Mirror

All the analog and mixed signal circuits require the generation of current sources whose values are identical with a reference current (I_{ref}). For example, this requirement is important for the design of OPAMPs, comparators, and DACs. Figure 4.13a shows the simplest current mirror with unmatched MOS transistors. Considering device and geometry mismatch, we obtain the relative current error compared to the reference current

$$\frac{\Delta I_D}{I_D} = \frac{\Delta S}{S} - \frac{\Delta V_{th}}{(1/2)V_{ov}} \quad (4.25)$$

The first term depends on geometry mismatch (e.g., LER) and it is independent on the circuit bias conditions. The second term depends on bias conditions (V_{ov}) and device mismatch introduced by RDF and BTI. Using the proposed subcircuit and simulating the current mirror at elevated temperatures and for different voltages, transistor sizes, and keeping the same device ratio that we designed, we obtain the degradation curves as shown in Figure 4.13b.

4.3.5 Ring Oscillator

Ring oscillator is the integral part for most of the integrated circuits and can be found in many applications, for example, clocks in mixed signal circuits,

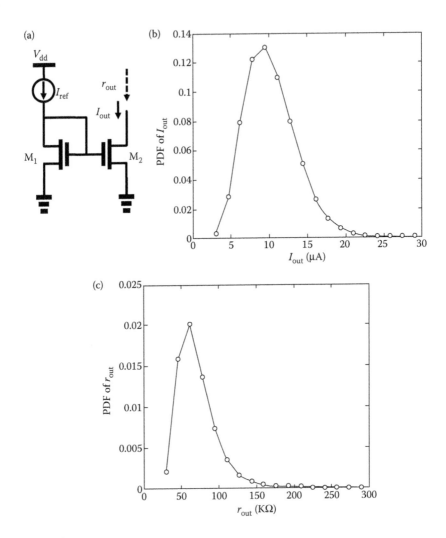

FIGURE 4.13
Basic current mirror used to show how mismatch between M_1 and M_2 results in I_{out} and r_{out} shift from the reference value. (a) Current mirror schematic, (b) mismatch shift to I_{out}, and (c) mismatch shift to r_{out}.

frequency synthesizers in RF systems, and voltage control oscillators in PLLs. Ring oscillator is a feedback system consisting of an odd number of inverter stages in a loop. The conditions for oscillation can be found in [43]. Oscillation frequency and power dissipation are two of the key performance parameters of the oscillator. In the following paragraphs, we analyze how RDF, BTI, and TDDB affect the performance parameters.

The oscillation frequency of an oscillator consisting of *n* number of inverters in the loop is given by the following formula:

$$f_{osc} = \frac{1}{2nt_d} \qquad (4.26)$$

where t_d is the delay of each stage. We have already seen that the delay of a stage is a random variable due to RDF and hence f_{osc} is also a random variable. In addition, t_d increases due to BTI leading to f_{osc} reduction. In Figure 4.14, we show the variability of oscillation frequency due to RDF.

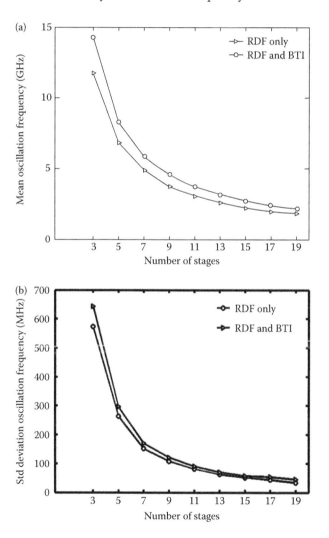

FIGURE 4.14
Deviation of oscillation frequency of ring oscillators from 3 up to 19 stages in the loop. Increasing the number of stages both frequency and variability decrease while BTI results in frequency decrease and variability increase. (a) Mean deviation and (b) standard deviation.

4.4 Conclusions

In summary, we have presented the most important variability and reliability issues, which are faced in the most advanced CMOS technologies. To address these issues, we have developed compact physics-based models to perform simulations at the circuit level. In this chapter, we present in detail the effect of RDF, BTI, HCI, and TDDB on MOS devices and we propose a subcircuit that can be used to simulate digital, analog, RF, and mixed-signal circuits. This subcircuit can be extended or can be modified depending on which model is available (physical or empirical).

It is shown that at scaled technologies, variability and reliability should be considered during the circuit design phase to ensure proper functionality over a large set of wafers (high yield) and over a long time (long lifetime). However, to overcome these issues we need to trade them with power, energy, area, as well as design time.

References

1. Moore, G. E. Progress in digital integrated electronics. *IEDM Tech. Dig.*, 11–13, 1975.
2. Keyes, R. W. Physical limits in digital electronics. *Proc. IEEE*, 63(5), 740–767, 1975.
3. Hoeneisen, B., Mead, C. A. Fundamental limitations in microelectronics—I, MOS technology. *Solid State Electron.*, 15(7), 819–829, 1972.
4. Panagopoulos, G. On variability and reliability of CMOS and spin-based devices, PhD Thesis, August 2012.
5. Asenov, A. Random dopant induced threshold voltage lowering and fluctuations in sub-0.1 μm MOSFETs: A 3-D 'atomistic' simulation study. *IEEE Trans. Electron Devices*, 45(12), 2505–2513, 1998.
6. Frank, D., Taur, Y., Leong, M., Wong, H. Monte Carlo modeling of threshold variation due to dopant fluctuations. *VLSI Symp. Tech. Dig.*, 169–170, 1999.
7. Borkar, S. Designing reliable systems from unreliable components: The challenges of transistor variability and degradation. *Micro IEEE*, 2005.
8. Frank, D. J., Dennard, R. H., Nowak, E., Solomon, P. M. Device scaling limits of Si MOSFETs and their application dependencies. *Proc. IEEE*, March 2001, 259–288, August 2002.
9. Tang, X., Bowman, K. A., Eble, J. C., De, V. K., Meindl, J. D. Impact of random dopant placement on CMOS delay and power dissipation. *Proceedings of 29th ESSDERC*, 1999, pp. 184–187.
10. Panagopoulos, G., Roy, K. A physical 3-D analytical model for the threshold voltage considering RDF. *Proceedings of DRC*, 2009, pp. 81–82.
11. Panagopoulos, G., Roy, K. A Three-dimensional physical model for Vth variations considering the combined effect of NBTI and RDF. *IEEE Trans. Electron Devices*, 58(8), 2337–2346, 2011.

12. Li, Y., Yu, S., Hwang, J., Yang, F. Discrete dopant fluctuations in 20-nm/15-nm gate planar CMOS. *IEEE Trans. Electron Devices*, 55(6), 1449–1455, 2008.
13. Xiong, S., Bokor, J. A simulation study of gate line edge roughness effects on doping profiles of short-channel MOSFET devices. *IEEE Trans. Electron Devices*, 51(2), 228–232, 2004.
14. Wong, H., Taur, Y. Three-dimensional atomistic simulation of discrete dopant distribution effects in sub-0.1 µm MOSFETs. *IEDM Tech. Dig.*, 705–708, 1993.
15. Ye, Y., Liu, F., Nassif, S., Cao, Y. Statistical modeling and simulation of threshold variation under dopant fluctuations and line edge roughness. *Proc. DAC*, 4, 900–905, 2008.
16. Hwang, C., Li, T., Han, M., Lee, K., Cheng, H., Li, Y. Statistical analysis of metal gate work function variability, process variation, and random dopant fluctuation in nano-CMOS circuits. *Proceedings of SISPAD*, 2009.
17. Keyes, R. W. Effect of randomness in the distribution of impurity ions on FET thresholds in integrated electronics. *IEEE J. Solid State Circuits*, SSC-10(4), 245–247, 1975.
18. Stolk, P. A., Widdershoven, F. P., Klaasen, D. B. M. Modeling statistical dopant fluctuations in MOS transistors. *IEEE Trans. Electron Devices*, 45(9), 1960–1971, 1998.
19. Stolk, P. A., Klaasen, D. B. M. The effect of statistical dopant fluctuations on MOS device performance. *IEDM Tech. Dig.*, 627–630, 1996.
20. Takeuchi, K. Channel size dependence of dopant-induced threshold voltage fluctuations. *VLSI Symp. Tech. Dig.*, 72–73, 1998.
21. Reid, D., Millar, C., Roy, G., Roy, S., Asenov, A. Analysis of threshold voltage distribution due to random dopants: A 100 000-sample 3-D simulation study. *IEEE Trans. Electron Devices*, 56(10), 2255–2263, 2009.
22. Tang, X., De, V. K., Meindl, D. Intrinsic MOSFET parameter fluctuations due to random dopant placement. *IEEE Trans. Very Large Scale Integr. (VLSI) Syst.*, 5(4), 369–376, 1997.
23. Lakshmikumar, K., Hadaway, R., Copeland, M. Characterization and modeling of mismatch in MOS transistors for precision analog design. *IEEE J. Solid-State Circuits*, SSC-21(6), 1057–1066, 1986.
24. Takeuchi, K., Tatsumi, T., Furukawa, A. Channel engineering for the reduction of random-dopant-placement-induced threshold voltage fluctuation. *IEDM Tech. Dig.*, 841–844, 1997.
25. Toriyama, S., Hagishima, D., Matsuzawa, K., Sano, N. Device simulation of random dopant effects in ultra-small MOSFETs based on advanced physical models. *Proceedings of SISPAD*, 2006, pp. 111–114.
26. Nobuyuki, S., Masaaki, T. Random dopant model for three-dimensional drift-diffusion simulations in metal oxide semiconductor field-effect-transistors. *Appl. Phys. Lett.*, 79(14), 2267–2269, 2001.
27. Roy, S., Lee, A., Brown, A. R., Asenov, A. Applicability of quasi-3D and 3D MOSFET simulations in the 'atomistic' regime. *Kluwer J. Comput. Electron.*, 2(2–4), 423–426, 1998.
28. Yin, J., Shi, X., Huang, R. A new method to simulate random dopant induced threshold voltage fluctuations in sub-50 nm MOSFETSs with non-uniform channel doping. *Solid State Electron.*, 50(9/10), 1551–1556, 2006.
29. Mukhopadhyay, S., Kim, K., Jenkins, K. A., Chuang, C.-T., Roy, K. An on-chip test structure and digital measurement method for statistical characterization of local random variability in a process. *IEEE J. Solid-State Circuits*, 43(9), 1951–1963, 2008.

30. Li, Y., Hwang, C.-H. High-frequency characteristic fluctuations of nano-MOS-FET circuit induced by random dopants. *IEEE Trans. Microw. Theory Tech.*, 56(12), 2726–2733, 2008.

31. Yeung, J., Mahmoodi, H. Robust sense amplifier design under random dopant fluctuations in nanoscale CMOS technologies. *Proceedings of SOC Conference*, September 2006, pp. 261–264.

32. Burnett, D., Erington, K., Subramanian, C., Baker, K. Implications of fundamental threshold voltage variations for high-density SRAM and logic circuits. *VLSI Symp. Tech. Dig.*, 15–16, 1994.

33. Li, Y., Hwang, C.-H., Yeh, T.-C., Li, T.-Y. Large-scale atomistic approach to random-dopant-induced characteristic variability in nanoscale CMOS digital and high-frequency integrated circuits. *Proceedings of ICCAD*, November 2008, pp. 278–285.

34. Millar, C., Reid, D., Roy, G., Roy, S., Asenov, A. Accurate statistical description of random dopant-induced threshold voltage variability. *IEEE Trans. Electron Device Lett.*, 29(8), 946–948, 2008.

35. Jiang, X.-W., Deng, H.-X., Luo, J.-W., Li, S.-S., Wang, L.-W. A fully three-dimensional atomistic quantum mechanical study on random dopant-induced effects in 25-nm MOSFETs. *IEEE Trans. Electron Devices*, 55(7), 1720–1726, 2008.

36. Asenov, A., Slavcheva, G., Brown, A. R., Davies, J. H., Saini, S. Increase in the random dopant induced threshold fluctuations and lowering in sub-100 nm MOSFETs due to quantum effects: A 3-D density gradient simulation study. *IEEE Trans. Electron Devices*, 48(4), 722–729, 2001.

37. Markovic, D., Nikolic, B., Brodersen, R. Analysis and design of low-energy flip-flops. *Proceedings of the 2001 International Symposium on Low Power Electronics and Design*, 2001, pp. 52–55.

38. Sze, S., Ng, K. *Physics of Semiconductor Devices*, 3rd edition. Hoboken, NJ: Wiley, 2008.

39. Pierret, R. F. *Fundamental of Solid State Devices*, 2nd edition. Addison-Wesley, 1996.

40. Taur, Y., Ning, T. *Fundamentals of Modern VLSI Devices*. Cambridge, UK: Cambridge University Press, 1998.

41. Balanis, C. A. *Advanced Engineering Electromagnetics*. Hoboken, NJ: Wiley, 1989.

42. Ramo, S., Whinnery, J. R., van Duzer, T. *Fields and Waves in Communication Electronics*, 2nd edition. Hoboken, NJ: Wiley, 1984.

43. Papoulis, A., Pillai, S. U. *Probability, Random Variables and Stochastic Processes*, 4th edition. New York: McGraw-Hill, 2002.

44. Razavi, B, *Design of Analog CMOS Integrated Circuits*. McGraw-Hill, 2000.

45. Gray, P., Hurst, P., Lewis, S., Meyer, R, *Analysis and Design of Analog Integrated Circuits*, 5th edition. Wiley, 2009.

46. http://ptm.asu.edu.

47. Alam, M., Kufluoglu, H., Varghese, D., Mahapatra, S. A comprehensive model for PMOS NBTI degradation: Recent progress. *Microelectron. Reliab.*, 47(6), 853–862, 2007.

48. Yang J. B., Chen, T. P. Analytical reaction-diffusion model and the modeling of nitrogen-enhanced negative bias temperature instability. *Appl. Phys. Lett.*, 88(17), 1–3, 2006.

49. Alam, M. A critical examination of the mechanisms of dynamic NBTI for PMOSFETs. *IEDM Tech. Dig.*, 345–348, 2003.

50. Islam, A.E., Kufluoglu, H., Varghese, D., Mahapatra, S., Alam, M. A. Recent issues in negative-bias temperature instability: Initial degradation, field dependence of interface trap generation, hole trapping effects, and relaxation. *IEEE Trans. Electron Devices*, 54(9), 2143–2154, 2007.
51. Kufluoglu, H., Alam, M. A generalized reaction-diffusion model with explicit H–H2 dynamics for negative bias temperature instability degradation. *IEEE Trans. Electron Devices*, 54(5), 1101–1107, 2007.
52. Chakravarthi, S., Krishnan, A., Reddy, V., Machala, C., Krishnan, S. A comprehensive framework for predictive modeling of negative bias temperature instability. *Proceedings of IEEE IRPS*, 2004, pp. 273–282.
53. Grasser, T., Gos, W., Kaczer, B. Dispersive transport and negative bias temperature bias instability: Boundary conditions, initial conditions, and transport models. *IEEE Trans. Device Mater. Rel.*, 8(1), 79–97, 2008.
54. Ielmini, D., Manigrasso, M., Gattel, F., Valentini, M. G. A new NBTI model based on hole trapping and structural relaxation in MOS dielectrics. *IEEE Trans. Electron Devices*, 56(9), 1943–1952, 2009.
55. Grasser, T., Kaczer, B., Goes, W., Aichinger, T., Hehenberger, P., Nelhiebel, M. Understanding negative bias temperature instability in the context of hole trapping. *Microelectron. Eng.*, 86(7–9), 1876–1882, 2009.
56. Zafar, S. Statistical mechanics based model for negative bias temperature instability induced degradation. *J. Appl. Phys.*, 97(10), p. 103, 709, 2005.
57. Zafar, S. A model for negative bias temperature instability in oxide and high-kpFETs. *Proceedings of IEEE ICICDT*, 2007, pp. 1876–1882.
58. Agostinelli, M., Lau, S., Pae, S., Marzolf, P., Muthali, H., Jacobs, S. PMOS NBTI-induced circuit mismatch in advanced technologies. *Microelectron. Reliab.*, 46(1), 63–68, 2006.
59. Hu, C., Tam, S. C., Hsu, F., Ko, P., Chan, T., Terrill, K. W., Hot-electron-induced MOSFET degradation—Model, monitor, and improvement. *IEEE Trans. Electron Devices*, 32(2), 375–385, 1985.
60. Depas, M., Nigam, T., Heyns, M. M. Definition of dielectric breakdown for ultra-thin (<2 nm) gate oxides. *Solid-State Electron.*, 41(5), 725–728, 1997.
61. Wu, E., Nowak, E., Aitken, J., Abadeer, W., Han, L. K., Lo, S. Structural dependence of dielectric breakdown in ultra-thin gate oxides and its relationship to soft breakdown modes and device failure. *IEDM*, 1998, pp. 187–190.
62. Weir, B. E., Silverman, P. J., Monroe, D., Krisch, K. S., Alam, M. A., Alers, G. B., Sorsch, T. W. et al. Ultra-thin gate dielectrics: They break down, but do they fail? *IEEE IEDM*, 1997, pp. 73–76.
63. Kaczer, B., Degraeve, R., Rasras, M., Mieroop, K., Roussel, P., Groseneken, G. Impact of MOSFET gate oxide breakdown on digital circuit operation and reliability. *IEEE Trans. Electron Devices*, 49(3), 500–506, 2002.
64. Kuang, W., Cao, L., Yu, C., Yuan, J. PMOS breakdown effects on digital circuits: Modeling and analysis. *Microelectron. Eng.*, 1597–1600, 2008.
65. Monsieur, F., Vincent, E., Roy, D., Bruyere, S., Vildeuil, J., Pananakakis, G., Ghibaudo, G. A thorough investigation of progressive breakdown in ultra-thin oxides. Physical understanding and application for industrial reliability assessment. *IRPS*, 2002, pp. 45–54.
66. Sune, J., Wu, E., Tous, S. Failure-current based oxide reliability assessment methodology. *IRPS*, 2008, pp. 230–239.

67. Sune, J., Wu, E., Lai, W. Statistics of competing post-breakdown failure model in ultrathin MOS devices. *IEEE Trans. Electron Devices*, 53(2), 224–234, 2006.
68. Miranda, E., Sune, J. Analytical modeling of leakage current through multiple breakdown paths in SiO films. *IRPS*, 2001, pp. 367–379.
69. Miranda, E., Suñé, J. Electron transport through broken down ultrathin SiO layers in MOS devices. *Microelectron. Reliab.*, 44, 1–23, 2004.
70. Stathis, Percolation models for gate oxide breakdown. *J. Appl. Phys.*, 86(10), 5757–5766, 1999.
71. Schuegraf, K.F., Hu, C. Hole injection SiO2 breakdown model for very low voltage lifetime extrapolation. *IEEE Trans. Electron Devices*, 41(5), 761–767, 1994.
72. Alam, M. A., Smith, R. K., Weir, B. E., Silverman, P. J. Statistically independent soft breakdowns redefine oxide reliability specifications. *IEDM*, 2002, pp. 151–154.
73. Sune, J., Miranda, E., Nafria, M., Aymerich, X. Point contact conduction at the oxide breakdown of MOS devices. *IEDM*, 1998, pp. 191–194.

5

Mixed-Signal Circuit Testing Using Wavelet Signatures

Michael G. Dimopoulos, Alexios Spyronasios, and Alkis Hatzopoulos

CONTENTS

5.1 Introduction

Testing of electronic circuits is an active research topic in recent years [1–6] and it remains crucial, especially for the rapidly developed mixed-mode (analog and digital) circuits. This is mainly due to the difficulties inherent to the nature of analog signals. The conventional fault models [6] for analog circuits are the catastrophic or hard faults, where an analog component becomes open or shorted and the parametric faults, where the value of a component such as R, L, C, or the value of a transistor parameter (transconductance, V_{th}, etc.) changes sufficiently outside the tolerance limits and causes unacceptable circuit performance. Process imperfections [6] due to subwavelength lithography, etching, and variation of the number of dopants in the channel of short channel devices result in large variations of related circuit parameters and particularly of the threshold voltage (V_{th}). Statistical variations in device parameters lead to a statistical distribution of V_{th}.

Test methods must be sensitive enough in order to detect faults due to parameter variations. Apart from the output voltage (V_{OUT}) measuring-based techniques, power supply current (I_{PS}) testing techniques have been investigated for several years and various approaches have been proposed [7–9] showing promising results. For the measured signal processing and its application in fault detection, different approaches have been proposed, like root mean square (RMS) value calculation, fast Fourier transform (FFT), amplitude and phase, correlation functions, and others. The exploitation of the wavelet transform, which resolves a signal in both time and frequency simultaneously, has also been investigated [10–12]. Moreover, a comparative sensitivity analysis between the wavelet and Fourier transforms but without fault detectability analysis has been reported [10]. A wavelet test method based on a Euclidean test metric applied to the I_{PS} current waveforms has been presented [11]. A common question that needs further investigation is the fault detection efficiency of the above methods when they are applied to measurement data, since problems such as measuring inaccuracies and parameter deviations are arising from their practical application.

In this chapter, fault detection methods for parametric and catastrophic faults in analog and mixed-signal circuits are presented based on the wavelet analysis of the measured signal waveform be it I_{PS} or V_{OUT}. Two different metrics employing the wavelet energy values of the measured signal are analyzed. Statistical processing of data from fault-free circuits is suggested as a means for systematically defining the tolerance limits of the fault-free circuit.

The chapter is organized as follows. A brief introduction to wavelet analysis is given in Section 5.2. The test algorithm utilizing the wavelet signatures is described in Section 5.3. The selection of the appropriate tolerance limit based on a cost analysis is presented in Section 5.4. Experimental results from the application of the proposed test algorithm to measurement data from a commercial electronic product are given in Section 5.5.

5.2 Introduction to Wavelets

The wavelet transform [13–15] is a transform that provides both time and frequency representation. It passes the time-domain signal from various high-pass and low-pass filters, which filter out either high-frequency or low-frequency portions of the signal. This procedure, which is called decomposition, is repeated until a predefined decomposition level is attained. After this, a set of signals is produced which actually represents the original signal. The continuous wavelet transform of a function $x(t)$ is defined as follows:

$$CWT_{\psi}^{x}(\tau,s) = \Psi_{\psi}^{x}(\tau,s) = \frac{1}{\sqrt{|s|}} \int_{-\infty}^{\infty} x(t)\psi^{*}\left(\frac{t-\tau}{s}\right)dt \qquad (5.1)$$

The transformed signal is a function of two variables, τ and s, the translation and scale parameters, respectively. $\Psi(t)$ is the transforming function, and it is called the mother wavelet. The variable τ represents the time shift (translation) while the variable s represents the amount of time scaling or dilation. The mother wavelet is a prototype for generating the other window functions. Several functions can be used as mother wavelets as long as they have finite energy and they satisfy the admissibility condition [14]. Indicative wavelet mother functions are presented in Figure 5.1.

For sampled signals (as in our case), the discrete wavelet transform is used. The main idea remains the same as with the continuous wavelet transform; however, in the discrete transform, the scaling and translating operations are done in discrete steps. In the discrete wavelet transform, the resolution of the signal, which is a measure of the amount of detail information in the signal, is changed by the filtering operations, and the scale is changed by upsampling and downsampling (subsampling) operations. The wavelet transform decomposes a discrete signal into two subsignals of half of its length [12–15]. One subsignal is a running average, approximation or trend (a); the other subsignal is a running difference or fluctuation (b).

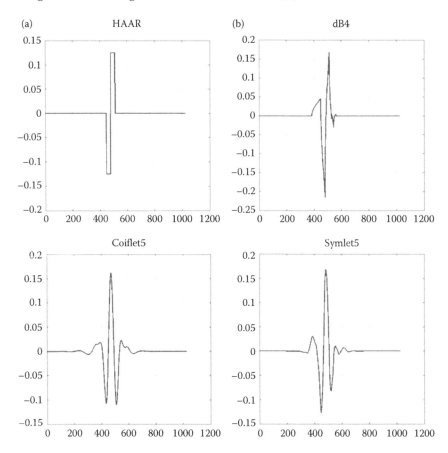

FIGURE 5.1
Indicative plots of mother wavelet functions. (a) Haar wavelet (HAAR) and Coiflet wavelet of order 5 (Coiflet5); (b) Daubechies wavelet of order 4 (dB4) and Symlet wavelet of order 5 (Symlet5).

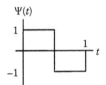

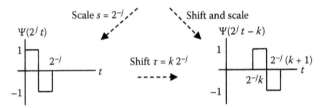

FIGURE 5.2
HAAR mother wavelet function.

The simplest of the mother functions is the Haar wavelet (Figure 5.2). It is a step function taking values 1 and −1, on [0, 1/2) and [1/2, 1), respectively. Without loss of generality in the following analyses the Haar mother wavelet is used. The Haar wavelet has also been proposed in literature [16] as an effective wavelet for mixed-signal test applications.

Thus, for a sampled signal $S = [S_1, S_2,..., S_M]$ with $M = 2L$ ($L \in N$) samples the first level decomposition subsignals T_1 and F_1 for the Haar mother wavelet are as follows:

$$T_1 = [T_{11},..., T_{1j},..., T_{1n}] = \left[\frac{S_1 + S_2}{\sqrt{2}} ,..., \frac{S_{2j-1} + S_{2j}}{\sqrt{2}} ,..., \frac{S_{2n-1} + S_{2n}}{\sqrt{2}} \right] \quad (5.2)$$

and

$$F_1 = [F_{11},..., F_{1j},..., F_{1n}] = \left[\frac{S_1 - S_2}{\sqrt{2}} ,..., \frac{S_{2j-1} - S_{2j}}{\sqrt{2}} ,..., \frac{S_{2n-1} - S_{2n}}{\sqrt{2}} \right] \quad (5.3)$$

where $j = 1, 2, ...,n = \dfrac{M}{2}$

The wavelet energy for the above subsignals T_1 (considering only the trend coefficients T_{1j}) and F_1 (utilizing the fluctuation or detail coefficients F_{1j}) are

$$E_{T1} = \sum_{j=1}^{n} T_{1j}^{2} \quad (5.4)$$

$$E_{F1} = \sum_{j=1}^{n} F_{1j}^{2} \quad (5.5)$$

So, the wavelet energy of the first level decomposition E_1 consists of two terms E_{T1} and E_{F1}.

It should be noted that the exact computation of coefficients T_{1j} and F_{1j} depends on the selected mother wavelet function.

In general, the wavelet energies for the m decomposition level are

$$E_{Tm} = \sum_{j=1}^{n} T_{mj}^2 \tag{5.6}$$

$$E_{Fm} = \sum_{j=1}^{n} F_{mj}^2 \tag{5.7}$$

5.3 Wavelet-Based Test Algorithm

Generally, a circuit test process involves three major steps: circuit stimulation, signal (response), measurement, and analysis of the results. Here, the circuit under test (CUT) is stimulated in normal operation by an externally generated signal and the I_{PS} or V_{OUT} signal is measured and stored. A wavelet analysis is performed to the stored I_{PS} or V_{OUT} signal in order to calculate the necessary metric values. The wavelet transform is utilized because it offers better approximation of a transient signal waveform than the Fourier transform [10] for a fixed limiting frequency of the measured signal. Also it allows the selection (in terms of minimization of the approximation error) of the appropriate basis function according to the measured signal encountered in the CUT.

It should be noted that since we are dealing with many known fault-free circuits with variations of their parameter values, the signals are specified by a nominal value along with an acceptable range of values around the nominal value. Thus, the notion of nominal circuit is replaced here by the notion of the reference circuit where the value of a parameter is defined as the mean value of the corresponding parameters of a set of known fault-free circuit instances.

Since the statistical distribution of analog faults [6] generally are not known with enough precision, the tolerance limits for the reference circuit, whose selection affects the circuit's fault detectability, are set after statistical preprocessing of the data collected from the fault-free circuits. These limits are set here systematically (Section 5.4) according to the chosen percentage of *fault-free CUT losses* [8]; that is, the percentage of free CUTs erroneously considered (and thus discarded) as faulty. So, when a tolerance limit results in *a% fault-free CUT losses*, this means that from a lot of 100 fault-free CUTs (100-*a*) pass the test process, whereas *a* fault-free CUTs are discarded as faulty.

In production line test, the tolerance limit computation may be dynamic, that is, periodically recompute and update the tolerance limits by including new data obtained from the production line.

Before describing the test algorithm, some notations will be given as follows:

z is the measured signal waveform ($z = \{I_{PS}, V_{OUT}\}$)

$E_{Xm,0-i}(z)$ is the energy value for the i fault-free circuit case using the trend ($X = T$) or detail ($X = F$) coefficients of the m decomposition level

$E_{Xm,0}(z)$ is the mean value of all $E_{Xm,0-i}(z)$ values for a set of fault-free circuit instances and $X = \{T, F\}$

$E_{Xm,k}(z)$ is the energy value for the k CUT using the trend ($X = T$) or detail ($X = F$) coefficients of the m decomposition level

$Df_0{}^i(z)$ is the discrimination factor for the i fault-free circuit case

$Df_0{}^{lim}(z)$ is the tolerance limit of the discrimination factor for a set of fault-free circuit instances

$Df_k(z)$ is the discrimination factor for the k CUT

$cdf_{Df}(z)$ is the cumulative distribution function computed from the values $Df_0{}^i(z)$

$MD_{0-i}(z)$ is the Mahalanobis distance value (actually a two-element array with two rows and one column) for the i fault-free circuit case

$MD_{Xm,0-i}(z)$ is the Mahalanobis distance value constituent of $MD_{0-i}(z)$ attributed to the trend ($X = T$) or detail ($X = F$) wavelet coefficients of the m decomposition level

$MD_0{}^{lim}(z)$ is the Mahalanobis tolerance limit value (actually it is a two-element array with two rows and one column)

$MD_{Xm,0}{}^{lim}(z)$ is the Mahalanobis tolerance limit value constituent of $MD_0{}^{lim}(z)$ attributed to the trend ($X = T$) or detail ($X = F$) wavelet coefficients of the m decomposition level for a set of fault-free circuit instances

$MD_k(z)$ is the Mahalanobis distance value of the k CUT. $MD_k(z)$ is a two-element array with two rows and one column

$MD_{Xm,k}(z)$ is the Mahalanobis distance value constituent of $MD_k(z)$ attributed to the trend ($X = T$) or detail ($X = F$) wavelet coefficients of the m decomposition level

$cdf_{MDXm,0}(z)$ is the cumulative distribution function computed from the values $MD_{Xm,0-i}(z)$

n is the number of known fault-free circuits

k is a circuit under test (CUT)

Num is the number of CUTs (i.e., the size of circuit test set)

n_f is the number of faulty circuit cases

n_W is the number of CUTs being detected as faulty by the discrimination factor $(W = Df)$ or Mahalanobis distance $(W = MD)$

FD_W is the percentage of total fault coverage utilizing the discrimination factor $(W = Df)$ or Mahalanobis distance $(W = MD)$

$S_{0-i,Xm-0}{}^{-1}$ is the inverse covariance matrix between the i energy value $(E_{Xm,0-i}(z))$ and the center of $E_{Xm,0}(z)$

It should be noted that, generally, the selected decomposition level m is signal dependent. Sometimes signal features at certain scales dominate and obscure certain signal details that may be of interest. In order to accentuate this "hidden" details, the scaling (decomposition level) has to be modified. Therefore, a preprocessing step may be necessary in order to choose the appropriate decomposition level to highlight certain signal details. For the examined test cases, two levels have found to be adequate to reveal the presented features.

The test algorithm which is presented below for a general metric function *fMetric* consists of two phases. In the first phase (*Training Phase*), statistical processing of the fault-free circuit data takes place in order to compute the wavelet energy values $E_{Tm,0}(z)$ and $E_{Fm',0}(z)$ (which correspond to the selected m and m' wavelet decomposition levels), generally with $m \neq m'$ for the reference circuit and the tolerance limit $fMetric_0{}^{\lim}(z)$.

In the second phase (*Main Test Phase*), the value $fMetric_k(z)$ for the k CUT is computed and compared with the corresponding limit $fMetric_0{}^{\lim}(z)$ of the reference circuit. The detection of a faulty circuit instance k will be successful when its $fMetric_k(z)$ value exceeds the tolerance limit $fMetric_0{}^{\lim}(z)$ (i.e., $fMetric_k(z) > fMetric_0{}^{\lim}(z)$).

The algorithm phases are described in the following:

Training Phase:

for each fault-free circuit $i = 1,...,n$ **do begin**
    A1)   Measure the waveform $z = \{I_{PS}, V_{OUT}\}$.
    A2)   Compute and store $E_{Tm,0-i}(z)$, $E_{Fm',0-i}(z)$.
end

A3)   Set: $E_{Tm,0}(z) = \dfrac{1}{n}\sum\limits_{i=1}^{n} E_{Tm,0-i}(z)$ and $E_{Fm',0}(z) = \dfrac{1}{n}\sum\limits_{i=1}^{n} E_{Fm',0-i}(z)$.   (5.8)

for each fault-free circuit $i = 1,...,n$ **do begin**

A4)   Compute $fMetric_0^i(z)$.   (5.9)

end
    A5)   From the values $x_i = fMetric_0^i(z)$ $(i = 1,...,n)$ compute their cumulative distribution function [8] $cdf_{fMetric}(z)$. The cumulative distribution

function $cdf_{fMetric}(z)\big|_{x=x_i}$ describes the probability that a random vari-
able x takes a value less than or equal to x_i.

A6)   For a selected value of fault-free CUT losses (for example 0%) obtain
from $cdf_{fMetric}(z)$ the value for $fMetric_0^{lim}(z)$.

So, for the reference circuit and for a predefined value of fault-free CUT
losses, the values stored are: $fMetric_0^{lim}(z)$, $E_{Tm,0}(z)$, $E_{Fm',0}(z)$.

It must be mentioned that the approach followed for the computation of
$cdf_{fMetric}(z)$ is by using the sample data and considering an *empirical cdf* [17].
For a sufficiently large sample size, the *empirical cdf* will approach the true
cdf. In this algorithm, the sample consists of the n stored (I_{PS} or V_{OUT}) signals
corresponding to the I_{PS} or V_{OUT} measurements of the n fault-free CUTs and
the *observed values* are the distances or limits from the mean value of the corresponding distribution. For example, for the $cdf_{fMetric}(z)$ distribution, if the
i observed value ($x_i = fMetric_0^i(z)$) is selected, then $fMetric_0^{lim}(z) = fMetric_0^i(z)$
and the percentage of *fault-free CUT losses* for $x_i = fMetric_0^i(z)$ is equal to the
value $[1 - cdf_{fMetric}(z)\big|_{fMetric_0^i(z)}] \cdot 100\%$.

Test Phase:

A7)   Set the counters of the number of detected faults equal to 0: $n_{fMetric} = 0$.
A8)   **For** a CUT k ($k = 1, ..., Num$) **do begin**
A8.1) Measure the waveform $z = \{I_{PS}, V_{OUT}\}$.
A8.2) Compute $E_{Tm,k}(z)$, $E_{Fm',k}(z)$.
A8.3) Compute $fMetric_k(z)$.
A8.4) **If** $fMetric_k(z) > fMetric_0^{lim}(z)$
then the k circuit case is declared faulty ($n_{fMetric} = n_{fMetric} + 1$)
else the k circuit case is considered fault-free.
end

In the following subsections, two different test metrics based on wavelet energy are described. The first test metric is named discrimination factor (*Df*) and is based on "Euclidean" distance while the second exploits the
Mahalanobis distance [18] and takes into account the correlation between
variables (measurements) when computing statistical distances.

5.3.1 The Discrimination Factor Metric Function *DF*

The discrimination factor *Df* metric is actually a normalized Euclidean
test metric that depends on the energy values [13–15] E_{T2} and E_{F1} of
the wavelet transform of the measured waveform (I_{PS}, V_{OUT}). For the wavelet
energy computation, the trend coefficients of the second level decomposition
and detail coefficients of the first level decomposition are considered.

For a circuit k and a waveform $z = \{I_{PS}, V_{OUT}\}$ the discrimination factor Df is defined as

$$Df_k(z) = \sqrt{\left(\frac{E_{T2,k}(z) - E_{T2,0}(z)}{E_{T2,0}(z)}\right)^2 + \left(\frac{E_{F1,k}(z) - E_{F1,0}(z)}{E_{F1,0}(z)}\right)^2} \qquad (5.10)$$

In the test algorithm of Section 5.3 by setting $fMetric \equiv Df$, we have a test algorithm using the discrimination factor Df metric. In the *Training Phase* (steps A1–A6), the wavelet energy values $E_{T2,0}(z)$ and $E_{F1,0}(z)$, and the discrimination factor limit Df_0^{lim} (from $cdf_{Df}(z)$) are computed for the reference circuit. In the *Main Test Phase* (steps A7–A8), the value of Df_k for the k CUT is computed and compared with the corresponding limit Df_0^{lim} of the reference circuit. The detection of a faulty circuit instance k will be successful when its Df_k value exceeds the tolerance limit Df_0^{lim}.

So, for the reference circuit and for a predefined value of fault-free CUT losses, the following values are stored: $Df_0^{lim}(z)$, $E_{T2,0}(z)$, $E_{F1,0}(z)$.

5.3.2 A Mahalanobis Distance-Based Metric Function MD

Generally, in the testing problem there is a need in computing the distance of an observation x (CUT observation or CUT measurement) from the center y (centroid) of a set (cluster) of points namely the fault-free cluster (fault-free or reference circuit). This distance helps in characterizing or better yet classifying the CUT as faulty or fault free. There are various ways of computing this distance based on the test metrics used. Here, a test metric is introduced based on Mahalanobis distance [18] which takes into account the correlation between variables (measurements) when computing statistical distances. The test metric exploits the first-level decomposition energy values [12–15] of the wavelet transform of the measured signal waveform (I_{PS}, V_{OUT}).

Given two points $x = (x_1,...,x_p)^t$ and $y = (y_1,...,y_p)^t$ in the p-dimensional space R^p the Mahalanobis distance between x and y (Figure 5.3a) is defined as

$$MD_{xy} = \sqrt{(x - y) \cdot S_{x,y}^{-1} \cdot (x - y)^T} \qquad (5.11)$$

where $S_{x,y}^{-1}$ the inverse of the covariance matrix between x and y.

In case the covariance matrix is the identity matrix, the Mahalanobis distance reduces to the Euclidean distance $d_{xy} = \sqrt{\sum_{i=1}^{p}(x_i - y_i)^2}$. For a diagonal covariance matrix, the resulting distance measure will be a normalized Euclidean distance (Figure 5.3b):

$$d'_{xy} = \sqrt{\sum_{i=1}^{p}\left(\frac{x_i - y_i}{\sigma_i}\right)^2} \qquad (5.12)$$

where σ_i is the standard deviation of the x_i over the sample set.

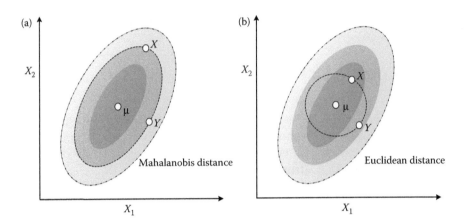

FIGURE 5.3
Diagrams showing the difference between Mahalanobis distance (a) and Euclidean distance (b).

For a circuit k and a waveform $z = \{I_{PS}, V_{OUT}\}$, the Mahalanobis distance-based metric is defined as a two-term function

$$MD_k(z) = \begin{bmatrix} MD_{T1,k}(z) \\ MD_{F1,k}(z) \end{bmatrix}$$ (5.13)

where

$$MD_{Xm,k}(z) = \sqrt{(E_{Xm,k}(z) - E_{Xm,0}(z)) \cdot S_{k,Xm-0}^{-1} \cdot (E_{Xm,k}(z) - E_{Xm,0}(z))^T},$$
$$(X = \{T, F\},\ m = 1)$$ (5.14)

In the test algorithm of Section 5.3 by setting *fMetric* $\equiv MD$, we have a test algorithm using the MD metric. In the *Training Phase* (steps A1–A6) the wavelet energy values $E_{T1,0}(z)$, $E_{F1,0}(z)$ and the tolerance limit $MD_0^{\lim}(z) = \begin{bmatrix} MD_{T1,0}^{\lim}(z) \\ MD_{F1,0}^{\lim}(z) \end{bmatrix}$ are computed for the reference circuit. As it can be seen, the tolerance limit $MD_0^{\lim}(z)$ consists of two terms (two different limits) $MD_{T1,0}^{\lim}(z)$ and $MD_{F1,0}^{\lim}(z)$. Each one of these limits is computed by the corresponding cumulative distribution function $cdf_{MD_{X1,0}}(z)$ $(X = \{T, F\})$. In the *Main Test Phase* (steps A7–A8), the value of $MD_k(z)$ for the k CUT is computed and compared with the corresponding limit $MD_0^{\lim}(z)$ of the reference circuit. The detection of a faulty circuit instance k will be successful when $MD_k(z) > MD_0^{\lim}(z)$.

For the reference circuit and for a predefined value of fault-free CUT losses, the following values are stored: $MD_0^{lim}(z)$, $E_{T1,0}(z)$, $E_{F1,0}(z)$.

After the *Test Phase*, the percentage fault detectabilities may be obtained by the following relation:

$$FD_W = \frac{n_W}{n_f} \cdot 100\%, \; W = \{Df, MD\} \tag{5.15}$$

The test algorithm of Section 5.3 can be applied for testing single catastrophic (opens and shorts) faults and parametric faults. The memory requirements are very low since for a given value of fault-free CUT losses, only three (for *Df* metric) or four (for *MD* metric) values need to be stored for the reference circuit. Finally, for each CUT, only one value ($Df_k(z)$ or $MD_k(z)$) needs to be computed and compared with the corresponding value of the reference circuit.

Both metrics (as it will be presented in Section 5.5) result in high fault coverage. Generally, the test algorithm with the Mahalanobis distance-based metric performs better than the one based on the discrimination factor metric mainly due to the more information obtained by capturing signal correlations but at an increased computational complexity. In Reference 12, the effect of the selection of various other mother functions is investigated. The presented test algorithm may be applied to any measured signal waveform.

In the next section, a systematic way for the computation of test tolerance limits based on cost-oriented analysis is described.

5.4 Cost-Oriented Statistical Analysis for Tolerance Limit Selection

The estimation of the tolerance limits when a testing method is applied is quite significant. According to the limits, there are faulty devices which are considered erroneously as fault-free (test-escape) and fault-free devices which are considered as faulty.

From Figure 5.4, we see that the placement of vertical line AB defines the tolerance limit $fMetric_0^{lim} = fMetric_0^i = x_i = x_B$. By moving the line AB further to the right, the tolerance limit value increases, the number of *fault-free CUT Losses* decreases and the number of *Test escapes* increases. Setting a tighter tolerance limit by shifting the line AB further to the left, decreases the number of *Test escapes* and at the same time increases the number of *fault-free CUT Losses*. From this discussion, it becomes apparent that an improper setting of the tolerance limits may result to increased total cost in the production line. Also, the yield Y of the production line is of great importance. The term "yield" refers to the average number of good devices in a production line. Higher yield means decreased number of faulty devices. In this

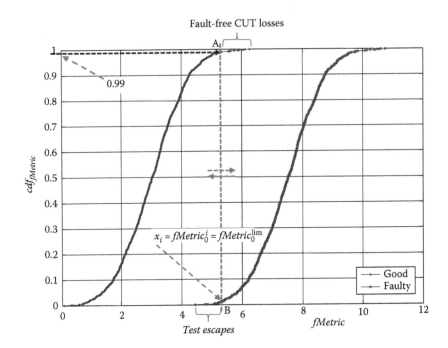

FIGURE 5.4
Indicative *cdf* diagrams for fault-free ("good") and faulty ("faulty") cases.

section, a cost oriented statistical analysis to properly set the tolerance limits
is presented.

The total test cost can be split into two terms namely direct and indirect
test cost. The direct test cost is defined by the chosen test setup and the
number of measurements which will be performed, while the indirect test
cost is defined by the tradeoff between the fault-free CUT losses and the test
escapes controlled by the setting of the test limits [19]. Assuming that the
cost of discarding a fault-free device is $c0$ and the cost of shipping a faulty
device (*cost factor*) is cf times more expensive ($cf \cdot c0$), if N CUTs are produced,
the total cost (indirect test cost) C_{Total} is

$$C_{\text{Total}} = N \cdot Y \cdot Losses_0^{\text{lim}} \cdot c0 + N \cdot (1 - Y) \cdot TE^{\text{lim}} \cdot c0 \cdot cf \tag{5.16}$$

where $Losses_0^{\text{lim}}$ and TE^{lim} are the chosen percentage of *fault-free CUT losses*
and the percentage of the *Test escapes*, respectively, both calculated for a certain test tolerance limit value *lim*.

The total normalized cost is

$$TCf = \frac{C_{\text{Total}}}{N \cdot c0} = Y \cdot Losses_0^{\text{lim}} + cf \cdot (1 - Y) \cdot TE^{\text{lim}} \tag{5.17}$$

The first term of (5.17) is the normalized cost of the fault-free instances that have been discarded, while the second term is the normalized cost of the test-escapes. The value of cf is not constant but may be varied according to the test application. The limit value may be obtained from Equation 5.17 where the objective is to minimize the total cost by making appropriate trade-offs between yield loss and *Test escapes*.

5.5 Experimental Results

The results presented in this section are based on measurement data from a commercial electronic circuit product, namely a nonmaintained emergency luminaire circuit [20–21]. The circuit consists of the following four major parts: (a) the rectifier circuit and the charging circuit, (b) the control circuit, which triggers the transition from normal operation mode to the emergency operation mode according to the input voltage level, (c) the lamp excitation circuit, which is responsible for producing proper voltage pulses of specific amplitude and duration, capable of igniting the lamp, and (d) rechargeable batteries. Emergency luminaire circuits need to be tested at different main power supply voltage levels (230 V AC, 180 V AC, 130 V AC, 0 V AC) in order to ensure proper operation [8,20,21]. The stimulus signal is a 50 Hz sinusoidal signal generated by a power supply circuit capable of generating all the required main AC power supply voltage levels (230 V AC, 180 V AC, 130 V AC). The measured circuit signal is the I_{PS} current. A microcontroller-based versatile test system [8,22] implementing the test procedure of Section 5.3 with the two proposed metric functions (the discrimination function Df (Section 5.3.1) and the Mahalanobis distance-based metric MD (Section 5.3.2)) has been used for the measurements. Moreover, the supply current spectrum test method of Reference 8 has been implemented in the same framework for comparison purposes.

From a set of $n = 500$ fault-free emergency luminaire circuits [8], the I_{PS} waveforms are measured and stored (256 samples per waveform). For the supply current spectrum test method of Reference 8, the following values are computed: $I_{PS}{}^{RMS0}$ (the mean RMS value of the n fault-free circuit instances), the first three harmonic magnitude components of the I_{PS}: C_1^0, C_2^0, C_3^0 $(C_j^0 = \frac{1}{n}\sum_{i=1}^{n} C_j^{0-i}$, where C_j^{0-i} is the j harmonic magnitude component of the i fault-free I_{PS} waveform and $j = 1,\ldots,3)$, $d^{RMS0-\text{lim}}$ (the tolerance limit of $I_{PS}{}^{RMS0})$, $dm\ 0_{,3}^0$ (the mean value of the set of $dm_{0,3}{}^i = \sqrt{\sum_{j=1}^{3}(C_j^{0-i} - C_j^0)^2 \Big/ 3}$, $i = 1,\ldots,n)$, $d_{0,3}^{\text{lim}}$ (the tolerance limit of the metric from Reference 8 using three harmonic magnitude components).

The set of CUTs consists of $Num = n_f = 70$ faulty emergency luminaire circuits taken from the production line [8]. These faulty circuits correspond to commonly encountered faulty cases in the production line: diode reversals, opens (absence) or short of a component, shorts and faulty fluorescent lamps. Functional testing has been performed for verifying the faulty circuits.

Indicative signal waveforms for the emergency luminaire circuit are presented in Figure 5.5. In this figure, we have the measured fault-free I_{PS} waveform and one faulty (Figure 5.5b) I_{PS} waveform. It should be noted that the test algorithm from Section 5.3 with either one of the metrics, DF or MD, succeeds in detecting the fault.

Results from the use of Df metric are depicted in the diagrams of Figure 5.6. These results, relate the tolerance limits with fault-free CUT losses and fault coverage FD_{Df} at various AC input voltages. The data values for the tolerance limits Df_0^{lim} are obtained after statistical processing of fault-free CUT measurement data for various selected percentage values of fault-free CUT losses, while the data for the fault detectabilities are obtained after processing the measurement data of CUTs and computing the percentage fault detectabilities FD_{Df} (Equation 5.14) for the various Df_0^{lim} values as described in Sections 5.3 and 5.3.1.

For a specific AC input voltage, the corresponding data in Figure 5.6 is generated by iteratively performing in sequence the following three steps which are explained in Section 5.3:

Step 1: Select the percentage a of fault-free CUT losses.

Step 2: Compute the corresponding (based on Step 1) tolerance limit Df_0^{lim}.

Step 3: Compute the corresponding (based on Step 2) percentage value of FD_{Df}.

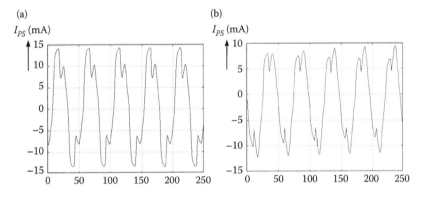

FIGURE 5.5
Indicative I_{PS} waveforms for (a) fault-free and (b) faulty emergency luminaire circuit. (Spyronasios, A., Dimopoulos, M., Hatzopoulos, A. Wavelet analysis for the detection of parametric and catastrophic faults in mixed-signal circuits. *Trans. Instrum. Meas.*, 60(6), 2025–2038. @ 2010. IEEE.)

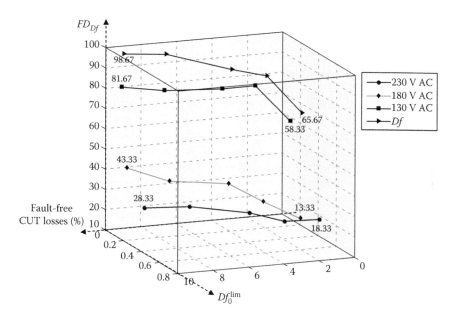

FIGURE 5.6

Results for emergency luminaire CUTs and discrimination factor *Df* metric. Measurements relating fault-free CUT losses with tolerance limits Df_0^{lim} and fault coverage FD_{Df} at various AC voltage levels.

For five different percentages of fault-free CUT losses (0%, 1.9%, 3.7%, 7%, and 9.3%) and a specific AC input voltage level, five tolerance limits Df_0^{lim} and five percentage values of FD_{Df} have been computed. The diagram data which corresponds to the *Df* label are the cumulative total results after testing the CUTs in all different AC input voltages. For example, by selecting a percentage value of fault-free CUT losses of 7% (i.e., from a lot of 100 fault-free CUTs seven of them do not pass the test process and are erroneously discarded as faulty due to the selected tolerance limits), the Df_0^{lim} values are 0.34 for the 230 V AC case, 0.12 for the 180 V AC case and 0.05 for the 130 V AC case and the succeeded total fault detectability will be $FD_{Df} = 96.67\%$.

Another observation which can be made in Figure 5.6 is that testing the emergency luminaire circuit under 130 V AC covers most of the faults with the fault coverage ranging from 58.33% (for 0% of fault-free CUT losses) to 81.67% (for 9.3% of fault-free CUT losses). When the AC input voltage is 230 V AC then the fault coverage is very low and ranges from 18.33% (for 0% of fault-free CUT losses) to 28.33% (for 9.3% of fault-free CUT losses). This is explained by the fact that some parts of the circuit are activated only when the circuit is in the emergency mode of operation [20,21] which happens when the input voltage is 130 V AC and 180 V AC. Therefore, the faults encountered in these parts cannot be observed when the circuit is in normal operation mode (input voltage of 230 V AC).

The percentages of the fault detectabilities obtained by the various test algorithms when fault-free CUT losses are considered are presented in Figure 5.7. In this figure, the wavelet test algorithm of Section 5.3 utilizing separately the *DF* metric (method "*DF*"), and the *MD* metric (method "*MD'*") is compared with other test methods from literature namely a test method based on the RMS value of I_{PS} (method "*RMS*" in Figure 5.7) and the test method of Reference 8 considering both the RMS value of I_{PS} and the first three magnitude components of the I_{PS} spectrum (method "I_{PS} *Spectrum*"). For example and for fault-free CUT loses of 7%, the RMS value alone detects a total of 83.33% of the faulty cases. The "I_{PS} *Spectrum*" detects a total of 95% of the faulty cases. The "*DF*" detects a total of $FD_{Df} = 96.67\%$ of faulty cases whereas the "*MD*" detects a total of $FD_{MD} = 97.33\%$ of faulty cases. As it is seen from Figure 5.7, the test algorithm with the *MD* metric obtains better results than the other methods.

It must be noted that the wavelet-based test method from Reference 11 implemented in the same test system and with a tolerance limit of 3σ, detects a total of 84.07% of the faulty cases but at a cost of 16.3% fault-free CUT losses. From Figure 5.7, we can see that both "*DF*" and "*MD*" and for a similar fault detection percentage (86.97% for "*DF*" and 88.33% for "*MD*") have only 1.9% fault-free CUT losses.

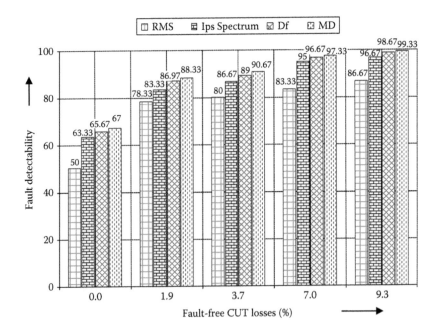

FIGURE 5.7
Comparative results for various test methods for emergency luminaire CUTs. (Spyronasios, A., Dimopoulos, M., Hatzopoulos, A. Wavelet analysis for the detection of parametric and catastrophic faults in mixed-signal circuits. *Trans. Instrum. Meas.*, 60(6), 2025–2038. @ 2010. IEEE.)

TABLE 5.1

Normalized Costs for MD Metric Function, Various Fault-Free
CUT Losses and Three Different Cost Factors for $Y = 0.95$

Fault-Free Losses (%)	Test Escapes (%)	TCf		
		$cf = 5$	$cf = 20$	$cf = 30$
0	33	0.0825	0.3300	0.4950
1.9	11.67	0.0472	0.1347	0.1931
3.7	9.33	0.0584	0.1284	0.1751
7	2.67	0.0731	0.0932	0.1065
9.3	0.67	0.0900	0.0950	0.0984

The appropriate tolerance limits for the presented methods may be selected
by applying the cost analysis of Section 5.4. Indicative results for the total
normalized cost of the described methods for three values of the cost factor
($cf = 5, 20, 30$) and for various values of fault-free CUT losses are presented in
Tables 5.1 and 5.2. It should be noted that the value of the total normalized
cost is calculated by using Equation 5.17.

From Table 5.1 and fifth row, we have for example: if the testing method
is applied with a tolerance limit which results in 7% fault-free CUT losses
and 2.67% test escapes (i.e., 97.33% fault detectability) and for a yield equal to
$Y = 95\%$, the total normalized cost from Equation 5.17 is

$$TCf\big|_{cf=5} = Y \cdot Losses_0^{\lim} + cf \cdot (1-Y) \cdot TE^{\lim} = 0.95 \cdot 0.07 + 5 \cdot (1-0.95) \cdot 0.0267 = 0.0731,$$

$$TCf\big|_{cf=20} = 0.1284 \text{ and } TCf\big|_{cf=20} = 0.1065$$

For the case of $cf = 5$, we see that the minimum cost is when we have 1.9%
fault-free CUT losses and 11.67% test escapes (i.e., 88.33% fault detectability).
Quite analogously are filled the entries in Table 5.2.

As it is observed from Table 5.1 and for the *MD* metric, the appropriate
limit which minimizes the total normalized cost for $cf = 5$ is the one with

TABLE 5.2

Normalized Costs for DF Metric Function, Various Fault-Free
CUT Losses and Three Different Cost Factors for $Y = 0.95$

Fault-Free Losses (%)	Test Escapes (%)	TCf		
		$cf = 5$	$cf = 20$	$cf = 30$
0	34.33	0.0858	0.3433	0.5149
1.9	13.03	0.0506	0.1483	0.2135
3.7	11	0.0626	0.1451	0.2001
7	3.33	0.0748	0.0998	0.1164
9.3	1.33	0.0916	0.1016	0.1083

1.9% *fault-free CUT Losses* and 11.67% *test escapes*. For $cf = 20$, the appropriate limit is the one with 7% *fault- free CUT losses* and 2.67% *test escapes* and when $cf = 30$, the minimum cost corresponds to the case of 9.3% *fault-free CUT losses* and 1.33% *test escapes*. Similar observations can be made for the *DF* metric and Table 5.2.

It should be noticed that when the cost factor *cf* increases, it is more preferable to have higher *fault-free CUT losses*. This *cf* value could be in some cases quite high in order to incorporate the high reliability requirements of certain applications, like automotive, space, etc.

Summary

The basic idea behind the described algorithm is to extract signatures which discriminate, as much as possible, the faulty circuits from the fault free. For this purpose, two metrics are analyzed. The wavelet transform, which is a representation of a signal in frequency and time domain, is used in order to exploit as many details as possible from the measured waveform. Comparative results with other methods from literature show the efficiency of the described test metrics. The presented test algorithm has a low computation cost and it is easy to apply using either the DSP of the ATE or a low-cost test system.

References

1. Singh, A., Plusquellic, J., Phatak, D. et al. Defect simulation methodology for i(DDT) testing. *J. Electron. Testing Theory Appl.*, 22(3), 255–272, 2006.
2. Bell, L., Spinks, S., Dasilva, J. Supply current test of analog and mixed-signal circuits. *IEE Proc. – Circuits Devices Syst.*, 143(6), 399–407, 1996.
3. Plusquellic, J., Singh, A., Patel, C. et al. Power supply transient signal analysis for defect-oriented test. *IEEE Trans. CAD Integr. Circuits Syst.*, 22(3), 370–374, 2003.
4. Krishnan, S., Doornbos, K., Brand, R. et al. Block-level Bayesian diagnosis of analogue electronic circuits. *Design, Automation & Test in Eur. Conf. Exhibit. (DATE)*, Dresden, Germany, 8–12, 1767–1772, 2010.
5. Fang, L., Nikolov, P., Ozev, S. Parametric fault diagnosis for analog circuits using a Bayesian framework. *24th IEEE Proc. VLSI Test Symp., (VTS)*, Berkeley, California, USA, 30 April–4 May 2006, pp. 272–277.
6. Bushnell, M.L., Agrawal. V.D. *Essentials of Electronic Testing for Digital, Memory and Mixed-Signal VLSI Circuits*. Kluwer Academic Publishers, Boston, MA, 2000.

7. Font, J., Ginard, J., Isern, E. et al. Oscillation-test technique for CMOS operational amplifiers by monitoring supply current. *Analog Integr. Circuits Signal Process.*, 33(2), 213–224, 2002.

8. Dimopoulos, M., Spyronasios, A., Papakostas, D. et al. Circuit implementation of a supply current spectrum test method. *IEEE Trans. Instrum. Meas.*, 59(10), 2660–2670, 2010.

9. Dragic, M., Margala, M. Power supply current test approach for resistive fault screening in embedded analog circuits. *Proceedings of the 18th IEEE International Symposium on Defect and Fault Tolerance in VLSI Systems*, Washington, DC, USA, 3–5 November 2003, pp. 124–131.

10. Bhunia, S., Roy, K. A novel wavelet transform-based transient current analysis for fault detection and localization. *IEEE Trans. Very Large Scale Integration (VLSI) Syst.*, 13(4), 503–507, 2005.

11. Dimopoulos, M., Spyronasios, A., Papakostas, D. et al. Wavelet energy-based testing using supply current measurements. *Sci. Meas. Technol., IET*, 4(2), 76–85, 2010.

12. Spyronasios, A., Dimopoulos, M., Hatzopoulos, A. Wavelet analysis for the detection of parametric and catastrophic faults in mixed-signal circuits. *IEEE Trans. Instrum. Meas.*, 60(6), 2025–2038, 2010.

13. Walker, J. *A Primer on Wavelets and their Scientific Applications.* Chapman & Hall, CRC Press, Boca Raton, FL, 1999.

14. Mallat, S. *A Wavelet Tour of Signal Processing.* Academic Press, San Diego, California, USA, 1999.

15. Daubechies, I. Ten Lectures on Wavelets, CBMS-NSF Lecture Notes nr. 61, SIAM, 1992.

16. Roh, J., Abraham, J. Subband filtering for time and frequency analysis of mixed-signal circuit testing. *IEEE Trans. Instrum. Meas.*, 53(2), 602–611, 2004.

17. Papoulis, A. *Probability, Random Variables and Stochastic Processes.* 3rd edition. McGraw-Hill, New York, 1991.

18. Mahalanobis, P. On the generalised distance in statistics. *Proc. Nat. Inst. Sci. India*, 2(1), 49–55, 1936.

19. Wegener, C., Kennedy, M. Test development through defect and test escape level estimation for data converters. *J. Electron. Testing Theory Appl.*, 22(4–6), 313–324, 2006.

20. Olympia Electronics S.A. home page. [Online]. Available at: http://www.olympia-electronics.gr.

21. Luminaires Part 2-22: Particular Requirements-Luminaires for Emergency Lighting. International Standard CEI/IEC 60598-2-22:1997 + A1:2002, 3.1 edition, 2002.

22. Dimopoulos, M., Papakostas, D., Hatzopoulos, A. et al. Design and development of a versatile testing system for analog and mixed-signal circuits. *Eur. Conf. Circuit Theory Design 2007 (ECCTD' 2007)*, Seville, Spain, August 2007, pp. 846–849.

6

Topological Investigations and Phase Noise Analyses in CMOS LC Oscillator Circuits

Ilias Chlis, Domenico Pepe, and Domenico Zito

CONTENTS

6.1 Introduction

Advances in wireless communications have a great impact on our societal and economic challenges [1,2]. One of the most critical circuits of modern radiofrequency transceivers is the local oscillator, that is, an autonomous circuit operating as the "pulsing heart" of such systems, in an analogy with a human body. As any other solid-state circuits, oscillators are affected by the inherent noise of the electronic devices. One of the major negative effects of noise in oscillators is given by the induced variations on the instantaneous oscillation frequency, leading to the degradation of the spectral purity of the output voltage, referred as phase noise (PN) [3,4]. Oscillator PN performance directly affects the bit-error rate (BER) of the overall communication system [5].

Understanding the generation mechanisms of PN has been a very intriguing challenge and consequently most of the efforts have been made in this direction for a number of circuit topologies [6]. However, no complete comparative studies have been carried out on how to choose the oscillator circuit topology that could potentially offer the best performance in terms of PN for the range of operating frequencies of modern telecommunication systems. Usually, thanks to its reliable startup, the common-source cross-coupled differential pair topology is chosen *a priori* without any further considerations.

Thereby, from a designer perspective, such a comparative analysis could be very helpful in focusing the design efforts toward specific directions.

The linear time-variant (LTV) oscillator model allows a quantitative understanding of oscillator PN through the impulse sensitivity function (ISF), represented as $\Gamma(x)$ [7]. Since the oscillator is assumed as a linear time-varying circuit, the phase sensitivity to noise perturbations can be described in terms of its (time-varying) impulse response.

The evaluation of the ISF involves a significant amount of transient simulations and data extractions, resulting in time-consuming calculations, potentially prone to inaccuracy. Recently, new efficient frequency-domain methods operating directly in the steady-state were proposed [8,9], allowing a consistent reduction of the simulation workload. Regardless of the methods, the analysis of the phase sensitivity can contribute significantly to a better understanding of the impact of noise sources to the oscillator PN in the most widespread circuit topologies.

Driven by the abovementioned motivations, in this chapter, we report comparative investigations of PN in the Colpitts, Hartley, and common-source cross-coupled pair oscillator topologies, with the main objective of bringing to the light the contributions of the inherent noise sources in the most widespread oscillator topologies reported in the literature. To get the complete picture of the PN performance, we will carry out analyses of both single-ended and differential Colpitts and Hartley oscillator topologies. In detail, all the steps for an accurate derivation of the ISF are summarized and the PN predictions for a wide set of amplitudes of the injected current pulse are compared with the results obtained by the direct plots obtained by means of SpectreRF-Cadence periodic steady-state (PSS) analysis. The oscillator circuit topologies are investigated under the common design conditions, such as (a) power consumption, (b) supply voltage, (c) transistor current density, (d) sizing (area, aspect ratio, and finger width), (e) inductance, (f) quality factor of the integrated spiral inductors, (g) coupling factor of the integrated transformers, and (h) considering the full models of the transistors available within the process design kit, including all their parasitic components related to their actual size, but excluding the layout interconnections, since the additional parasitic components introduced by the layout implementation could mask the results of the topological investigations which are the objective of this study extensively addressed in References 10 through 14 and summarized hereinafter. The ISF is used to quantify the impact of each noise source on the overall PN in each oscillator circuit topology, allowing the identification of the major contributions to the PN degradation versus the oscillation frequency. The results could drive the designer through the choice of the oscillator circuit topology that could potentially offer the best PN.

The chapter is organized as follows. Section 6.2 summarizes the key analytical expressions for PN predictions through the ISF and the key steps and settings for accurate evaluations. Section 6.3 reports the comparative results of PN for single-ended Colpitts and Hartley and common-source cross-coupled differential pair oscillator topologies. Quantitative and qualitative analyses of the PN contribution by each circuit component are carried out for each topology for a discrete set of oscillation frequencies ranging from 1 to 100 GHz. In Section 6.4, the comparative results of PN for differential Colpitts and Hartley and common-source cross-coupled differential pair oscillator topologies are reported. Analyses of the PN contributed by each circuit component are also carried out for each topology for a discrete set of oscillation frequencies ranging from 1 to 100 GHz. Finally, in Section 6.5, the conclusions are drawn.

6.2 Impulse Sensitivity Function

To get an insight of the noise contribution of each circuit component in each circuit topology, we make use of the ISF as a predictive tool for quantitative and qualitative PN evaluations.

A detailed procedure for computation of the ISF and PN prediction in a linear time-varying system in the case of a source-coupled CMOS multivibrator with operating frequency up to 2 MHz [15]. All the results were achieved only for a single amplitude value of the injected pulse. However, the time-domain evaluation of the ISF involves a number of transient simulations, resulting potentially prone to inaccuracy. Thereby, it is worth consolidating all the steps in order to achieve accurate results.

The impulse response from each current noise source to the oscillator output phase can be written as [7]

$$h_\phi(t, \tau) = \frac{\Gamma(\omega_0\tau)}{q_{max}} u(t - \tau) \tag{6.1}$$

where q_{max} is the charge injected into a specific circuit node of the oscillator at time $t = \tau$, $u(t)$ is the unity step function, and $\Gamma(\omega_0\tau)$ is a dimensionless periodic function that can be expressed as a Fourier series [7]

$$\Gamma(\omega_0\tau) = \frac{c_0}{2} + \sum_{n=1}^{\infty} c_n \cos(n\omega_0\tau + \theta_n) \tag{6.2}$$

The DC and root mean square (rms) values of $\Gamma(\omega_0\tau)$ are given by the following two equations [7,16]:

$$\Gamma_{DC} = \frac{c_0}{2} \tag{6.3}$$

$$\Gamma_{\text{rms}} = \sqrt{\frac{c_0^2}{4} + \frac{1}{2}\sum_{n=0}^{\infty} c_n^2} \tag{6.4}$$

All the simulations have been carried out by using the SpectreRF simulator in the Cadence design environment. The ISF of the oscillator topologies has been evaluated for an oscillation frequency of 10 GHz, which will be considered hereinafter as a reference for all the other cases. First we run a transient simulation in order to observe and record when the amplitude of the oscillation waveform reaches the steady state regime. In our case, this occurs with large margins after 5 ns. Afterward, we perform other transient simulations applying current impulsive sources acting in parallel with the actual inherent current noise sources of the LC tank and transistors, by activating only one noise source at one time. The current impulses are set to occur in the steady state regime starting from a given time reference for the unperturbed solution. The pulse width of each current source has been chosen equal to 1 ps (i.e., one hundredth of the oscillation period) with 0.1 ps rise and fall time, as shown in Figure 6.1.

The simulation has been repeated for amplitudes of the injected current of 1, 10, 100 μA and 1, 10 mA. Each transient analysis is performed using the conservative mode and a maximum time step of 10 fs (i.e., one 10-thousandth of the oscillation period), in order to have a good accuracy even in the case of the smallest injected current pulse (i.e., 1 μA). The charge q_{max} injected in each node corresponds to the area under each pulse, that is, the area of the trapezoid, shown in Figure 6.1.

$$q_{\text{max}} = I_{\text{pulse}} \times 1.1 \times 10^{-12} \text{ Coulombs} \tag{6.5}$$

where I_{pulse} is the amplitude value of each source pulse. This is repeated for all the N noise sources connected in parallel for all the M instants of time over one period of oscillation. Here, M was selected equal to 40. The time instants have been chosen to be equally spaced in an oscillation period. The time shift caused by the impulse injection can be extracted by comparing the perturbed and unperturbed waveforms. This means that when the

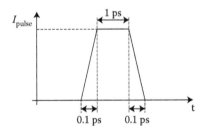

FIGURE 6.1
The injected current pulse.

oscillation has reached the steady state regime, the time shift Δt_i of the zero-crossing instant of the perturbed oscillation with respect to the unperturbed one, that is, when no impulse is applied, is calculated as shown in Figure 6.2.

Then, these time shifts are converted into phase shifts by using the following relation:

$$\Gamma(x = \omega_0 t) = 2\pi \frac{\Delta t_i(t)}{T} \tag{6.6}$$

To take into account the cyclostationary nature of the active device noise sources, $\Gamma(x)$ is multiplied with $\alpha(x)$, where $\alpha(x)$ is the absolute value of the unperturbed current flowing in the respective node in which the impulses are injected, and at the same time instant in which they are injected, normalized to its maximum value in the period. Then, the DC and root mean square (rms) components of the product $\Gamma(x) \times \alpha(x)$ can be calculated as follows:

$$\Gamma_{DC} = \frac{\sum_{i=1}^{40} [\Gamma(x)\alpha(x)]}{40} \tag{6.7}$$

$$\Gamma_{rms} = \sqrt{\frac{\sum_{i=1}^{40} \{[\Gamma(x)\alpha(x)]^2\}}{40}} \tag{6.8}$$

The total PN of the oscillator is computed by adding the contributions from all the noise sources acting in the circuit. In particular, the active devices are responsible for both flicker and thermal noise contributions to the oscillator PN, whereas the LC tank participates only in the thermal noise contribution to PN.

The total thermal noise contribution to the PN, the latter traditionally indicated with $\mathcal{L}$, from all m noise sources with a white power spectral density, can be expressed as follows [7]:

$$\mathcal{L}\{\Delta\omega\}\big|_{white} = \sum_{i=1}^{m} \left[\left(\frac{\Gamma_{rms}^2}{q_{max}^2} \right)_i \frac{(\overline{i_n^2}/\Delta f)_i}{2\Delta\omega^2} \right] \tag{6.9}$$

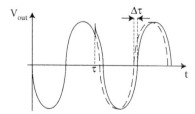

FIGURE 6.2
Time shift Δt caused by the impulse injection occurring at the time τ.

where $(\overline{i_n^2}/\Delta f)_i$ is the thermal noise generated from the ith noise source, q_{max} is the charge injected into a circuit node by the noise source i_n insisting in that node, Γ_{rms} is the root mean square (rms) value of the ISF, and $\Delta\omega$ is the offset from the oscillation angular frequency. The contribution of each noise source with white spectrum to the total thermal noise appearing at the output spectrum of the oscillator, the latter given by Equation 6.9, is independent of the angular frequency offset $\Delta\omega$ as seen from Equation 6.10.

$$\frac{[\mathcal{L}(\Delta\omega)\,|_{white}]_i}{\mathcal{L}\{\Delta\omega\}\,|_{white}} = \frac{(\Gamma_{rms}^2/q_{max}^2)_i(\overline{i_n^2}/\Delta f)_i}{\sum_{i=1}^{m}[(\Gamma_{rms}^2/q_{max}^2)_i(\overline{i_n^2}/\Delta f)_i]} \tag{6.10}$$

Moreover, the total flicker contribution to the PN from all n noise sources with a $1/f$ (flicker) spectrum can be expressed as follows [7]:

$$\mathcal{L}\{\Delta\omega\}\,|_{1/f} = \sum_{i=1}^{n}\left[\left(\frac{4\Gamma_{DC}^2}{q_{max}^2}\right)_i \frac{(\overline{i_n^2}/\Delta f)_i}{8\Delta\omega^2} \frac{(\omega_{1/f})_i}{\Delta\omega}\right] \tag{6.11}$$

where

$$\left(\frac{\overline{i_n^2}}{\Delta f}\right)_i \times \frac{(\omega_{1/f})_i}{\Delta\omega}$$

is the flicker noise generated from the ith noise source, Γ_{DC} is the DC value of the ISF, and $(\omega_{1/f})_i$ is the flicker noise corner of the ith active device. The contribution of each flicker noise source to the total flicker noise appearing at the output spectrum of the oscillator, the latter given by Equation 6.11, is independent of the angular frequency offset $\Delta\omega$ as noted from Equation 6.12.

$$\frac{[\mathcal{L}\{\Delta\omega\}\,|_{1/f}]_i}{\mathcal{L}\{\Delta\omega\}\,|_{1/f}} = \frac{(\Gamma_{DC}^2/q_{max}^2)_i(\overline{i_n^2}/\Delta f)_i(\omega_{1/f})_i}{\sum_{i=1}^{n}[(\Gamma_{DC}^2/q_{max}^2)_i(\overline{i_n^2}/\Delta f)_i(\omega_{1/f})_i]} \tag{6.12}$$

6.3 Single-Ended Colpitts, Hartley, and Common-Source Cross-Coupled Pair

Three LC oscillator topologies have been analyzed: single-ended Colpitts, single-ended Hartley, and top-biased common-source cross-coupled differential pair oscillator topologies, as shown in Figure 6.3. The same figure

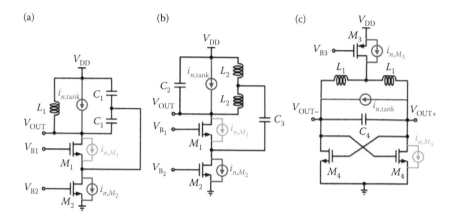

FIGURE 6.3
Schematic of the oscillator circuit topologies: (a) single-ended Colpitts; (b) single-ended Hartley; (c) top-biased common-source cross-coupled differential pair. V_{B1}, V_{B2}, and V_{B3} are DC bias voltages.

shows the current impulsive sources acting in parallel to the inherent current noise sources and used for the evaluation of the ISF. The three oscillator circuit topologies have been implemented in 28-nm bulk CMOS technology by STMicroelectronics by adopting the same criteria for a fair comparison as follows. The frequency of operation is 10 GHz. The sizes of the transistors and the value of the inductors and capacitors used are reported in Table 6.1. Despite this work is addressed to the investigations of the circuit topologies as such, rather than the circuit design and implementation, that is, regardless of the effects of parasitic components, we considered a reasonable quality factor for the LC tank in order to carry out the comparative study of the properties of each circuit topology under the same typical conditions. Thereby, a quality factor (Q) equal to 10 has been assumed for the inductors, considering a parasitic resistance (noisy) in series with the inductor, whereas the capacitors have been considered as ideal devices. In all cases, the power consumption is 6.3 mW.

Figure 6.4 reports $\Gamma(x)\alpha(x)$ for an injected current pulse amplitude of 1 µA versus the phase for the injected noise sources, during one oscillation period, for the three oscillator circuit topologies.

TABLE 6.1

Device Sizing

Devices	Transistor Width (µm)				Capacitor Value (pF)				Inductor Value (pH)	
	M_1	M_2	M_3	M_4	C_1	C_2	C_3	C_4	L_1	L_2
Values	30	30	30	15	0.97	0.495	0.8	0.229	500	250

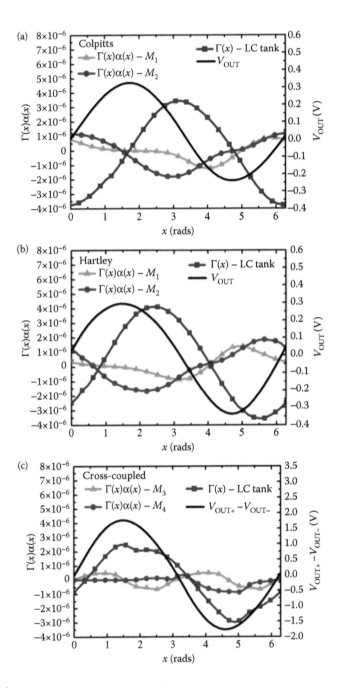

FIGURE 6.4

$\Gamma(x)\alpha(x)$ of the MOSFETs and $\Gamma(x)$ of the LC tank versus phase for a 1-μA amplitude current impulse, for an oscillation frequency of 10 GHz: (a) Colpitts topology; (b) Hartley topology; (c) common-source cross-coupled differential pair topology.

Figure 6.5 reports the comparison between the PN obtained through the ISF and the PN obtained by direct plots from PSS and periodic noise simulations, for the three oscillator circuit topologies. It is worth noting that the PN predicted by the ISF is very close to the values obtained by means of SpectreRF simulations. Table 6.2 provides the PN results for all the current impulse amplitude values for a 1-MHz frequency offset from the carrier.

The agreement degrades for higher pulse amplitudes, when the current-to-phase transfer function starts becoming nonlinear. The amplitude in which this occurs is slightly different for each oscillator topology, but for the injected current impulse of 1 μA, the difference between the PN predicted by the ISF method and the one given by PSS and periodic noise (Pnoise) analysis is lower than 1% at a 1-MHz frequency offset.

The investigations through the ISF can provide a better understanding of the PN in each oscillator topology. To be able to extract further useful considerations about the devices and topologies, the previous analyses have been reiterated also for other oscillation frequencies. In detail, the three oscillator topologies have been implemented also for 1 and 100 GHz operations, by keeping the quality factor of 10 for the LC tank and preserving the same power consumption of 6.3 mW as in the case of the 10-GHz oscillation frequency. The transistor sizes were also kept the same as in the previous case. As a consequence of the results reported in the previous section, we injected noise current impulses with the amplitude of 1 μA.

Table 6.3 reports the values of the individual circuit components for the topologies of Figures 6.3a–c used for the oscillation frequencies of 1 and 100 GHz.

Table 6.4 reports the PN values at a 1-MHz offset predicted by the ISF along with the values obtained by means of SpectreRF simulations for the oscillation frequencies of 1 and 100 GHz.

Figures 6.6a and b report the relative contributions of M_1, M_2, and LC tank to the overall PN versus the oscillation frequency for the Colpitts topology, in both the flicker and thermal noise contributions to PN.

Figures 6.7a and b and 6.8a and b report the results for the Hartley and common-source cross-coupled differential pair topologies, respectively.

The results suggest considering additional oscillation frequencies. For this reason, the three topologies have been designed also for the additional oscillation frequencies of 30, 50, and 70 GHz, according to the same criteria reported at the beginning of Section 6.3.

Table 6.5 reports the values of the circuit components for each topology for the oscillation frequencies of 30, 50, and 70 GHz.

Figure 6.9 reports the PN results obtained by SpectreRF for 1, 10, 30, 50, 70, and 100 GHz at a 1-MHz frequency offset from the carrier.

These results allow us to identify the following four main frequency regions: 1–20, 20–30, 30–80, and 80–100 GHz. They offer the opportunity to carry out further comparative analyses and derive a number of observations.

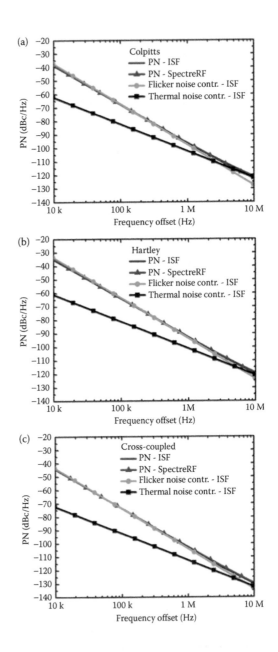

FIGURE 6.5
Phase noise versus frequency offset for the three oscillator circuit topologies, obtained through the ISF for a 1-μA current impulse and direct plot from PSS and periodic noise (Pnoise) SpectreRF simulations for an oscillation frequency of 10 GHz. The flicker and thermal noise contributions to the overall phase noise are also plotted in order to identify the $1/f^3$ phase noise frequency corner. (a) Colpitts. The $1/f^3$ phase noise corner is at a frequency offset of 3.1 MHz. (b) Hartley. The $1/f^3$ phase noise corner is at the frequency offset of 5.7 MHz. (c) Common-source cross-coupled differential pair. The $1/f^3$ phase noise corner is at a frequency offset of 7.5 MHz.

TABLE 6.2

Summary of the Phase Noise Results Obtained by SpectreRF and ISF

		Phase Noise (dBc/Hz) @ 1-MHz Frequency Offset				
			ISF			
Topology	SpectreRF	1 µA	10 µA	100 µA	1 mA	10 mA
Colpitts	−96.25	−96.20	−98.33	−98.49	−98.50	−98.45
Hartley	−92.75	−92.79	−95.18	−94.36	−94.85	−95.29
Cross-coupled	−102.66	−102.69	−102.84	−102.83	−102.84	−102.94

TABLE 6.3

Device Sizing for Oscillation Frequencies of 1 and 100 GHz

	Transistor Width (µm)				Capacitor Value (fF)				Inductor Value (pH)	
Devices	M_1	M_2	M_3	M_4	C_1	C_2	C_3	C_4	L_1	L_2
Values @ 1 GHz	30	30	30	15	10^4	5×10^3	10^4	2.5×10^3	5×10^3	2.5×10^3
Values @ 100 GHz	30	30	30	15	51.5	23	100	5.7	50	25

TABLE 6.4

Summary of the Phase Noise Results Obtained by SpectreRF and ISF

	Phase Noise (dBc/Hz) @ 1-MHz Frequency Offset			
	1 GHz		100 GHz	
Topology	SpectreRF	ISF (1 µA)	SpectreRF	ISF (1 µA)
Colpitts	−115.31	−116	−77.06	−77.69
Hartley	−114.52	−114.85	−81.18	−81.38
Cross-coupled	−123.7	−124.06	−74.78	−75.59

Region 1 (1–20 GHz): According to Figure 6.9, the common-source cross-coupled differential pair topology exhibits the lowest PN with respect to the other two topologies. The highest PN is exhibited by the Hartley topology.

Region 2 (20–30 GHz): In this region, we note from Figure 6.9 that the common-source cross-coupled differential pair topology still maintains the best PN performance, but unlike the case mentioned above, we can observe an inversion between the Hartley and Colpitts topologies. The latter exhibits the worst PN at 30 GHz.

Region 3 (30–80 GHz): In Figure 6.9, we register an inversion for the best PN performance, given now by the Hartley topology, whereas the Colpitts topology still exhibits the worst PN as in the previous case.

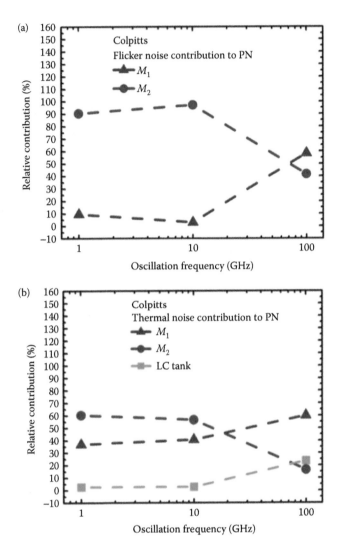

FIGURE 6.6
Relative contributions of M_1, M_2, and the LC tank for the Colpitts topology versus oscillation frequency. The contribution of each noise source to the flicker and thermal noise appearing at the output spectrum of the oscillator is independent of the angular frequency $\Delta\omega$ as seen from Equations 6.12 and 6.10, respectively. (a) Flicker noise contribution to phase noise. (b) Thermal noise contribution to phase noise.

Region 4 (80–100 GHz): Figure 6.9 indicates that Hartley continues to exhibit the lowest PN. However, with respect to the previous case, here we can observe an inversion of performance between the Colpitts topology and the common-source cross-coupled differential pair topology, which now exhibits the highest PN. The operation in the triode region for some part of

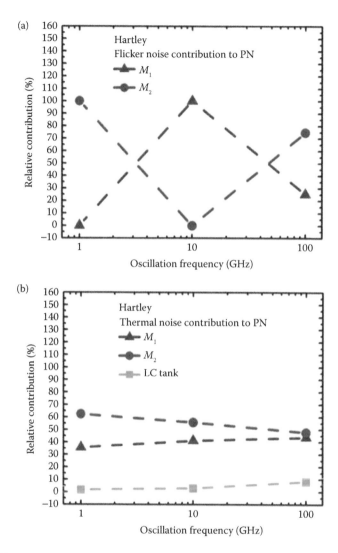

FIGURE 6.7
Relative contributions of M_1, M_2, and the LC tank for the Hartley topology versus oscillation frequency. The contribution of each noise source to the flicker and thermal noise appearing at the output spectrum of the oscillator is independent of the angular frequency $\Delta\omega$ as seen from Equations 6.12 and 6.10, respectively. (a) Flicker noise contribution to phase noise. (b) Thermal noise contribution to phase noise.

the oscillation period is the main reason for this noise performance degradation at the highest frequencies in the common-source cross-coupled differential pair topology. Indeed, our design operates in the voltage-limited regime, thus causing the active devices enter in the triode region at the peaks of the differential output node voltage.

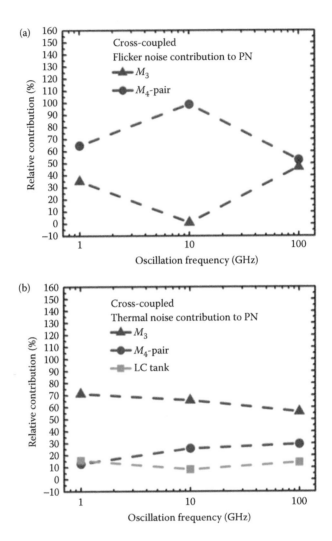

FIGURE 6.8
Relative contributions of M_3, M_4, and the LC tank for the common-source cross-coupled differential pair topology versus oscillation frequency. The contribution of each noise source to the flicker and thermal noise appearing at the output spectrum of the oscillator is independent of the angular frequency $\Delta\omega$ as seen from Equations 6.12 and 6.10 respectively. (a) Flicker noise contribution to phase noise. (b) Thermal noise contribution to phase noise.

6.4 Differential Colpitts, Hartley, and Common-Source Cross-Coupled Pair

Despite the results reported in Section 6.3 provide a first interesting perspective, this is limited by the comparison between a differential topology,

TABLE 6.5

Device Sizing for Oscillation Frequencies of 30, 50, and 70 GHz

Devices	Transistor Width (μm)				Capacitor Value (fF)				Inductor Value (pH)	
	M_1	M_2	M_3	M_4	C_1	C_2	C_3	C_4	L_1	L_2
Values @ 30 GHz	30	30	30	15	286	138.5	400	62.7	166.7	83.35
Values @ 50 GHz	30	30	30	15	150	71	250	29.7	100	50
Values @ 70 GHz	30	30	30	15	92.8	43.9	150	15.8	71.4	35.7

that is, cross-coupled common-source differential pair, and two single-ended topologies, Colpitts and Hartley. In fact, assuming a perfect symmetry, common-mode noise sources (e.g., noise coming from the common bias circuitry) do not produce effects in a differential topology; thereby, in principle, this aspect may play a significant role in determining the PN performance, then leading to a less effective comparison. Moreover, despite Colpitts and Hartley single-ended topologies have long been used in discrete circuit design, the advances in silicon integration have led to the implementation of differential versions as well, which have shown the potential for superior performance, typically at the expenses of larger area occupancy on silicon.

Consequently, extending the comparative analyses of PN to the Colpitts and Hartley differential topologies is in order for a comparison under peer conditions. Figures 6.10a and b show the oscillator circuit topologies designed in 28-nm bulk CMOS technology, operating from a 1-V supply voltage. All the circuit topologies operate in the voltage-limited regime. The same figure shows the current impulsive sources acting in parallel to the inherent current noise sources and used for the evaluation of the ISF. Based on the findings in Section 6.3, transient simulations were performed for an injected current amplitude of 1 μA.

The sizes of the active and passive devices are reported in Table 6.6. Capacitors are considered ideal, whereas a quality factor (Q) of 10 is assumed for the spiral inductors, that is, a feasible value for the oscillation frequencies in the range of interest from 1 to 100 GHz [17,18]. A coupling factor k of 0.85 is assumed for the transformers. As in the previous case, for all the investigated differential circuit topologies the total power consumption is 6.3 mW.

In particular, the comparative analysis takes into account the common-source cross-coupled differential pair, Colpitts, and Hartley differential circuit topologies shown in Figures 6.3c and 6.10a and b, which have shown the best PN performances with respect to other design variations.

Thereby, these topologies allow an effective comparison based on the actual needs and opportunities, rather than a comparison between basic topologies and their variations which are known from the literature to provide worse PN performance with respect to those considered in this comparative

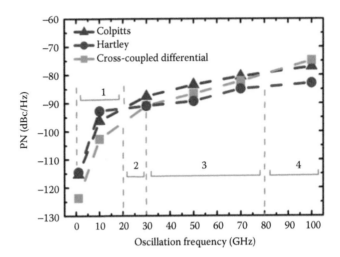

FIGURE 6.9
Phase noise at a 1-MHz frequency offset from carrier versus oscillation frequency for Colpitts, Hartley, and common-source cross-coupled differential pair topologies by SpectreRF.

analysis. In other words, the investigated topologies are the most promising in their category. The common-source cross-coupled pair in Figure 6.3c provides the negative resistance needed for the oscillation start-up. A p-MOSFET is chosen as a current source since it exhibits lower flicker noise. The transformer coupling in the Colpitts topology of Figure 6.10a contributes to the suppression of common-mode oscillations. Moreover, for lower PN, two separate tail current transistors are used for biasing. As for the Hartley topology in Figure 6.10b, the transformer coupling is used in order to reduce the area occupied by the inductors [19].

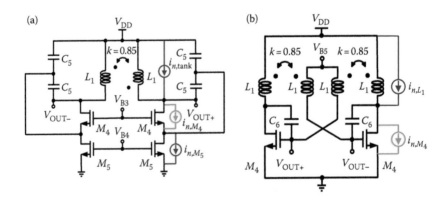

FIGURE 6.10
Schematic of the oscillator circuit topologies: (a) differential Colpitts; (b) differential Hartley; V_{B3}, V_{B4}, and V_{B5} are DC bias voltages.

TABLE 6.6

Device Sizing

Oscillation Frequency (GHz)	Transistor Width (µm)		Capacitor Value (fF)		Inductor Value (pH)
	M_4	M_5	C_5	C_6	L_1
1	15	15	5.55×10^3	1.41×10^3	5×10^3
10	15	15	527	132.5	500
100	15	15	29	4.72	50

Figures 6.11 through 6.13a–c report the results in terms of PN obtained through the ISF and direct plots from periodic steady state (PSS) and periodic noise (Pnoise) simulations by SpectreRF in Cadence.

The ISF allows us to determine the flicker and thermal noise contributions to the overall PN, as reported in Figures 6.11 through 6.13a–c [10–14]. Consequently, the $1/f^3$ corner of the PN can be identified in each case. Table 6.7 provides the results for a 1-MHz frequency offset from the carrier. The results show that the PN predicted by ISF matches well (within 1.7 dB) with the values obtained directly by means of SpectreRF simulations.

To gain a better understanding of the performances in between the initial discrete set of frequencies, the topologies have been designed also for the additional operating frequencies of 30, 50, and 70 GHz. The PN at a 1-MHz frequency offset from the carrier frequency obtained by direct plots from PSS and Pnoise simulations is shown in Figure 6.14. By inspection, Figure 6.14 reveals three distinct regions in which the topologies rank unevenly in terms of best PN performance.

Region 1 (1–10 GHz): The common-source cross-coupled topology exhibits the lowest PN, whereas differential Colpitts exhibits the worst one. In between, differential Hartley shows PN performance very close to that given by the common-source cross-coupled topology.

Region 2 (10–70 GHz): The differential Hartley topology shows the lowest PN. Its superior performance with respect to the other topologies, improves as the oscillation frequency increases. In particular, at 70 GHz, the differential Hartley topology shows a PN about 10 dB lower with respect to the others.

Region 3 (70–100 GHz): The differential Hartley is still characterized by the best PN. Here, differential Colpitts exhibits a better PN with respect to the common-source cross-coupled topology.

Thereby, it can be concluded that for oscillation frequencies between 1 and 10 GHz, the common-source cross-coupled oscillator topology considered here could be potentially the best choice. Outside this range, between 10 and 100 GHz, its performance dramatically deteriorates. For this oscillation frequency range, the differential Hartley topology considered in this study appears to be potentially the best choice.

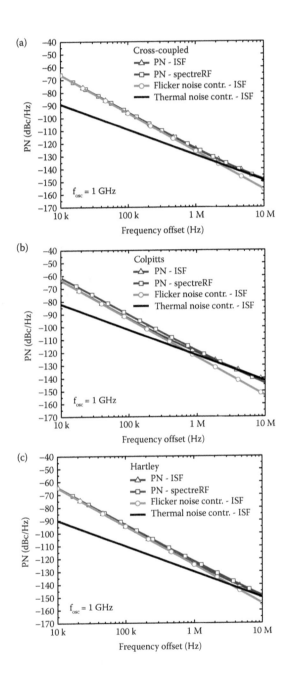

FIGURE 6.11
Phase noise versus frequency offset obtained through the ISF for a 1μA current impulse and direct plot from PSS and Pnoise SpectreRF simulations, for the oscillation frequency of 1 GHz for (a) common-source cross-coupled differential pair. The $1/f^3$ phase noise corner is at the frequency offset of 2 MHz; (b) differential Colpitts. The $1/f^3$ phase noise corner is at a frequency offset of 0.74 MHz; (c) differential Hartley. The $1/f^3$ phase noise corner is at a frequency offset of 3.7 MHz.

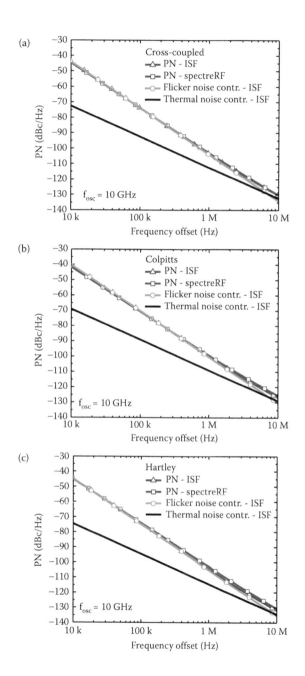

FIGURE 6.12

Phase noise versus frequency offset obtained through the ISF for a 1 μA current impulse and direct plot from PSS and Pnoise SpectreRF simulations, for the oscillation frequency of 10 GHz for (a) common-source cross-coupled differential pair. The $1/f^3$ phase noise corner is at the frequency offset of 7.5 MHz; (b) differential Colpitts. The $1/f^3$ phase noise corner is at a frequency offset of 8 MHz; (c) differential Hartley. The $1/f^3$ phase noise corner is at a frequency offset of 10 MHz.

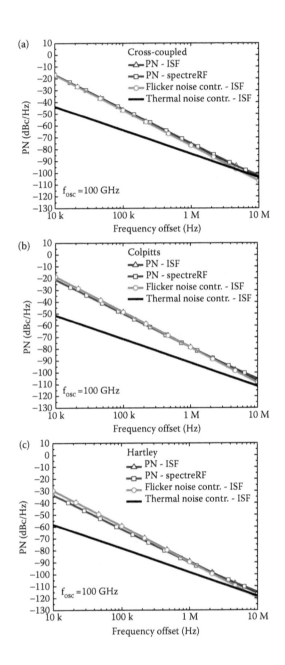

FIGURE 6.13
Phase noise versus frequency offset obtained through the ISF for a 1 μA current impulse and direct plot from PSS and Pnoise SpectreRF simulations, for the oscillation frequency of 100 GHz for (a) common-source cross-coupled differential pair. The $1/f^3$ phase noise corner is at the frequency offset of 5.5 MHz; (b) differential Colpitts. The $1/f^3$ phase noise corner is at the frequency offset of 20 MHz; (c) differential Hartley. The $1/f^3$ phase noise corner is at the frequency offset of 7.4 MHz.

TABLE 6.7

Summary of Phase Noise Performance

| | Phase Noise (dBc/Hz) @ 1 MHz Frequency Offset | | | | | |
| | 1 GHz | | 10 GHz | | 100 GHz | |
Topology	SpectreRF	ISF	SpectreRF	ISF	SpectreRF	ISF
Cross-coupled	−123.7	−124.06	−102.66	−102.69	−74.78	−75.79
Colpitts	−118.34	−119.71	−99.56	−100.26	−78.45	−78.27
Hartley	−121.69	−123.38	−102.59	−104.15	−89.78	−89.00

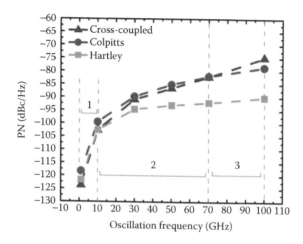

FIGURE 6.14

Phase noise at a 1-MHz frequency offset from the carrier frequency obtained by direct plots from PSS and Pnoise simulations versus oscillation frequency for the common-source cross-coupled differential pair, as well as for the differential Colpitts and Hartley oscillator circuit topologies.

Figures 6.8a and b and 6.15 through 6.17 report the percent contributions of the active and passive device noise sources to the flicker and thermal noise components of the PN. They show also the total contributions of each device to PN at a frequency offset of 1 MHz from the oscillation frequency. These results allow us to derive several important observations.

For the 1-GHz oscillation frequency, in the common-source cross-coupled pair topology, the flicker noise sources of the cross-coupled pair contribute for about 65% to the flicker noise component of the PN, as shown in Figure 6.8a. From Figure 6.8b, it can be observed that only 13% of the thermal noise component of PN comes from the thermal noise of the cross-coupled pair. In spite of the latter observation, Figure 6.17a shows that the total noise from the cross-coupled pair is the major contribution to the PN at a 1-MHz frequency offset. This can be explained by noticing from Figure 6.11a that the $1/f^3$ corner of the PN is at a frequency offset of 2 MHz.

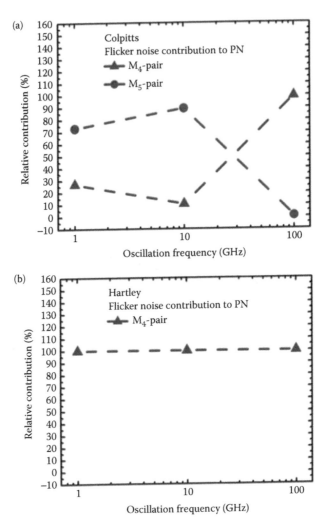

FIGURE 6.15
Relative flicker noise contributions to phase noise from active devices in the oscillator circuit topologies for (a) differential Colpitts; (b) differential Hartley. The contribution of each noise source to the flicker noise appearing at the output spectrum of the oscillator is independent of the angular frequency offset $\Delta\omega$ as seen from Equation 6.12.

In the Colpitts topology, from Figures 6.15a and 6.16a both the flicker noise and thermal noise components of the PN are mainly due to the tail current transistor pair M_5. This explains why at a 1-MHz frequency offset, the tail current transistor pair M_5 takes the largest portion of the total PN as reported in Figure 6.17b. With respect to the differential Hartley, the active device pair M_4 is the only source of flicker noise. In addition, the thermal noise generated by M_4 pair is mainly responsible for the white noise affecting the PN. This is

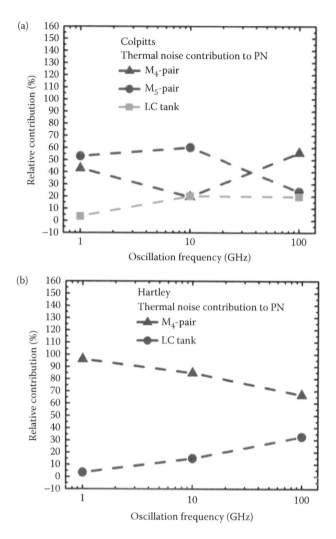

FIGURE 6.16
Relative thermal noise contributions to phase noise from active and passive devices in the oscillator circuit topologies for (a) differential Colpitts; (b) differential Hartley. The contribution of each noise source to the thermal noise appearing at the output spectrum of the oscillator is independent of the angular frequency offset $\Delta\omega$ as seen from Equation 6.10.

why at a 1-MHz frequency offset, the total noise contribution to PN is dominated by the M_4 common-source crossed-coupled pair.

For the oscillation frequency of 10 GHz, in the common-source cross-coupled pair topology, from Figure 6.8a the cross-coupled pair M_4 is the major contributor to the flicker noise component of PN. Also, at a 1-MHz frequency offset the PN spectrum is still at the $1/f^3$ region according to Figure 6.12a.

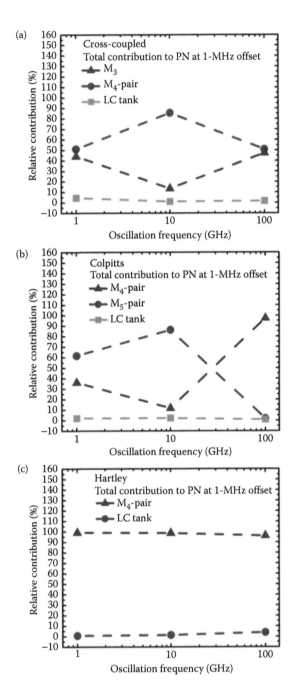

FIGURE 6.17
Relative total noise contributions to phase noise, from active and passive devices in the oscillator circuit topologies at a 1-MHz offset from the carrier frequency for (a) common-source cross-coupled differential pair; (b) differential Colpitts; (c) differential Hartley.

Thereby, the cross-coupled pair M_4 is expected to be the dominant source of the overall PN as confirmed from Figure 6.17a.

As for the Colpitts topology, the flicker contribution of the tail current transistor pair M_5 is predominant. Thereby, for the same reason as above, the tail current noise is the major contributor to PN at a 1-MHz frequency offset.

The pair of transistors M_4 in differential Hartley is almost the sole source of noise. This is because the noise generated by the parasitic resistance of the inductors is at least one order of magnitude lower than the flicker and thermal noise of the active device pair M_4.

For the oscillation frequency of 100 GHz, the $1/f^3$ frequency corners of PN in the three topologies under investigation are beyond 5 MHz, according to Figures 6.13a–c. Consequently, at a 1-MHz frequency offset the PN is mostly due to the flicker noise. In particular, from Figures 6.8a and 6.17a, in the common-source cross-coupled topology, the cross-coupled pair M_4 and the p-MOSFET current source M_3 show similar contributions to the flicker noise component of PN and to the overall PN. This happens in spite of the higher thermal noise component of PN from the tail current transistor M_3, reported in Figure 6.8b.

In the Colpitts topology, flicker and thermal noise contributions to PN are mainly due to the M_4 transistor pair as shown in Figures 6.15a and 6.16a. Thereby, the pair of transistors M_4 represents the noise source which is mainly responsible for the PN at a 1-MHz frequency offset.

In the Hartley topology, the active device pair M_4 is the only flicker noise source, which dominates the PN at a 1-MHz frequency offset.

From all above, we can conclude that, for all three topologies studied in this section and for the oscillation frequencies of 10 and 100 GHz, the $1/f^3$ region of the PN extends above 1 MHz. At an oscillation frequency of 1 GHz, this is true for the common-source cross-coupled pair and for the Hartley topologies. This is a consequence of the adoption of nano-scale CMOS technologies characterized by flicker noise corners of several tens or hundreds of MHz, which lead to flicker noise up-conversion being responsible for most of PN even at large offsets from the carrier frequency. This means that the devices with the highest contribution to the flicker noise present at the output spectrum will also dominate PN.

Hence, design efforts should be made in minimizing the flicker noise sources as much as possible, as well as flicker noise up-conversion mechanisms in each topology. For example, increasing the width of the cross-coupled devices shown in Figure 6.3c would reduce the flicker noise produced by the transistor pair M_4. On the other hand, flicker noise up-conversion gain would be increased. This is due to the increased small-signal loop gain which would in turn cause a higher distortion of the voltage output. As another example, a filtering technique adopting a resonant filter could also be adopted for reducing flicker noise up-conversion [20].

For the oscillation frequency of 1 GHz, in the Colpitts topology the $1/f^3$ PN corner is below the 1-MHz frequency offset. In this case, the active or passive devices with the highest contribution to the thermal noise component of the PN will be dominant. Short channel effects such as velocity saturation and channel length modulation are responsible for the significant increase in the thermal noise excess factor γ in deep submicron CMOS technologies. Thus, thermal noise from the active devices is usually the principal source of white noise in the output spectrum as already observed from Figures 6.8b and 6.16a and b.

It is worth noting that at lower frequency, that is, below 10 GHz, the common-source cross-coupled differential pair topology of Figure 6.3c shows superior PN performance than both single-ended and differential versions of the Colpitts and Hartley topologies considered in this study. This is consistent with the current trend in literature of adopting a common-source cross-coupled topology without any further investigations. However, the results presented here also suggest the opportunity to explore circuit design implementations also of other topologies, the potential of which may have been perhaps underestimated until today, especially at very high frequencies.

Moreover, a comparison between the PN performances of the single-ended and differential Colpitts and Hartley topologies studied in this chapter based on the SpectreRF results reported in Tables 6.2, 6.4, and 6.7 reveals the following. The differential versions of those topologies show a superior PN performance, at every examined oscillation frequency, thereby confirming the discussion reported in Section 6.4.

6.5 Conclusions

In this chapter, we reported comparative analyses of phase noise (PN) for three oscillator circuit topologies: common-source cross-coupled differential pair, Colpitts, and Hartley. To get a complete picture of the PN performance, we investigated both single-ended and differential versions of Colpitts and Hartley oscillator topologies. The oscillator circuit topologies were designed in a 28-nm bulk CMOS technology, for a set of operating frequencies in the range of 1–100 GHz. All the topologies were investigated under the same common design conditions, such as power consumption, supply voltage, transistor current density and sizing, inductance and quality factor of the integrated spiral inductors, coupling factor of the integrated transformers, and considering the full models of the transistors available within the process design kit, including all their parasitic components related to their actual size. Furthermore, the PN results from PSS and Pnoise simulations were compared with PN predictions obtained by the impulse sensitivity function. Finally, we evaluated and discussed the noise contributions from each active or passive device to the flicker and thermal PN components and to the overall PN at a 1-MHz frequency offset from the carrier frequency.

The results show that, under the adopted design conditions, the PN of the three topologies degrades unevenly over the considered oscillation frequency range. In particular, the comparative analyses show the existence of several distinct frequency regions. Thereby, the results presented here suggest that the identification of the best oscillator circuit topology in terms of PN is related to the operating frequency range. Consequently, these results suggest the opportunity to address further investigations on the Hartley topology considered in this study. The investigations could allow us to extend the range of possibilities beyond the common practice of choosing the common-source cross-coupled differential pair topology, traditionally selected for its reliable start-up, but without further topological considerations.

Finally, the investigations through the impulse sensitivity function allowed the identification of the dominant noise contributions for each oscillator circuit topology. Despite a few exceptions, the results show that the flicker noise from the active devices is the component with the most significant effect on the oscillator PN at a 1-MHz frequency offset from the carrier frequency, confirming its rising role in nano-scale CMOS technology.

References

1. Zito, D., Pepe, D., Mincica, M., Zito, F., Tognetti, A., Lanata, A., De Rossi, D. SoC CMOS UWB pulse radar sensor for contactless respiratory monitoring. *IEEE Tran. Biomed. Circuits Systems*, 5(6), 503–510, 2011.
2. Zito, D., Pepe, D., Mincica, M. A 90 nm CMOS SoC UWB pulse radar for respiratory rate monitoring. *IEEE International Solid-State Circuits Conference (ISSCC), Digest of Technical Papers*, San Francisco, CA, USA, pp. 40–41, 2011.
3. Zito, D., Pepe, D., Fonte, A. 13 GHz CMOS LC active inductor VCO. *IEEE Microwave Wireless Compon. Lett.*, 22(3), 138–140, 2012.
4. Voicu, M., Pepe, D., Zito, D. Performance and trends in millimetre-wave CMOS oscillators for emerging wireless applications. *Int. J. Microwave Sci. Technol.*, 2013, 6, 2013, article ID 312618.
5. Pepe, D., Zito, D. System-level simulations investigating the system-on-chip implementation of 60GHz transceivers for wireless uncompressed HD video communications. *Applications of MATLAB in Science and Engineering*, InTech Open Access, Vienna, Austria, pp. 181–196, 2011.
6. Maffezzoni, P., Pepe, F., Bonfanti, A. A unified method for the analysis of phase and amplitude noise in electrical oscillators. *IEEE Trans. Microwave Theory Tech.*, 61(9), 3277–3284, 2013.
7. Hajimiri, A., Lee, T. H. A general theory of phase noise in electrical oscillators. *IEEE J. Solid-State Circuits*, 33(2), 179–194, 1998.
8. Maffezzoni, P. Analysis of oscillator injection locking through phase-domain impulse-response. *IEEE Tran. Circuits Syst. I: Reg. Papers1*, 55(5), 1297–1305, 2008.

9. Levantino, S., Maffezzoni, P., Pepe, F., Bonfanti, A., Samori, C., Lacaita, A.L. Efficient calculation of the impulse sensitivity function in oscillators. *IEEE Tran. Circuits Syst. II: Exp. Briefs 2*, 59(10), 628–632, 2012.

10. Chlis, I., Pepe, D., Zito, D. Comparative analysis of phase noise in 28nm CMOS LC oscillator circuit topologies: Hartley, Colpitts and common-source cross-coupled differential pair. *Sci. World J.*, 2014, 13, 2014, article ID 421321.

11. Chlis, I., Pepe, D., Zito, D. Phase noise comparative analysis of LC oscillators in 28nm CMOS through the impulse sensitivity function. *IEEE Proceedings of Ph.D. Research in Microelectronics and Electronics (PRIME)*, Villach, Austria, pp. 85–88, 2013.

12. Chlis, I., Pepe, D., Zito, D. Comparative analyses of phase noise in differential oscillator topologies in 28nm CMOS technology. *IEEE Proceedings of PhD Research in Microelectronics and Electronics (PRIME)*, pp. 1–4, 2014.

13. Chlis, I., Pepe, D., Zito, D. Comparative analysis of phase noise in 28nm CMOS LC oscillator differential topologies: Armstrong, Colpitts, Hartley and common-source cross-coupled pair. *Sci. World J.*, 2014, 13, 2014, article id 421321.

14. Chlis, I., Pepe, D., Zito, D. Comparison on phase noise in common-source cross-coupled pair and Armstrong differential topologies. *IEEE Proceedings of Irish Signals and Systems Conference*, Limerick, Ireland, pp. 1–4, 2014.

15. Paavola, M., Laiho, M., Saukoski, M., Kamarainen, M., Halonen, K. A. I. Impulse sensitivity function-based phase noise study for low-power source-coupled CMOS multivibrators. *Analog Integr. Circuits Signal Process.*, 62(1), 29–41, 2010.

16. Lu, L., Tang, Z., Andreani, P., Mazzanti, A., Hajimiri, A. Comments on 'A general theory of phase noise in electrical oscillators'. *IEEE J. Solid-State Circuits*, 43(9), 2170, 2008.

17. Zito, D., Fonte, A., Pepe, D. Microwave active inductors. *IEEE Microwave Wireless Components Lett.*, 19(7), 461–463, 2009.

18. Pepe, D., Zito, D. 50 GHz mm-wave CMOS active inductor. *IEEE Microwave and Wireless Compon. Lett.*, 24(4), 254–256, 2014.

19. Bao, M., Li, Y. A compact 23 GHz Hartley VCO in 0.13 μm CMOS technology. *IEEE Proceedings of 3rd European Microwave Integrated Circuits Conference*, Amsterdam, Netherlands, pp. 75–78, 2008.

20. Hegazi, E., Sjoland, H., Abidi, A. A. A filtering technique to lower LC oscillator phase noise. *IEEE J. Solid-State Circuits*, 36(12), 1921–1930, 2001.

7

Design of an Energy-Efficient ZigBee Transceiver

Antonio Ginés, Rafaella Fiorelli, Alberto Villegas,
Ricardo Doldán, Manuel Barragán, Diego Vázquez,
Adoración Rueda, and Eduardo Peralías

CONTENTS

7.1 Introduction

ZigBee transceivers are being encouraged in this last decade for ultra-low power, low-cost low-data-rate wireless applications. In most cases, these devices should be able to operate from a battery with a life time of years. Typical applications for this standard include diverse areas in the industry or home automation. Of relevant interest, we can highlight its specific use to create and to control the smart grids which distribute the utilities (as electricity, gas, and water). In order to achieve the power targets, it is critical not only to have low power consumption during transmission and reception, but also to be able to have a fast wake up time, so that the devices can be most of the time in sleep mode.

This chapter tries to summarize our experience in the development of the analog front-end part of 2.4 GHz ZigBee transceivers with the main objective of optimizing power consumption during normal operation in both reception (described in Section 7.3) and transmission modes (at Section 7.4). Other interesting design aspects, such as optimizing the transceiver protocol, the design of the digital subsystems, or managing the sleep modes, have not been included due to space limitation. To gather together the presented design ideas, the chapter concludes in Section 7.5 with an example of a complementary metal-oxide semiconductor (CMOS) integrated transceiver analog front-end. The competitive experimental performances for this integration endorse the employed design flow, procedures, and analysis.

7.2 The ZigBee Standard

ZigBee® is a standard developed by the ZigBee Alliance (www.zigbee.org), built in 2002, that defines a communication protocol for short-range wireless networking mainly covering applications where low-data-rate, low cost, and long battery life are required. ZigBee is compliant in its physical layer and medium access control protocols with the IEEE 802.15.4 standard [1]. Operation frequency bands are 868 MHz, 915 MHz, and 2.4 GHz with a maximum data rate of 250 kb/s. ZigBee uses spreading methods to improve the receiver sensitivity level, to increase the jamming/interference resistance, and to reduce the effect of the multipath. The spreading method used in the 2.4 GHz-band is the direct sequence spread spectrum (DSSS). Data bit streams going at 250 kbps, after the spreading, are transformed in pseudo random sequences of bits, named chips, at a rate of 2 Mbps. Number of channels in this band are 16 (of 5 MHz-wide each one) and distributed between frequencies of 2402.5 and 2482.5 MHz. Sensitivity of the receiver in each channel has been specified from −85 to −20 dBm input power for detection error rates less than 1% with 208-bit information packets in a range up to 100 m. The transmitter output level must reach at least −3 dBm with a variable adjustment of 30 dB.

7.2.1 ZigBee Transceiver Architectures

The choice of the receiver and transmitter architectures in a ZigBee transceiver strongly determines the performance of the whole system, especially concerning power consumption. Architecture selection has been widely analyzed in many recent works—a good survey of them appears in Reference 2. The work in this chapter uses the one shown in Figure 7.1, which is the widest accepted in the state-of-the-art of 2.4 GHz transceivers. It comprises a low-IF chain with complex filters [3] for the receiver (RX) and a Direct-Up Conversion architecture [4] for the transmitter (TX). More details of selected architectures are given in Sections 7.3 and 7.4, respectively. Both receiver and transmitter share a PLL frequency synthesizer as local oscillator (LO), which is based on a simple integer-N scheme with the VCO running at double frequency of the RF band (~5 GHz) [5]. The clock frequency of the receiver chain (f_{LOD}) has an off-set from the RF central frequency (f_{RF}) of the reception channel, in an amount equal to the desired low-intermediate frequency (f_{IF}) in baseband (BB) for the down-converted signal. In transmitter chain, the clock frequency (f_{LOU}) coincides with f_{RF} frequency.

7.2.2 Transceiver Requirements for ZigBee

The electrical characteristics for the receiver and transmitter—noise figure (NF), input level for the third-order intermodulation product (IIP$_3$), output compression point (OP$_{1\,dB}$), image rejection ratio (IRR), and error vector magnitude (EVM), are not specified in the ZigBee standard, so they have to be obtained from system specifications, which are in physical layer: the data rate, the modulation and spreading types, the receiver sensitivity and jamming resistance, and the mask of output power spectral density. In this chapter, only a summary with the most representative specifications are given (see Table 7.1), but the interested reader can find a detailed procedure to obtain them in Reference 6.

7.3 Proposed Receiver Synthesis

7.3.1 RX: Architecture Selection

The first step in the receiver synthesis is the choice of the optimum architecture for a fully integrated low-power design. The decision was restricted between the direct conversion and the low-IF topologies, since the heterodyne receiver [4] would generally require external components and it uses two different frequency down-conversion steps with the consequent area, cost, and power consumption penalties. Among the two candidates, and taking into account that the image rejection specifications in the ZigBee standard are not very restrictive (image rejection ratio IRR > 30 dB is safe enough), the low-IF approach was finally selected because it relaxes the DC offset and flicker noise in the receiver BB section.

Figure 7.1 shows the details of the resulting low-IF receiver (RX). A fully differential implementation is considered to improve noise immunity and

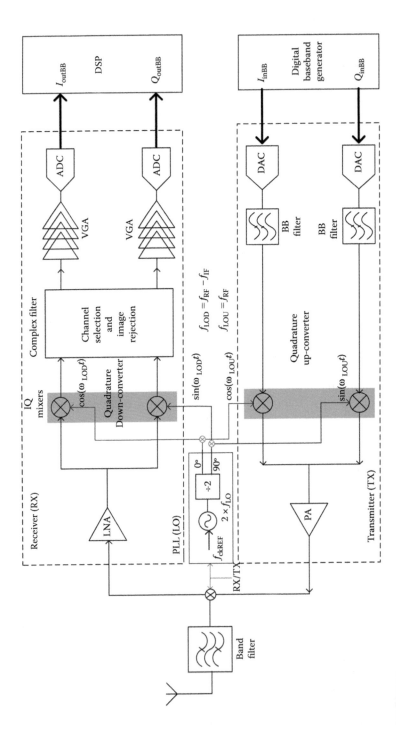

FIGURE 7.1
Proposed ZigBee architecture: Low-IF receiver and direct conversion transmitter.

TABLE 7.1

Summary of the Transceiver Electrical Specifications and Constraints

Receiver Specification	Value	Comments
Input power range	$(-85, -20)$ dBm	
Noise figure	$< 21 + (P_{in} + 85)$ dB/Hz	$P_{in} > -85$ dBm
IIP_3	$> -35 + (P_{in} + 85)$ dBm	$P_{in} > -85$ dBm
	> -10 dBm	$P_{in} > -60$ dBm
IIP_2	$> -20 + (P_{in} + 85)$ dBm	$P_{in} > -85$ dBm
	> 5 dBm	$P_{in} > -60$ dBm
Spurious free dynamic range	> 32 dBm	
Channel selectivity requirements	< 0 dB	adjacent channels (± 5 MHz)
	< 30 dB	alternate channels (± 10 MHz)
Transmitter specification		
Output power range	$(-35, -3)$ dBm	
OP_{1dB}	> 0 dBm	
EVM	$< 35\%$	for 1000 chips at -3 dBm output power
Dynamic range	> 30 dBm	
PLL specification		
Turnaround time	< 192 μs	
Frequency offset	< 40 ppm	

common-mode rejection. The RF signal is amplified by the front-end low-noise amplifier (LNA) and down-converted to the intermediate frequency (IF) by two mixers in quadrature (IQ-mixers). Channel selection and image rejection are performed by an active complex filter. Finally, the IF in-phase and quadrature (I/Q) signals are passed through the variable-gain amplifiers (VGAs) to adequate their amplitude to the input ranges of the back-end analog-to-digital converters (ADCs).

The choice of the intermediate frequency requires a trade-off between power consumption and noise. A widely accepted compromise is to use an IF frequency equal (or slightly larger) than the signal bandwidth (2 MHz in ZigBee). Considering extra safety margin for relaxing the DC-coupling networks between stages and flicker noise requirements, a 2.5 MHz IF was finally set as a good compromise. This selection has the additional advantage of simplifying the PLL frequency divider establishing an integer relationship between the input clock reference ($f_{ckREF} = 2.5$ MHz) and the synthesizer output frequencies, $f_{LOU} = N \cdot f_{ckREF}$, $f_{LOD} = (N-1) \cdot f_{ckREF}$.

7.3.2 RX: Building Blocks Specifications

When compared to the transmitter channel, the task of determining the building block specifications in the receiver path is especially complex since the input signal power level will cover several decades—in the transmitter counterpart, the difficulty is not found in the size of design space, but in the close coupling between up-converter and PA specifications which require a codesign at circuit level, as will be shown in Section 7.4.

In the ZigBee standard, the input power specification must be verified within a nominal power input range from −85 dBm to −20 dBm. In practice, extra safety margins at the upper level should be also allocated for robustness against possible interferences. A maximum power level of $P_{in} = -16$ dBm could be a good choice, this meaning a sensitivity range of 78 dB for the complete channel.

Once the architecture has been selected, we have to determine the circuit-level realization and specifications for the different building blocks. This implies finding noise and distortion error budget distribution among the receiver stages to satisfy the target performance, while minimizing power consumption. The key specifications include the average power consumption of each building block ($P_{AVG,i}$), input-referred noise voltage or noise figure (NF$_i$), input-referred intermodulation products of third and second orders (IIP$_{3,i}$, IIP$_{2,i}$), power gains (G_i), and the internal power levels (P_i) at the input of the different stages in the receiver queue (STG$_i$), where the subindex $i \in [1, NSTG]$ specifies the order of a particular block in the chain from the antenna {STG$_1$ = Band Filter, STG$_2$ = LNA, … , STG$_{NSTG}$ = ADC} in the implementation of Figure 7.1 and $P_1 \equiv P_{in}$ is the input power to the receiver.

Due to its difficulty, this task has been traditionally addressed by a high qualified engineer, the so-called *RF cook*, who often performs a manual channel sizing based on his previous expertise and knowledge of technology limits from different integrations and assisted by more or less complex distributing tools (as ADS™ from Agilent Technologies). As shown in the RF receiver model of Figure 7.2, critical aspects—such as distribution of power consumption per stage ($P_{AVG,i}$), number (N_G) of programmable gain modes to satisfy the target requirements at all internal input levels (P_i), input

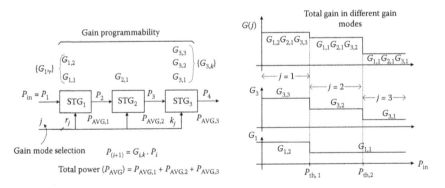

FIGURE 7.2
Conceptual example of a RF receiver with $N_G = 3$ different gain modes, $j \in [1, 3]$.

signal power thresholds between modes ($P_{\text{th},j}$), and specific gain distribution set $\{G_{i,k}\}_j$, where $j \in [1, N_G]$ specifies a particular gain mode—could lead to a tedious iterative process which eventually conduce to no solution, especially when complex trade-offs or target specifications close to the technological limits are present. Even if a solution is found, manual sizing usually leads to nonoptimum designs in terms of power consumption.

In Reference 7, a guided synthesis procedure for RF receiver channels was proposed considering semi-analytical expressions. The main limitation of this approach is the lack of models that incorporate gain as a design parameter. Hence, it needs some manual adjustment when different gain-programmability modes are defined in the power range of input signal.

As alternative to overcome these drawbacks, we have introduced a systematic procedure for automatic synthesis of RF receivers with emphasis in power optimization. Following a top-down bottom-up design flow, the procedure provides an energy efficient distribution of gain and power between stages, and hence, an optimum allocation of noise and distortion error budget. As will be seen in Section 7.3.3, the proposed design flow is suitable for most of the RF chains and other specific transistor-level realizations of the building blocks. In Section 7.3.4, it is applied for power optimization of the low-IF receiver in the ZigBee transceiver of Figure 7.1 as demonstrator.

Figure 7.3 shows the simplified schematics of the main building blocks in this case of study. The front-end LNA (Figure 7.3a) uses a differential common source stage with inductive degeneration [8]. Both IQ-mixers share the same active implementation in Figure 7.3b [9] inspired on a Gilbert multiplier, but the differential pair associated with the local oscillator (LO) is substituted by a CMOS digital buffer [10]. An impedance adaptation network between LNA and IQ-mixers has been used for their coupling.

The channel filter in Figure 7.3c [11] comprises a preamplifier followed by an eighth-order bandpass complex filter core (two second-order bandpass Gm-C sections in cascade). The preamplifier sets the common-mode of all used transconductors in the filter. The last key block in the receiver chain is the VGA in Figure 7.3d which provides signal conditioning for the ADCs. A multistage VGA architecture is considered. Each stage core is formed by a gain boosted amplifier with resistive degeneration, since this topology presents an optimum trade-off between nonlinear response and power consumption for ZigBee receivers [12]. The specific number of stages in the VGA has also been included in the optimization process. In the integrated example (see Section 7.5), two stages were implemented.

7.3.3 RX: Gain, Noise, and Linearity Distribution Procedure

This subsection introduces the fundamental concepts of the proposed design flow. The synthesis is formulated as an optimization problem with multiple constrains, generally described in the form,

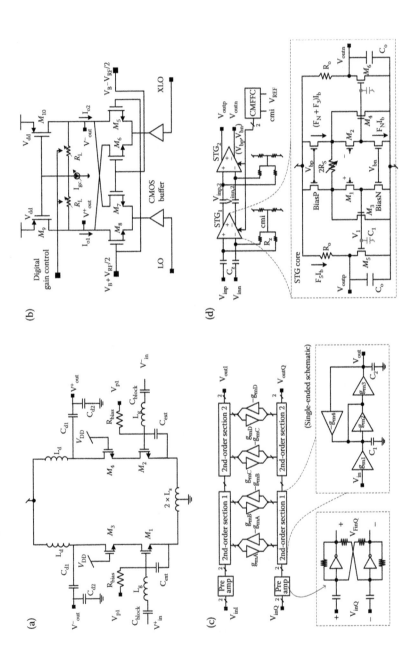

FIGURE 7.3
Simplified schematic of the main building blocks in the receiver chain: (a) LNA, (b) IQ-mixers, (c) complex filter, and (d) VGA.

$$\text{find } \underset{\sim}{x} = \{N_G, \{P_{th,j}\}_{j=1...N_G-1}, \{G_{i,k_j}\}_{j=1...N_G}, \{P_{AVG,i}\}_{i=1...NSTG}\}$$

$$\text{which minimizes : } P_{AVG}(\underset{\sim}{x}) = \sum_{i=1}^{N_{STG}} P_{AVG,i}$$

$$\text{constrained by: } NF(P_{in}) \leq NF^{spec}(P_{in}); \quad IIP_3(P_{in}) \geq IIP_3^{spec}(P_{in});$$

$$IIP_2(P_{in}) \geq IIP_2^{spec}(P_{in})$$

(7.1)

Taking into account the target specifications of the receiver chain (see Table 7.1)—that is, the total NF^{spec} and intermodulation products (IIP_3^{spec}, IIP_2^{spec}) as function of the input signal power (P_{in}), the objective of the synthesis procedure is to find the number of gain modes (N_G) and their distribution along the input power range ($P_{th,j}$), the specific gain distribution set $\{G_{i,k}\}_j$ in each mode and the power consumption per stage $P_{AVG,i}$, which minimize the total power consumption, $P_{AVG} = \Sigma P_{AVG,i}$, once all targets are satisfied.

Considering this formulation of the problem, that can be obviously generalized with other target constrains, a top-down bottom-up design flow has been developed for automatic synthesis of the receiver channel. Its flow diagram is illustrated in Figure 7.4. Starting from specifications, the topology generator (see label step1 in the figure) defines which stages in the receiver queue have the possibility of gain programmability and determines the number of gain levels (L_j) for these blocks. This phase in the design flow requires a close interaction with block designers to assure that the inclusion of programmability does not affect the block performance. In the example of Figure 7.2, programmability was included in first and third stages (STG$_1$, STG3) with $L_1 = 2$ and $L_3 = 3$. The topology generator could also incorporate critical design parameters (DP$_i$) in some specific building blocks, when there exist complex trade-offs between specifications. As an example, the DP$_i$ could be the biasing conditions (g_m/I_D) of the main transducer transistors (these transistors determine in each circuit implementation the main trade-off between noise and distortion for a given power consumption, $P_{AVG,i}$).

Taking into account the outputs of the topology generator, the synthesis procedure performs an optimization loop to minimize power consumption while achieving specifications in each design (step 2). In our case, the whole synthesis loop is supported by MATLAB® and the external optimization tool Fridge [13], which is the engine that processes performance results and generates new iterations seeds following a simulated annealing algorithm. In each iteration, the design variables (step 3) are reduced to the average power consumption ($P_{AVG,i}$) and the specific values of the gain in each stage (including programmability). In the conceptual example of Figure 7.2, the set of gains were $\{G_{1,1}, G_{1,2}, G_{2,1}, G_{3,1}, G_{3,2}$, and $G_{3,3}\}$.

From these values, and taking into account technology look-up tables (LUTs) which characterize the performance of the building blocks at transistor levels, the high-level parameters of each stage are derived in step 4 as a

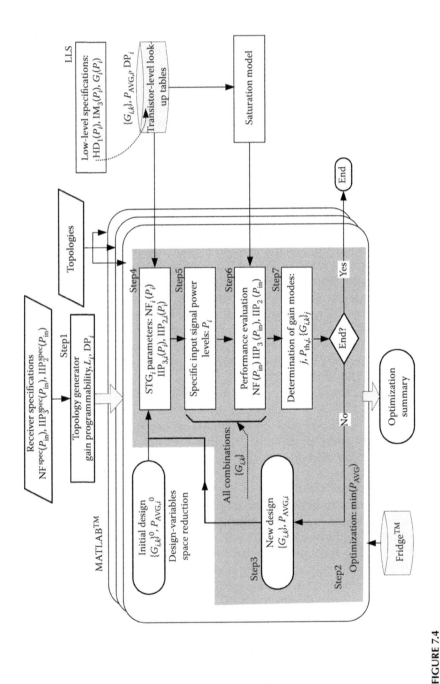

FIGURE 7.4
Basic flow diagram of the receiver synthesis procedure.

function of P_i. In this step, the parameter P_i should be understood just as an independent variable which defines the mathematical relationship between the high-level parameters and the signal power at the input of each stage. The specific distribution of signal power along the channel queue will be later evaluated at step 5, as described next in detail. In our ZigBee case, the effective noise figure NF_i and intermodulation products ($IIP_{3,i}$ and $IIP_{2,i}$) are inferred from the low-level specifications tabulated in the LUTs.

Depending on the building block, these LUTs could incorporate different specifications, such as the amplitude of the first harmonic (HD_1), the effective gain (G_i, from which the 1 dB compression point can be extracted) or the third-order intermodulation products IM_3, all of them evaluated as a function of P_i based on transistor-level simulations with one-tone and/or two-tone input stimuli. As an example of the information capture in the LUTs, Figure 7.5 shows the transistor-level simulations of a two-tone test for an implementation of the IQ-mixers in Figure 7.3b in a 1.2 V 90 nm CMOS process. Details on the amplitude of one of the tones HD_1 and the corresponding intermodulation product IM_3 for two possible gain values are depicted.

It is worth noticing that in order to reduce the design effort and characterization time during the LUT generation, the determination of the high-level parameters in step 4 could be supported by analytical expressions which allow expanding the covered design space as function of the power consumption on each block ($P_{AVG,i}$). In our ZigBee case, the information in the

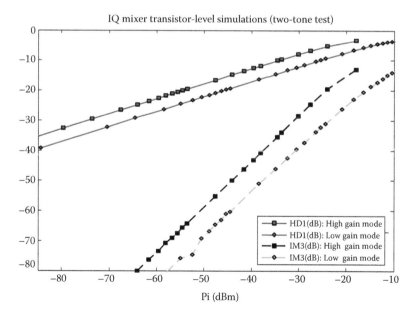

FIGURE 7.5
Transistor-level simulations of HD1 and IM3 versus input signal power of an implementation of the IQ-mixers in Figure 7.3b in a 1.2 V 90 nm CMOS process.

LUT has been processed by means of two analytical expressions, denoted by $NF_i = f_i(P_i, P_{AVG,i})$ and $IIP_{x,i} = g_{x,i}(P_i, P_{AVG,i}, G_{i,k})$ which depends on the specific building block realization (similar expressions could be derived for other additional targets). The agreement between these analytical predictions and transistor-level simulations for different requirements in term of average power consumption of the block ($P_{AVG,i}$) and gain selection (G_i) were verified by multiple designs in each building block. Although the detailed descriptions of these expressions are out of the scope of this chapter, we would like to highlight that they capture the natural design knowledge in terms of noise and distortion for a given design biasing conditions; these are (a) the noise or equivalently NF_i can be reduced incrementing power consumption and (b) given an input signal power P_i, the input referred intermodulation products ($IIP_{3,i}$ and $IIP_{2,i}$) have an inverse relationship with gain since nonlinearities increase with the signal level.

In step 5, the amplitude signal levels at the different stages are evaluated for the specified power input range at the antenna (from −85 to −16 dBm in our ZigBee demonstrator), and for all the possible gains combinations. In the example of Figure 7.2, the set of gains has six possibilities, given by $\{G_{1,1}G_{2,1}G_{3,1},\ G_{1,2}G_{2,1}G_{3,1},\ G_{1,1}G_{2,1}G_{3,2},\ G_{1,2}G_{2,1}G_{3,2},\ G_{1,1}G_{2,1}G_{3,3},\ \text{and}\ G_{1,2}G_{2,1}G_{3,3}\}$. For each gain distribution, the specifications of the receiver channel are evaluated—in step6—as a function of the input power: $f(P_i, P_{AVG,i})$, $g_3(P_i, P_{AVG,i}, G_i)$, and $g_2(P_i, P_{AVG,i}, G_i)$. This evaluation is based on the classical Frii's formula for the determination of the noise factor (NF) in a RF channel with m stages in cascade, given by

$$NF = 1 + (NF_1 - 1) + \frac{(NF_2 - 1)}{G_1} + \frac{(NF_3 - 1)}{G_1 \cdot G_2} + \cdots + \frac{(NF_m - 1)}{G_1 \cdot G_2 \cdots G_{(m-1)}} \quad (7.2)$$

and the equivalent expression for the intermodulation products,

$$\frac{1}{a_{IIP3}^2} = \frac{1}{a_{IIP3,1}^2} + \frac{A_1^2}{a_{IIP3,2}^2} + \frac{A_1^2 \cdot A_2^2}{a_{IIP3,3}^2} + \cdots + \frac{A_1^2 \cdot A_2^2 \cdots A_{(m-1)}^2}{a_{IIP3,m}^2} \quad (7.3)$$

$$\frac{1}{a_{IIP2}} = \frac{1}{a_{IIP2,1}} + \frac{A_1}{a_{IIP2,2}} + \frac{A_1 \cdot A_2}{a_{IIP2,3}} + \cdots + \frac{A_1 \cdot A_2 \cdots A_{(m-1)}}{a_{IIP2,m}} \quad (7.4)$$

where $a_{IIPx,i}$ stands for the input referred to as *rms* amplitude (in Volts) of the intermodulation products, and A_i is the linear voltage gain of the ith block.

In these traditional formulas, two aspects were particularized to take into account that (1) impedance matching is not necessary and (2) the signal level excursions at the input of each block cannot be arbitrarily defined, since the available voltage range is limited depending on the circuit-level realization. According to the first aspect, the input referred voltage noise of the BB section has been translated to an effective NF which considers degradation in the signal-to-noise ratio. Regarding the second consideration, a soft saturation model has been included in such a way that situations with excessive gains result penalized in the optimization process, and therefore, it leads to noncompetitive designs in terms of power and performance (discarded by the search engine). The key idea of this soft saturation model is to introduce a degradation of the NF and intermodulation products when the amplitude signal levels exceed a safety region boundary in each design.

The final step in each iteration (step 7) consists of determining the optimum number (N_G) of gain distributions modes and power thresholds $P_{th,j}$ between gain modes. To carry out this selection, the channel performance evaluated in step 6 is used for all the possible gain combinations to determine the gain subset $\{G_{i,k}\}_j$ which provides the best performance for each input power level (P_{in}). In the example of Figure 7.2, the six gain possibilities are finally reduced following this criterion to just three cases ($N_G = 3$): $\{G_{1,1}G_{2,1}G_{3,1}, G_{1,1}G_{2,1}G_{3,2}$ and $G_{1,2}G_{2,1}G_{3,3}\}$.

The optimization of each topology finishes when all specifications are met and minimum power consumption is found by the optimization tool. This process usually takes less than half an hour per topology. If no solution is found for a specific case, the maximum number of iterations is achieved and the incidence will be back-annotated in the final optimization summary.

7.3.4 RX: Optimization and Verification

In this section, the final optimization results for the ZigBee receiver design is presented, as well as the corresponding verification simulations. These results have been carried out using the synthesis procedure presented in previous sections with multiple iterations to refine the topology and critical design parameter (DP_i) constrains captured by the LUTs. The convergence of the whole process was quite fast, requiring just three iterations at the top level (including the refinement of LUTs) to achieve the optimum topology with minimum power consumption (17 mW for the receiver channel, excluding the PLL 6.9 mW, and mixer drivers 7.4 mW).

Table 7.2 shows a summary of the derived specifications with details on the gain programmability for each building block. The optimization procedure introduced programmability in all stages excluding the front-end LNA—the inclusion of gain programmability in the LNA was also considered in the optimization problem, but these solutions resulted not efficient in terms of power consumption, and therefore, they were automatically discarded in the process. In Table 7.3, the optimum gain distribution as function of the input power (P_{in}) is presented. In total, 13 different gain modes were selected for the best performance ($N_G = 13$).

Considering the above optimum distribution of gains, Figure 7.6 shows the overall performance results based on MATLAB simulations. They include the total NF and third intermodulation products (IIP_3 and IIP_2) as a function of

TABLE 7.2

Main Optimization Results for the Specifications of the Main Building Blocks in the Channel Receiver with Details on the Gain Programmability Modes

Receiver Block	Specification	Gain Programmability per Building Block				
		1	2	3	4	5
LNA	Power gain (dB)	6				
	Power (mW)	1.44				
	NF (dB)	5				
	IIP$_3$ (dBm)	−7				
	IIP$_2$ (dBm)	15				
Mixer	Conversion gain (dB)	6	15			
	Power (mW)	2 × 4.92				
	SSB NF (dB)	18	18			
	IIP$_3$ (dBm)	3	1			
	IIP$_2$ (dBm)	25	20			
Channel	Voltage gain (dB)	0	12			
Complex Filter	Power (mW)	3.60				
	Input noise (V/√Hz)	100	25			
	IIP$_3$ (dBV)	−3	−8			
	IIP$_2$ (dBV)	12	7			
PGA Stage	Voltage gain (dB)	0	6	12	18	24
	Power (mW)	4 × 0.56				
	Input noise (V/√Hz)	36	33	26	18	13
	IIP$_3$ (dBV)	10	10	3	−4	−10
	IIP$_2$ (dBV)	27	22	17	12	7
ADC	Power (mW)	2 × 0.2				
	Full scale (V)	1				
	Resolution (effective bits)	6				

input power. In each case, limits imposed by the ZigBee standard (see Table 7.1) have been highlighted, clearly showing that all specifications are satisfied.

The agreement of the specifications assignment with the standard has been alternatively validated using Agilent Advanced Design System (ADS™) simulator. Figure 7.7 shows the simulation setup for the ZigBee receiver. In this environment, behavioral simulations include relevant nonideal effects such as channel noise, distortion, phase noise, and I/Q imbalance.

The performance of the receiver was exhaustively characterized in terms of error rate. As an example, Figure 7.8 shows the sensitivity simulation by varying the input power level. For the required chip error rate (CER) of 6.89% (i.e., 1% packet error rate PER) [14], the input power of the receiver is about −88 dBm with a system NF of 21 dB. This simulation not only includes the noise effects but also the distortion in the receiver blocks and phase noise in the LO. As it can be observed, the chip error rate is below 6.89% in the whole

TABLE 7.3

Optimization Results for the Gain Distribution among the Receiver Chain

Gain Modes	Input Range (dBm)		Voltage Gain (dB)						RX Gain (dB)
	From	To	Band Filter	LNA	Mixer	Channel Filter	VGA STG 1	VGA STG 2	
1	−90	−85	−3	6	15	12	24	24	78
2	−84	−79	−3	6	15	12	24	18	72
3	−78	−73	−3	6	15	12	24	12	66
4	−72	−67	−3	6	15	12	24	6	60
5	−66	−64	−3	6	6	12	24	12	57
6	−63	−58	−3	6	6	12	18	12	51
7	−57	−52	−3	6	6	12	18	6	45
8	−51	−46	−3	6	6	12	12	6	39
9	−45	−40	−3	6	6	12	6	6	33
10	−39	−34	−3	6	6	12	6	0	27
11	−33	−28	−3	6	6	12	0	0	21
12	−27	−22	−3	6	6	0	6	0	15
13	−22	−16	−3	6	6	0	0	0	9

standard input range (from −85 to −20 dBm), corroborating that the chosen distribution of specifications among receiver building blocks meets the standard requirements.

Figures 7.9 and 7.10 show some details of the eye diagram and the I/Q constellation in two extreme input power level. Some effects, as the phase noise and amplitude saturation over the receiver output signals, can be appreciated clearly in these figures.

7.4 Proposed Transmitter Synthesis

7.4.1 TX: Architecture Selection

The low-power and completely-integrated transmitter required for the desired application only admits certain types of architectures. The one chosen in this implementation is the direct up-conversion transmitter (DUCT) of Figure 7.1, which, despite its known drawbacks due to high DC-offset and flicker noise values [2], is a good choice because of the reasonable transmitter input levels. To make this architecture effectively feasible, some efforts in reducing offset and mismatch should be done since they cause LO feedthrough, which overlaps with the modulated carrier at frequency f_{RF}. This way, it is possible that the DUCT correctly achieves EVM ZigBee specifications (EVM $\leq$ 35% at ZigBee 2.4 GHz), as shown afterward.

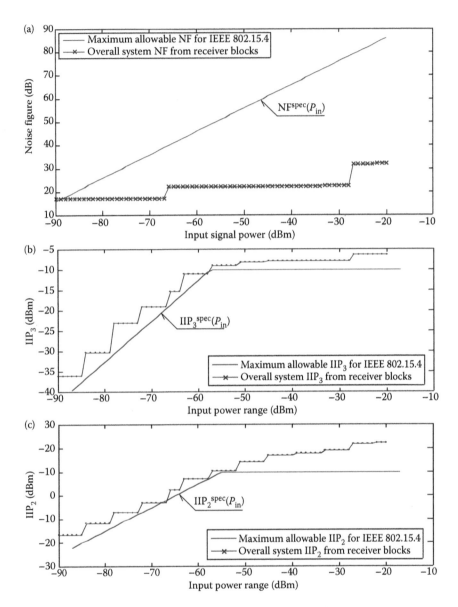

FIGURE 7.6
Optimization results for the overall receiver specifications: (a) NF, (b) IIP_3, and (c) IIP_2.

7.4.2 TX: Building Blocks Specifications

Both the ZigBee physical layer and the transmitter architecture, in this case
the DUCT of Figure 7.1, determine the specifications of its building blocks.
For the DAC, its minimum number of conversion bits is derived from the

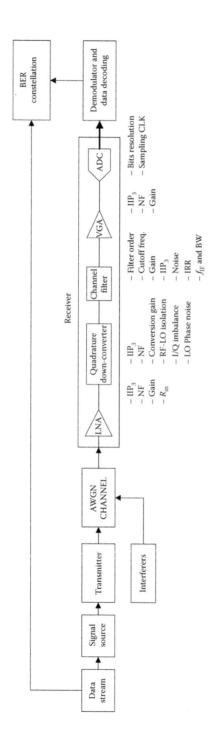

FIGURE 7.7
Simulation scheme for the ZigBee receiver in Agilent Advanced Design System (ADS).

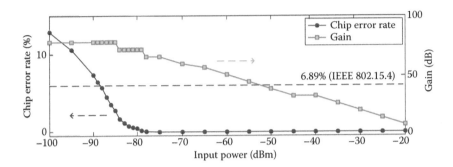

FIGURE 7.8
System simulation results in the receive path: sensitivity performance.

ZigBee SNR as follows. As the absolute limits of spurious outside the band must be below 30 dB [1] and each bit of resolution improves 6.02 dB, then the DAC should have, at least, 5 bits. Also, to improve the SNR, the signal is oversampled eight times the chip rate, that is, $f_{CLK,DAC} = 16$ MHz [6].

The BB filter is employed to reduce the power of the DAC output images. Its corner frequency can be as low as the bandwidth of ZigBee signal (~1.5 MHz), and the use of DAC oversampling relaxes its design up to very low-order

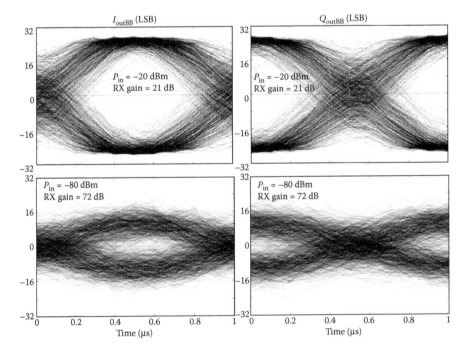

FIGURE 7.9
System simulation results in the receive path: eye diagram for two different input power levels.

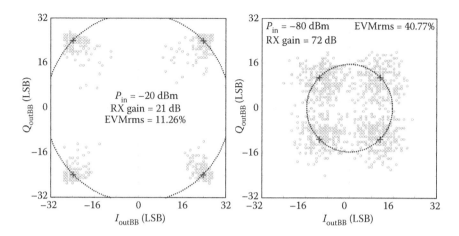

FIGURE 7.10
System simulation results in the receive path: I/Q constellation for two input power levels.

type (~second-order), resulting in EVM less than 5% [15]. The variation in the transmitter's output power can be completely gathered in the BB filter, by means of a variable gain (e.g., having an attenuation of up to 30 dB in steps of 1 dB) to relax the amplitude requirements of following blocks.

For the DAC and the BB filter, the I/Q mismatch should be reduced, in order to avoid crosstalk between the two data streams modulated on the quadrature signals of the carrier, as it would create an undesired sideband signal at the output, placed symmetrically respect to f_{RF}. Those mismatches are described in terms of rejection of the unwanted single-sideband (SSB). Considering an SSB rejection of 30 dB [1], the I/Q mismatch contribution in EVM should be below 3.2%.

The up-conversion mixer should have a double-balanced structure so as to achieve high isolation between LO and RF ports. The typical carrier leakage specifications in this block achieve –30 dBc, resulting in an EVM contribution of less than 3.2%. Also, mixer nonlinearities generate in-band spurs. For ZigBee, a 30 dB attenuation of these spurs respect to the output signal means a reasonable EVM of 3.2%. Moreover, mixer linearity requirements are derived as follows: the $OP_{1\,dB}$ for the DUCT is generally fixed to at 0 dBm, above the nominal P_{out} (= –3 dBm); hence, in a first approximation, 10 dB can be considered the minimum mixer OIP3 [4]. Finally mixer's conversion gain is adjusted to achieve the desired input amplitude of the PA, as it will be explained in Section 7.4.3.

The final block of the chain, the power amplifier (PA), must provide a maximum P_{out} of, at least, 0 dBm for a standard output impedance as 50 Ω (linearity constraints are the same as in the mixer). As ZigBee applications have to be low power consumption, PA efficiency is a concern. To achieve the required maximum output power with the best efficiency, the election of the

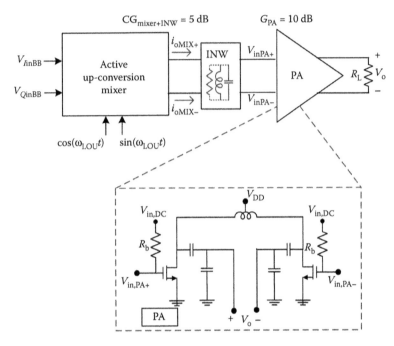

FIGURE 7.11
Scheme of the transmitter blocks and PA differential topology (detail).

PA sizing as well as its bias and input signal amplitude is made carefully [16]. To transform the output amplitude of the mixer into the required PA input signal amplitude, without high losses, an interface network INW is needed between both blocks, as seen in Figure 7.11. The DUCT PA carrier frequency f_{RF} is equal to that of the LO, hence the former might represent an interference in the LO, that is, LO pulling [4]. It is worse, if the system integrates the voltage-controlled oscillator (VCO) and the PA on the same chip. To reduce this effect, it is desirable to use a VCO working with twice the channel frequency ($f_{VCO} = 2f_{RF}$). This approximation has been implemented in our transceiver.

To evaluate the total DUCT EVM, it is assumed that the EVM contributors above mentioned are uncorrelated, hence the total DUCT EVM is the root-mean square (RMS) sum of the individual components [15], resulting in a total EVM of 7.5% for the transmitter chain up to the PA. Compared to 35% from ZigBee requirements, this leaves a wide safe margin for the PA inherent nonlinearity.

7.4.3 TX: Gain, Noise, Linearity, Power Distribution Procedure, and Other Specifications

After the presentation of the main specifications of the DUCT blocks, the final distribution of power consumption budget, gain, and linearity are discussed here. When compared to the receiver counterpart in Section 7.3.3, the design space is now more reduced in terms of input range and channel selectivity. However, its implementation presents multiple interrelations between blocks at circuit level, especially at the interface network INW between up-converter mixer and PA, making necessary the developing of a different optimization approach. To do so, we use a combination of a bottom-up methodology together with a top-down methodology which incorporates the simultaneous codesign between critical blocks at transistor level. Initially, the distribution of blocks specifications is chosen by considering both state-of-the-art designs and our previous experience, and then optimization procedures are followed in each block, for example, the PA design is done following an optimization procedure to obtain the most efficient of the Class-AB architecture used, as discussed in Section 7.4.4. Next, high-level simulations are carried out in MATLAB and ADS software to distribute optimally the budget of power consumption, gain and linearity, and to check the achievement of specifications during all the design procedure.

The reduction of transmitter power consumption is a great concern in the whole ZigBee DUCT design. Its reduction strongly depends on the PA output power, P_{out}, on the appropriate distribution of gain and IIP3 between DUCT blocks, and on the correct power transfer between the mixer and the PA. There is an inherent trade-off between power and the other characteristics that have to be considered, since reducing power consumption degrade the linearity of the blocks and gain characteristics.

The DUCT chain begins with the DAC, whose output swing is generally set to full-scale to achieve the best SNR. However, BB signals are, in general, too large for the DUCT analog-RF section. Hence, proper signal attenuation is necessary before the quadrature modulator, being this performed in the BB filter. This attenuation increases the overall NF but it is insignificant because the DAC signal level is much higher than that of the input equivalent noise of the filter. Moreover, to relax the complexity of the up-conversion mixer and PA sections, the ZigBee required programmable gain/attenuation of 30 dB is incorporated in this block. Gain, $OP_{1\,dB}$ and OIP3 values of the DUCT are derived from their respective formulas of a cascade of nonlinear stages [15], where no clipping exists at the PA output signal and blocks ahead the up-conversion mixer are highly linear. For an $OP_{1\,dB}$ of about 0 dBm for the mixer and the PA, a power gain G_{PA} of 10 dB has to be assigned to the PA. The mixer conversion gain is set to shift its input voltage to a certain value at PA input. For example, for a mixer with 300 mV input signal and for a PA input signal of 540 mV (this last value will be justified in Section 7.4.4), the mixer and the INW should have a voltage conversion gain CG of about 5 dB. The OIP3 is set about +10 dBm for both blocks, considering that OIP3 is about 10 dB higher than $OP_{1\,dB}$.

Once the transmitter planning has been performed and the specifications for each building block have been assigned, behavioral system-level simulations were carried out to evaluate the expected performance including relevant nonideal effects using ADS simulator. Figure 7.12 shows the simulation

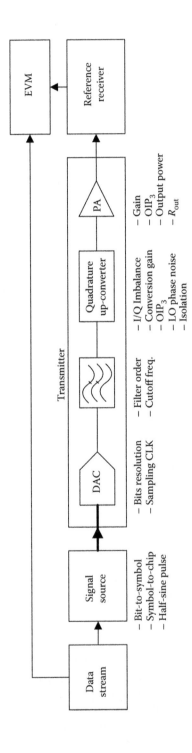

FIGURE 7.12
System ADS simulation scheme for ZigBee transmitter.

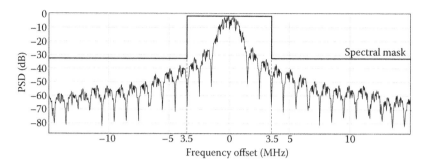

FIGURE 7.13
DUCT output spectrum obtained by means of ADS simulations.

scheme for the DUCT, whose performance is evaluated in terms of EVM. For simplicity, the output signal is assumed to propagate through an additive white Gaussian noise channel with no fading, frequency selectivity, interference, nonlinearity, or dispersion [4]. An example of the PSD of the obtained output signal is shown in Figure 7.13.

The modulation accuracy of the transmitter is determined by its EVM, which should be less than 35% when measured over 1000 chips for ZigBee transmitters. The EVM is measured on BB I and Q chips after recovery through a reference (ideal) receiver. In Figure 7.14, the simulated EVM performances when varying the DAC bit resolution, 1 dB compression power of the output PA and the corner frequency of the DAC reconstruction filter are shown. Simulation results show that increasing the DAC resolution beyond 5 bits does not further improve EVM performance. Moreover, it is not strongly affected by the 1 dB compression point of the PA due to the use of the constant envelope modulation scheme, DSSS-OQPSK.

Mismatches between the transmitter I and Q path cause impairments in the signal constellation, thereby degrading the EVM. In this case, simulations

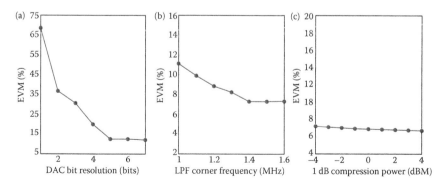

FIGURE 7.14
ADS system simulation results in the transmission path. EVM performances are simulated with different (a) DAC bit resolution, (b) BB filter corner, and (c) $OP_{1\,dB}$ values.

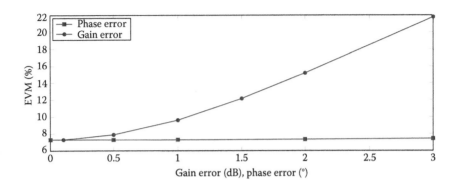

FIGURE 7.15
ADS system simulation results in the transmission path. EVM performance with different I/Q mismatch.

are performed with a transmitter and a reference receiver by introducing gain and phase errors in the transmitter I/Q paths. As shown in Figure 7.15, gain errors affect the transmitter performance significantly. Then, based on these simulations, a gain error of less than 0.5 dB and a phase error of less than 3° should be required for an optimal transceiver implementation.

7.4.4 TX: Power Amplifier Optimization

This section discusses the procedure followed to design the monolithic Class-AB PA used in this DUCT implementation and presented in Reference 16. The single-ended PA comprises a MOS transistor (MOST) and an RLC passive output network ONW [4], shown as a differential structure in detail in Figure 7.11.

The applied design procedure finds the best set of PAs in the sense of efficiency and harmonics filtering for a maximum required output power P_{out} at the carrier frequency. To derive this set, a sinusoidal voltage $v_{in,PA}(t) = V_{G,DC} + V_{G,RF} \sin(\omega_{RF}t)$ is injected at the gate of MOST and by means of the nonlineal MOST model of Reference 16, the MOST normalized drain current $\hat{i}(t)$ is found. Then, for each feasible pair $(V_{G,DC}, V_{G,RF})$, it is derived the output network ONW which filters $\hat{i}(t)$ achieving the best PA efficiency and the required harmonics level at the output voltage.

This study considers three hypothesis: (1) when maximum efficiency is reached, due to the ONW, the MOST drain voltage is almost sinusoidal (centered in a chosen drain voltage $V_{D,DC}$) and the load resistance R_L is transformed into a higher pure resistance R_{NW} seen from the MOST drain; (2) the instantaneous normalized drain current $\hat{i}(t)$ is the normalized static DC-current $I_D(v_G(t), v_D(t))/(W/L)$ (with W and L the MOST width and length, respectively) corresponding to the instantaneous MOST gate and drain voltages; and (3) the MOST is working under the quasi-static condition.

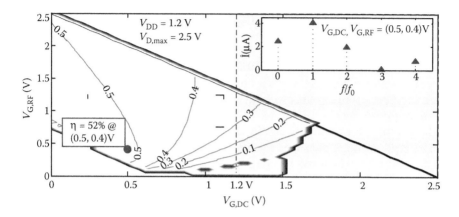

FIGURE 7.16
Efficiency contours versus $V_{G,DC}$ and $V_{G,RF}$, where $V_{G,DC}$ above 1.2 V are discarded (zone in gray). In the inset, harmonics of the chosen design.

This methodology is implemented in a set of MATLAB computational routines that generates maps of the PA characteristics versus the pair ($V_{G,DC}$, $V_{G,RF}$). To give an example, the PA efficiency map is shown in Figure 7.16, considering a thick oxide transistor in a 90 nm bulk RF CMOS technology.

The final election of the PA design cannot be made solely by choosing, for example, the PA design with the best efficiency. It is at this point of the DUCT synthesis where the design limitations in the mixer and in the INW arise. Several designs have been discarded despite their good efficiency and low power due to the lack of a feasible on-chip inductor to be used in that network. Also, in the proposed DUCT, the supply voltage is 1.2 V, so designs with $V_{G,DC}$ above this value are discarded in order to avoid additional external voltage regulators.

For example, for the mixer architecture used [10] and the technology limitations, the $V_{G,DC}$ voltage is limited; hence the zone of $V_{G,DC}$ is constrained. A good compromise is found around ($V_{G,DC}$, $V_{G,RF}$) = (0.5, 0.4) V, where the efficiency is 52% and the third harmonic is very low as shown in the inset of Figure 7.16 (the values of even harmonics are disregarded because a differential architecture is used).

7.5 Example of Integrated Transceiver

This section summarizes the design and implementation of a prototype of analog front-end for a 2.4 GHz ZigBee transceiver, which was developed by the authors especially for ultra-low power consumption in low-cost applications. Its design was carried out following the system planning and procedures exposed along this chapter. So, the front-end architecture is depicted in Figure 7.1. Although the design was optimized to reduce the consumption of RX and TX chains in its normal operation, special care was put in achieving a negligible consumption in the front-end off-state. Moreover, several states of selective shutdown of blocks were implemented to minimize the start-up time of each chain in the front-end, according to the application requirements.

The prototype was integrated in the TSMC 90 nm RF CMOS technology with supply voltage between 1.0 and 1.2 V for the whole core, while digital I/O and RF transmission outputs extend to 2.5 V (a microphotograph is shown in Figure 7.17). No external passive elements (as bypass/blocking capacitors or choke devices) are necessary for the front-end operation, only the RF switch is left out of this integration.

The key blocks in reception path are: an LNA [8], an active quadrature down-conversion mixer [9], an image-rejection complex bandpass channel filter [11], and a two-stage programmable gain amplifier (PGA) [12]. The main blocks in transmission path are an active up-conversion mixer and a Class-AB one-stage PA [17]. The prototype also includes a PLL frequency synthesizer whose core is a 5 GHz LC-tank VCO [5], and a design-for-testability (DfT) circuitry to increase the functional monitoring of each integrated block. Both paths, RX and TX, were implemented with fully-differential topologies to enhance common-mode noise immunity. Although data converter blocks are not integrated in this prototyping phase, their design was completely defined together with all previous blocks to have the best performance.

Integrated prototype was boxed in a QFN48 package and the characterization was performed on soldered samples to a specific home-made PCB. Commercial SMA antennas and baluns were used to interact with the devices. No other filters were used in the I/O RF paths. RX output IF-band signals were digitized with the ADCs in the AD9201 part at 10.0 Msps and configured with 2.0 V of differential full-scale. TX input BB signals were generated with the DACs in the AD9761 part. Measured results prove that the sensitivity of RX chain is about −90 dBm for a ~0% PER (800-bit per packet)

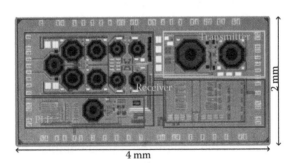

FIGURE 7.17
Microphotograph of the integrated transceiver analog front-end.

with an average consumption of 31 mW (including the PLL). The output power of TX chain achieves +2 dBm, exhibiting an *rms* error vector magnitude (EVM) less than 7% when the output power is 0 dBm and the averaged power consumption is about 22 mW (including the PLL). Basic characteristics of the integrated prototype are summarized in Table 7.4.

Some results of characterization of normal operation modes of the RX and TX paths are depicted in Figures 7.18 and 7.19.

After an exhaustive characterization of the integrated prototype and a postanalysis of the synthesis procedure, we can state that certain readjustment over some blocks and topologies, as for example changing the preamplifier in the receiver complex filter, could lead to an overall reduction in the power consumption up to 25%. Notwithstanding, by comparing the achieved consumption with those of the most popular commercial transceivers in the year 2010 (see Table 7.5), when our prototype was designed, the result was still very competitive.

TABLE 7.4

Main Characteristics of the Integrated Front-End

Technology	TSMC 90 nm LP-RF CMOS
Supply voltage	1.0–1.2 V
Operation band	2.4 GHz
Temperature range	0–100°C
RX-mode	
Power consumption	31 mW
(ON mode)	PLL Freq. Synthesizer included and 6 bit 8 Msps ADCs estimated power included
Sensitivity	≥–90 dBm (0% PER, 100-octet packets)
Gain range	12–81 dB (16 levels)
Input power range	[–95, –15] dBm
TX-mode	
Power consumption (ON mode at 0 dBm)	23 mW (PLL Freq. Synthesizer included and 6 bit 8 Msps DACs estimated power included)
Output power	≤2 dBm (in antenna)
EVMrms	<7% (0 dBm output power)
OFF-mode	
Power consumption	~50 nA
Reference clock	2.5 MHz
RX-TX turnaround	<100 μs
Sleep mode control	all blocks

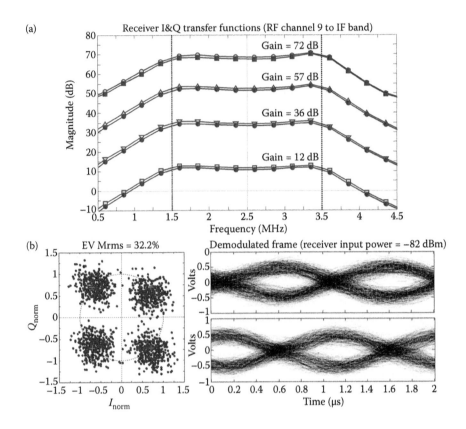

FIGURE 7.18
Reception mode characteristics: (a) Transfer functions in reception chain and (b) Constellation and eye diagram of a low-energy received frame.

7.6 Conclusion

Wireless networks with low-data-rate and low-power have significantly increased their presence in the market since they provide efficient development and cost in the smart-control of many application areas of the economy and technology. A remarkable example of this type of network is based on the ZigBee standard. The search for the high efficiency in the power consumption of their transceiver devices (long battery life) is what actually surrounds its technological and scientific development. This chapter summarizes the approach followed to optimize the power consumption of the transceiver analog front-ends in normal operation. Guidelines are given for the optimal distribution of the specifications for fundamental blocks in ZigBee receivers and transmitters for the 2.4 GHz band. The quality of the approximation

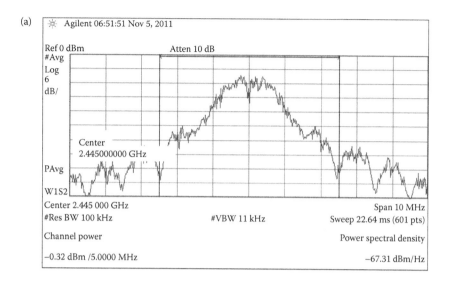

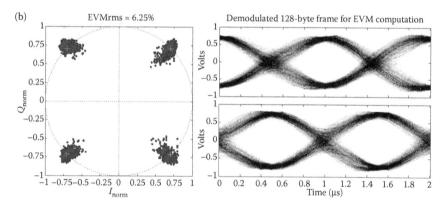

FIGURE 7.19
Transmission mode characteristics: (a) Output power spectrum and (b) Constellation and eye diagram of a transmitted 0 dBm frame.

is shown by the experimental results in several integrated prototypes in a 90 nm RF CMOS technology.

Acknowledgments

We would like to thank Nicolás Barabino and Professor Fernando Silveira of the Universidad de la República (Uruguay), for providing us the

TABLE 7.5

Comparative between Commercial Transceivers and the Integrated Example

	Authors	ATMEL	Texas Instruments			Ember		Microchip	Freescale				Jennic	
	ZigBee Front-End	ATRF 230	CC 2420	CC 2430	CC 2431	EM 250	EM 260	MRF 24J40	MC 13193	MC 13203	MC 13213	MC 13225	JN 5121	JN 5139
Vsupply (V)	1.2	1.8	1.8	1.8	1.8	1.8	1.8	2.4	2.7	2.7	2.7	2.7	2.7	2.7
RX (mW)	31.0	28.8	36	48.6	48.6	64.8	64.8	43.2	113.4	113.4	113.4	54	135	91.8
TX (mW)	22.0	30.6	32.4	48.6	48.6	64.8	64.8	52.8	94.5	94.5	94.5	54	121.5	91.8

computational support for the PA design, and Richard Jansen of European Space Agency (ESA) for his interest in promoting this publication. Thanks also to the funding support from Andalusian Government (project P09-TIC-5386), the Spanish Government (project TEC2011-28302), the project SR2 (TSI-020400-2010-55/Catrene2A105), and European FEDER program.

References

1. IEEE. Standard 802.15.4: Wireless MAC and PHY Specifications for Low-Rate Wireless Personal Area Networks (WPANs), 2006.
2. Mak, P.-I., Seng-Pan, U., Martins, R.P. Transceiver architecture selection: Review, state-of-the-art survey and case study. *IEEE Circuits Syst. Mag.*, 7(2), 6–25, 2007.
3. Crols, J., Steyaert, M.S.J. Low-IF topologies for high-performance analog front ends of fully integrated receivers. *IEEE Trans. Circuits Syst. II: Analog Dig. Sig. Process.*, 45(3), 269–282, 1988.
4. Razavi, B. *RF Microelectronics*. Upper Saddle River, NJ, USA: Prentice-Hall, 2012.
5. Gines, A., Doldan, R., Villegas, A. et al. A 1.2 V 5.14 mW quadrature frequency synthesizer in 90 nm CMOS technology for 2.4 GHz ZigBee applications. *IEEE Asia Pacific Conference on Circuits and Systems (APCCAS)*, Macao, China, 2008, pp. 1252–1255.
6. Oh, N.-J., Lee, S.-G. Building a 2.4 GHz radio transceiver using IEEE 802.15.4. *IEEE Circuits Devices Mag.*, 21(6), 43–51, 2005.
7. Shen, S., Emira, A., Sánchez-Sinencio, E. CMOS RF receiver system design: A systematic approach. *IEEE Trans. Circuits Syst. I: Regular Papers*, 53(5), 1023–1034, 2006.
8. Fiorelli, R., Silveira, F., Peralias, E. MOST moderate–weak-inversion region as the optimum design zone for CMOS 2.4-GHz CS-LNAs. *IEEE Trans. Microwave Theory Tech.*, 62(3), 556–566, 2014.
9. Fiorelli, R., Villegas, A., Peralias, E. et al. 2.4-GHz single-ended input low-power low-voltage active front-end for ZigBee applications in 90 nm CMOS. *20th European Conference on Circuit Theory and Design (ECCTD)*, Linkoping, Sweden, 2011, pp. 829–832.
10. Klumperink, E.A.M., Louwsma, S.M., Wienk, G.J.M. et al. A CMOS switched transconductor mixer. *IEEE J. Solid-State Circuits*, 39(8), 1231–1240, 2004.
11. Villegas, A., Vazquez, D., Peralias, E. et al. A 3.6 mW @ 1.2 V high linear 8th-order CMOS complex filter for IEEE 802.15.4 standard. *37th European Solid-State Circuits Conference (ESSCIRC)*, Helsinki, Finland, 2011, pp. 99–102.
12. Gines, A., Doldan, R., Rueda, A. et al. Power optimization of CMOS programmable gain amplifiers with high dynamic range and common-mode feed-forward circuit. *IEEE International Conference on Electronics, Circuits, and Systems (ICECS)*, Athens, Greece, 2010, pp. 45–48.
13. Medeiro, F., Fernandez, F.V., Dominguez-Castro, R. et al. A statistical optimization-based approach for automated sizing of analog cells. *IEEE/ACM International Conference on Computer-Aided Design (ICCAD)*, San Jose, CA, USA, 1994, 594–597.

14. Goyal, M., Prakash, S., Xie, W. et al. Evaluating the impact of signal to noise ratio on IEEE 802.15.4 PHY-level packet loss rate. *13th International Conference on Network-Based Information Systems (NBiS)*, Takayama, Gifu, Japan, 2010, 279–284.
15. Gu, Q. *RF System Design of Transceivers for Wireless Communications.* New York, USA: Springer, 2005.
16. Barabino, N., Fiorelli, R., Silveira, F. Efficiency based design for fully-integrated class C RF power amplifiers in nanometric CMOS. *IEEE International Symposium in Circuits and Systems (ISCAS)*, Paris, France, 2010, pp. 2223–2226.
17. Fiorelli, R., Peralias, E., Barabino N. et al. A fully differential monolithic 2.4 GHz PA for IEEE 802.15.4 based on efficiency design flow. *IEEE International Conference on Electronics, Circuits, and Systems (ICECS)*, Athens, Greece, 2010, 603–606.

8

Simulation Techniques for Large-Scale Circuits

Nestor Evmorfopoulos, Sotiris Bantas, and George Stamoulis

CONTENTS

8.1 Introduction

Computational methods have their own version of Moore's law, which states that the size of problems that can be solved in a given amount of time doubles every few years. However, it is also a historical fact that the size of problems that *need* to be solved always increases faster than our ability to solve them, necessitating algorithmic breakthroughs and paradigm shifts beyond the raw increase of computational power offered by Moore's law. One of these problems is the simulation of electrical and electronic circuits.

Circuit simulation at the device and electrical level is indispensable for the design of a broad range of large-scale integrated circuits such as digital, mixed-signal, or memory circuits. It is also equally essential in the verification of the electrical models of important integrated circuit subsystems such as power distribution networks, clock networks, or multiconductor buses.

Ever since the emergence of the ground-breaking Berkeley SPICE program [1], many SPICE-like clones have appeared that perform various improvements mainly toward computational efficiency, to render feasible simulation of ever larger circuits. This chapter will attempt to give an assessment of the current status of large-scale circuit simulation, as well as offer a perspective on how relevant trends might shape up in the future. To keep the text at an acceptable length, we will only talk about exact, full-circuit, and large-scale simulation. We will not cover efficient simulation of small- or medium-scale circuits, for example, by efficient selection of the number of time or frequency points needed for simulation, nor will we go into the simulation of only an input/output subset of circuit nodes, for example, by reducing the internal circuit by model order reduction (MOR) techniques. We will likewise not discuss circuit partitioning or other forms of domain decomposition (such as the historically important relaxation techniques) that are not used in exact—or direct—SPICE simulation but only in the so-called fast-SPICE approaches, as they compromise accuracy by individually simulating smaller circuit blocks and attempting to combine the results at the full circuit by a relaxation-type process. For these—nonetheless important—topics there exist excellent reviews [2–4].

The rest of this chapter is organized as follows. The next section presents an overview of circuit simulation and the basic associated types of analyses. Section 8.3 examines the simulation of large-scale nonlinear circuits (incorporating transistors and other nonlinear devices). Section 8.4 discusses the important subcase of large-scale linear circuits (comprising only RLC elements and active sources, and usually being the result of electrical modeling of parasitic effects). Overall conclusions are drawn in Section 8.5.

8.2 Overview of Circuit Simulation

The main types of analyses in circuit simulation are static or DC analysis, time-domain or transient analysis, and frequency-domain or AC analysis. In this section, we will describe briefly the equations that are formulated for each type of analysis and the main steps that are executed for their solution, with the intention to identify the relative importance of each step (from a computational point of view) and suggest the appropriate course of action to be taken when dealing with large-scale circuits. For more details on the derivation of the equations for each type of analysis, and the subject of circuit simulation in general, the reader is referred to Reference 5.

The equations for circuit simulation are formulated by combining the Kirchhoff's current and voltage laws that arise from the circuit connectivity with the current–voltage relationships of each circuit element. There are various possibilities for such a combination, but almost all modern SPICE-like circuit simulators use the modified nodal analysis (MNA) framework, in which the branch currents are eliminated wherever possible (specifically for all elements whose current–voltage relationships can be written in admittance form), and the final set of variables are the node voltages—with respect to a reference node called datum or ground—and the branch currents that cannot be eliminated (corresponding to either inductances or voltage sources). If the number of nodes (excluding the ground node) is n and the number of branches whose current is retained in the equations is b, then this procedure results in a system of $N = n + b$ equations with $N = n + b$ unknowns. All unknown variables are combined in a vector of unknowns $\mathbf{x}$, which for DC analysis is a vector of scalars (real numbers) $\mathbf{x} \in \mathfrak{R}^N$, while for transient analysis it is a vector-valued function $\mathbf{x}(t)$ of time t and for AC analysis a vector-valued function $\mathbf{x}(s)$ of (complex) frequency s.

8.2.1 Static (DC) Analysis

The DC analysis computes the operating point of the circuit under constant or static inputs. In DC analysis, the dynamic or energy storage elements like capacitances and inductances are zeroed out (capacitances are opened and inductances are shorted) and the input sources participate only with their DC parts. In a circuit that generally contains nonlinear elements such as semiconductor devices, this results in a nonlinear system of N equations with N unknowns, of the form

$$\mathbf{F(x)} = 0 \qquad (8.1)$$

where $\mathbf{F(x)}$ is a vector-valued function of the vector variable $\mathbf{x}$ (i.e., $\mathbf{F} : \mathfrak{R}^N \to \mathfrak{R}^N$).

The only general way to solve this nonlinear system of equations is by a numerical method. The most widely used method for solving nonlinear systems is the Newton–Raphson (or simply Newton) method, in which an iterative procedure is formulated generating a sequence of approximate solutions $\mathbf{x}^{(k)}$, $k = 0,1,\dots$ that (hopefully) converge to the true solution $\mathbf{x}^*$ (starting from an initial guess $\mathbf{x}^{(0)}$). At each iteration, a local linearization of $\mathbf{F(x)}$ is performed around the current approximation $\mathbf{x}^{(k)}$, and the update $\Delta\mathbf{x}^{(k)}$, which gives the new approximation $\mathbf{x}^{(k+1)} = \mathbf{x}^{(k)} + \Delta\mathbf{x}^{(k)}$, is computed by the solution of the following linear system of equations:

$$\mathbf{J}(\mathbf{x}^{(k)}) \cdot \Delta\mathbf{x}^{(k)} = \mathbf{F}(\mathbf{x}^{(k)}) \tag{8.2}$$

where $\mathbf{J}(\mathbf{x}^{(k)})$ is the Jacobian matrix of the partial derivatives of $\mathbf{F}(\mathbf{x})$ evaluated at $\mathbf{x}^{(k)}$, that is, the matrix with elements $\left[\mathbf{J}(\mathbf{x}^{(k)})\right]_{ij} = (\partial F_i(\mathbf{x})/\partial x_j)\big|_{\mathbf{x}=\mathbf{x}^{(k)}}$. Effectively, the application of the Newton method in DC analysis generates a sequence of linear systems to be solved until convergence.

8.2.2 Time-Domain (Transient) Analysis

The transient analysis computes the transient response of the circuit to a set of time-varying input excitations. The dynamic elements of capacitance and inductance introduce time derivatives into the MNA formulation, and the resulting system of equations is a system of differential-algebraic equations (DAE) of the form

$$\mathbf{f}(\mathbf{x}(t)) + \frac{d\mathbf{q}(\mathbf{x}(t))}{dt} = \mathbf{b}(t) \tag{8.3}$$

where $\mathbf{f}:\mathfrak{R}^N \to \mathfrak{R}^N$ and $\mathbf{q}: \mathfrak{R}^N \to \mathfrak{R}^N$ are vector-valued functions and $\mathbf{b}(t)$ is the vector of input excitations (from independent current and voltage sources).

The only general way to solve this system is by numerical integration, which amounts to discretizing time and replacing the continuous time derivative at every discrete time point t_i by a finite-difference approximation such as the backward-Euler $d\mathbf{q}(\mathbf{x}(t_i))/dt \approx [\mathbf{q}(\mathbf{x}(t_i)) - \mathbf{q}(\mathbf{x}(t_{i-1}))]/h_i$ or the trapezoidal $[d\mathbf{q}(\mathbf{x}(t_i))/dt + d\mathbf{q}(\mathbf{x}(t_{i-1}))/dt]/2 \approx [\mathbf{q}(\mathbf{x}(t_i)) - \mathbf{q}(\mathbf{x}(t_{i-1}))]/h_i$, where $h_i \equiv t_i - t_{i-1}$ is the time-step at time point t_i, $i = 0,1,....$ Other choices of approximation exist, and each one has each own advantages and disadvantages [6], but they all lead to a nonlinear system of equations to be solved at every time point t_i. For example, the backward-Euler approximation leads to the following sequence of nonlinear systems of equations:

$$\mathbf{f}(\mathbf{x}(t_i)) + \frac{\mathbf{q}(\mathbf{x}(t_i))}{h_i} - \left(\frac{\mathbf{q}(\mathbf{x}(t_{i-1}))}{h_i} + \mathbf{b}(t_i)\right) = 0 \tag{8.4}$$

Each of the above nonlinear systems is solved for $\mathbf{x}(t_i)$ (with known $\mathbf{x}(t_{i-1})$ and $\mathbf{b}(t_i)$) by the same iterative Newton procedure described in the previous section to compute the DC operating point. Thus, from a computational perspective, transient analysis is equivalent to the solution of a DC problem at every point in time. A DC analysis must still be performed prior to transient

---

Inputs: Circuit netlist and simulation interval $[t_0, t_f]$
Output: Circuit response $x(t_i)$ at time points $t_i \in [t_0, t_f]$

1: Calculate DC operating point (Solve (8.1) for **x** by Newton's method)
2: Set $x(t_0) = x$; $i = 0$; $h_1 = h_{min}$
3: **while** $(t_i < t_f)$
4: $i = i + 1$; $t_i = t_{i-1} + h_i$
5: Solve (8.4) for $x(t_i)$ by Newton's method
6: Calculate next time-step h_{i+1} by step control algorithm
7: **end**

FIGURE 8.1
Overall flow of transient analysis.

analysis to determine the DC operating point, which is to be used as the vector of initial conditions $x(t_0)$ for the solution of (8.4).

The time discretization is not performed *a priori* and is not generally uniform, that is, the step $h_i = t_i - t_{i-1}$ is not constant, because it is inefficient to enforce needlessly small time-steps (and thus more simulation points) during intervals of low activity where the response changes very slowly. An adaptive step control mechanism (through monitoring of a local error measure) is always introduced in the integration of the system of differential equations, with the purpose of acquiring the fewest possible simulation points that supply acceptable accuracy [6].

The overall flow of transient analysis is shown in Figure 8.1.

8.2.3 Frequency-Domain (AC) Analysis

The AC analysis computes the sinusoidal steady-state response of the circuit to small-signal sinusoidal inputs, as a function of frequency. Since the sinusoidal inputs are small-signal ones, all nonlinear devices are linearized around the DC operating point and substituted by small-signal linear models. The dynamic elements (also linearized, if nonlinear) are then substituted by complex impedances, each capacitance C by a complex impedance with value $Z_C(s) = 1/sC$ and each inductance L by a complex impedance with value $Z_L(s) = sL$. A complex linear system of equations is formulated for each complex frequency $s_i = j\omega_i$, of the following form:

$$A(s_i) \cdot x(s_i) = b(s_i) \tag{8.5}$$

The frequency discretization is performed in advance (since the solution at one frequency point does not affect the solution at the next point) and is usually uniform or logarithmic.

The overall flow of AC analysis is shown in Figure 8.2.

Inputs: Circuit netlist and frequency points $\omega_1 < \omega_2 < \cdots < \omega_f$
Output: Circuit response $\mathbf{x}(s_i)$ at frequency points $s_i = j\omega_i$

1: Calculate DC operating point (Solve (8.1) for $\mathbf{x}$ by Newton's method)
2: Linearize current–voltage relationships for nonlinear elements around $\mathbf{x}$
3: **for** $(i = 1, \ldots, f)$
4: Substitute each capacitance C by complex impedance $1/s_iC$
5: Substitute each inductance L by complex impedance s_iL
6: Solve complex linear system (8.5)
7: **end**

FIGURE 8.2
Overall flow of AC analysis.

8.3 Large-Scale Nonlinear Circuit Simulation

This section examines the basic operations involved in the simulation loop from a large-scale computational perspective, with emphasis on the solution of linear systems which is at the core of circuit simulation (and indeed, almost any other kind of simulation) and, as we will elaborate, accounts for the bulk of the computational cost.

8.3.1 Evaluation of Device Model Equations

To solve the nonlinear systems (8.1) and (8.4), the corresponding linear systems involving the Jacobian matrices must be formulated at every Newton iteration. This involves evaluation of the semiconductor model equations for each device at every discrete time point and every Newton iteration per time point. The semiconductor devices and especially MOS transistors with very fine nanometer-scale geometries have very sophisticated models, for example, BSIM3 or BSIM4 [7], which involve very complex equations with hundreds of parameters describing the current–voltage relationships between terminals. Evaluating these models so many times constitutes a significant fraction of the total computation time, which for small to medium circuits can reach as much as 70% in a serial implementation. Fortunately, there is a large degree of data-level parallelism involved, since model evaluations for different devices are independent to one another and can be performed in parallel. The parallel evaluation of device equations has been a recurring subject over the years, with implementations ranging from traditional multiprocessors [8] to modern massively parallel and heterogeneous platforms [9]. Its large degree of parallelism notwithstanding, device model evaluation scales only linearly with problem size and its impact on total computation time deteriorates for large circuits, causing the solution of linear systems to become dominant.

8.3.2 Solution of Nonlinear Systems of Equations

It is well known that Newton's method exhibits a quadratic rate of convergence [10], which practically means that it only takes a handful of iterations to converge to the true solution, *if* convergence is guaranteed. Indeed, the greatest problem in solving nonlinear systems is convergence and not efficiency of the algorithm. There are many techniques devised to facilitate convergence of Newton's method, but they are beyond the scope of this chapter and the interested reader is referred to [11].

8.3.3 Solution of Linear Systems of Equations

Since even for nonlinear systems the solution is performed by linearization around the current approximation in Newton's method, the simulation inevitably and always reduces to some kind of linear system of equations to be solved at every nonlinear iteration and every time point (in transient analysis) or frequency point (in AC analysis). Therefore, the solution of linear systems of equations is the central step in circuit—and almost any other—simulation process. This is reinforced by the fact that the linear system solution is naturally an $O(N^3)$ process, and for large circuits it takes up the bulk of the total computation time. Here, in contrast to the solution of nonlinear systems, the efficiency of the algorithm is the primary concern, in regard to both computational cost and memory consumption. A great deal of research has been conducted in the past toward improving the efficiency of the solution of linear systems, for various kinds of applications, and all efforts generally strive to exploit two key factors, namely matrix sparsity and parallelism.

A matrix is called sparse if the number of nonzero elements is $O(N)$, as opposed to $O(N^2)$, which holds for fully populated or dense matrices. From a computational point of view, a matrix is considered as sparse if only its nonzero elements are stored and manipulated in matrix operations, leading to substantial savings in memory footprint and execution time. The linear systems arising in circuit simulation are typically very sparse, since for nearly all practical circuits (excluding those with lots of mutual inductances) each node is connected to no more than four to five other nodes and gives an analogous number of contributions to the system matrix. Therefore, it is generally of great advantage to employ sparse matrix techniques for the solution of linear systems encountered in circuit simulation. In fact, for very large circuits it constitutes the only option that enables the system matrix to fit into the available memory.

Parallelization of the solution of linear systems is a fairly old topic, with many past efforts targeting traditional multiprocessors and vector architectures. However, its importance is now greater than ever due to the apparent end in the upscaling of clock frequency associated with Moore's law, which makes parallel processing the only available possibility for improving software performance. The landscape of parallel processing has also changed, with the abundance of inexpensive desktop and portable PCs equipped with multicore processors and graphic processing units (GPUs). This has gradually moved parallel machines from arcane scientific labs to the mainstream arena and, more importantly, has created a paradigm shift by incorporating two different types of parallel architectures under the same hood. Multicore processors are representatives of multiple-instruction multiple-data (MIMD) parallel systems, in which there is a small number of relatively powerful processing elements, each one capable of performing its own computations (i.e., executing its own program) on its available data. This model is well-suited to general-purpose parallel computation. On the other hand, GPUs are representatives of single-instruction multiple-data (SIMD) parallel systems, in which there is a large number of relatively weak pro-cessing elements, each capable of performing only specific scalar and vector operations. The processing elements in such systems are usually arranged in a two-dimensional array, and are all executing the same computations (i.e., the same program imposed by a central control), each on its own avail-able data. Such a model is ideal for specialized problems with some kind of structural or operational regularity (GPUs may have originated as dedi-cated hardware for accelerating graphics applications, but they are perfectly capable of handling other problems in scientific computing). Many practical problems exhibit regularity at various levels and degrees, and the challenge is to deploy the right amount of regular computations on the GPU and the more irregular ones on the multicore processor. This extraordinary blend of MIMD and SIMD parallelism, together with deep and multilevel memory hierarchies, offers many unique opportunities and challenges for scientific computing in the future.

From an algorithmic standpoint, there are two broad classes of linear sys-tem solution methods, or "linear solvers" as they are known, namely direct methods and iterative methods. We will next examine and contrast both of these classes in the context of large-scale circuit simulation, and explore for each one the issues of sparsity and parallelism.

8.3.3.1 Direct Linear Solvers

Direct methods for linear systems employ the factorization of the system matrix into a lower triangular matrix L and an upper triangular matrix U (called LU factorization), which is followed by the solution of the triangular systems (called forward solve and backward solve, respectively). These meth-ods are referred to as "direct" because they terminate in a fixed and predeter-mined number of operations, which for the factorization part is $O(N^3)$ and for the forward and backward solve part is $O(N^2)$. Direct methods are generally preferred in industrial circuit simulators since they are robust for most types of matrices and they are adequately fast for small- and medium-sized problems (up to a few hundred thousands of unknowns). Unfortunately, they do not fare very well with sparsity and parallelism, which has serious impact in their applicability to large-scale problems. Also, the potential reusability of the LU factorization across a sequence of linear systems does not hold for circuit simulation, since the matrices in (8.2), (8.4), or (8.5) are not constant but dependent on the current approximation $x^{(k)}$, the time-step h_i, or the frequency s_i (all of which are generally variable).

The biggest problem with direct methods is that the triangular factors L and U are in no way as sparse as the system matrix itself. This translates in to significantly more memory to store the extra nonzero elements (called "fill-ins") that are introduced in L and U as part of the factorization process, and more importantly, significant additional computation time (beyond the $O(N)$ elements of the sparse matrix itself) in both to calculate these fill-ins during factorization and to solve the triangular systems afterward. The increased memory requirements also affect indirectly the computational cost, due to the need to resort to larger memory banks which are substantially slower.

The number of fill-ins generated during factorization can be reduced considerably by appropriate ordering of the equations and the unknowns (or equivalently, by permutation of the rows and columns of the matrix). The optimal ordering that produces the minimum number of fill-ins cannot be determined, since the problem is NP-complete [12], but powerful heuristics have been developed that do a reasonable job in practice. The most successful among them are the methods of approximate minimum degree (AMD) and nested dissection (ND). Both methods have their origins in graph theory and exploit the graph–matrix connection, in which a $N \times N$ sparse matrix **A** is represented by a graph with N vertices where the location of the nonzeros or "sparsity pattern" of **A** specify the edges of the graph (i.e., the vertices i and j are linked by an edge if and only if $a_{ij} \neq 0$). The AMD method [13] orders the rows and columns by repeatedly eliminating vertices with small—approximately minimum—degree (i.e., with few neighbors), and is the most widely used method for reduction of fill-ins in general sparse matrices. The ND method [14] is based on recursive graph partitioning (via techniques like METIS [15]), and is superior for special types of matrices like those represented by planar graphs and their higher dimensional extensions, for which there also exist theoretical results for the asymptotic upper bound of fill-ins (e.g., for matrices corresponding to planar graphs the fill-ins are $O(\log N)$, leading to complexity of $O(N^{3/2})$ for the factorization and $O(N \log N)$ for the forward and backward solve).

To give a concrete example of the effectiveness of proper ordering to produce sparser LU factors, we experimented with a set of large-scale circuit benchmarks arising from the electrical modeling of on-chip power distribution networks (see Section 8.4.1). The smallest of them had $N = 127{,}238$ nodes (there were no extra inductive branches) and furnished a system matrix of dimension $127{,}238 \times 127{,}238$ having 334,900 nonzero elements. A naive implementation of LU factorization produced 824,497,514 nonzeros in each of the factors L and U, while the application of AMD ordering resulted in only 4,570,337 nonzeros. Still, even with AMD ordering, a circuit with size $N = 2{,}627{,}445$ and 10,172,376 nonzeros generated 1,242,192,768 nonzeros in the LU factors, while all larger circuits could not fit the LU factors into the 24 GB memory of the target machine. This demonstrates beyond any doubt that direct methods scale very badly with problem size.

Direct methods are not very amenable to parallelization either. Although much research has been carried out in this direction, parallelization of direct methods is very difficult due to many sequential dependencies in the factorization process as well as in the forward and backward solves [16], and progress appears to have slowed down if not stagnated.

There are several software packages for general-purpose unsymmetric sparse direct linear system solution for a variety of target architectures. The most well-known are UMFPACK [17] for serial (not parallel) execution, PARDISO [18] for multicore processors, and more general shared-memory multiprocessor systems, and SuperLU_DIST [19] for distributed-memory clusters. Each of them also has a version for complex linear systems that is suitable for AC analysis. There also exists a circuit-specific (serial-only) sparse direct linear solver, KLU [20], which performs a different kind or ordering known as block-triangular or Dulmage–Mendelsohn decomposition [21] but further experiments are needed to verify its efficiency across a wide range of circuits, and for truly large-scale problems iterative methods become more efficient.

8.3.3.2 Iterative Linear Solvers

Iterative methods for linear systems work—like Newton's method for non-linear systems—by iteratively improving on approximate solutions until convergence is reached in some metric. They do not terminate in a fixed number of operations, unlike direct methods, because the number of iterations required for convergence is not known in advance and can vary among different matrices of equal size. However, they do not modify the system matrix (potentially destroying its sparsity) but only employ it in matrix–vector products, and thus are more efficient for large-scale problems in all three important factors, namely execution time, memory footprint, and parallel potential.

Traditional iterative methods like the Jacobi, Gauss–Seidel, or successive over-relaxation (SOR) are of relaxation type, because they operate by relaxing some of the components of the approximate solution vector in every iteration. More recent and advanced iterative methods are of projection type, as they construct each approximate solution by a process of projection on a subspace of $\mathfrak{R}^N$ with increasing dimension in every iteration. The subspace involved is usually of a specific kind known as Krylov subspace, and thus modern projection-type iterative methods are also referred to as "Krylov-subspace" methods. There are quite a few projection or Krylov-subspace methods [22], but for general unsymmetric linear systems like (8.2) the preferred method in industrial solvers is the method of generalized minimum residuals (GMRES), due to its robustness (other popular methods like Bi-CG or QMR can break down occasionally) and despite its larger memory requirements (as it has to keep the entire history of approximate solutions in memory).

As hinted previously, the operation with the largest complexity inside the iteration loop is the matrix–vector product, which is $O(N^2)$ for dense matrices and $O(N)$ for sparse matrices with $O(N)$ nonzeros multiplying a dense vector of size N (thus the serial complexity of a single iteration is $O(N^2)$ and $O(N)$, respectively). The other operations of the loop are inner products and scalar–vector products with vector additions (so-called saxpy), all of which are serially $O(N)$. A lot of parallel potential also exists, since all three types of products can be straightforwardly mapped to parallel hardware by splitting the vectors into segments and the matrix into corresponding strips, and assigning each processing core to compute the local product related to one segment.

The main problem of iterative methods is their rate of convergence which is not predictable up front. In principle, the maximum number of iterations for Krylov-subspace methods is equal to N, since then the N-dimensional subspace coincides with the entire space. In practice, however, it is common to converge to the true solution in much fewer iterations, but the actual number depends greatly on the properties of the system matrix. Specifically, it can be shown [23] that the required number of iterations—for a given convergence tolerance—is bounded in $O\left(\sqrt{\kappa_2(\mathbf{A})}\right)$, where $\kappa_2(\mathbf{A}) = ||\mathbf{A}||_2||\mathbf{A}^{-1}||_2$ is the spectral condition number of the matrix $\mathbf{A}$. This can be written as

$$\kappa_2(\mathbf{A}) = \frac{\sigma_{max}(\mathbf{A})}{\sigma_{min}(\mathbf{A})} = \frac{\sqrt{\lambda_{max}(\mathbf{AA}^T)}}{\sqrt{\lambda_{min}(\mathbf{AA}^T)}}$$

where $\sigma_{max}(\mathbf{A})$ and $\sigma_{min}(\mathbf{A})$ are the largest and smallest singular values of $\mathbf{A}$ ($\lambda_{max}(\mathbf{AA}^T)$ and $\lambda_{min}(\mathbf{AA}^T)$ are the largest and smallest eigenvalues of $\mathbf{AA}^T$), and it is always $\kappa_2(\mathbf{A}) \geq 1$. The condition number represents the distance of $\mathbf{A}$ to being singular (noninvertible), with $\kappa_2(\mathbf{A}) \approx 1$ when $\mathbf{A}$ is close to the identity matrix $\mathbf{A} \approx \mathbf{I}$ (well-conditioned matrix) and $\kappa_2(\mathbf{A}) \gg 1$ when $\mathbf{A}$ is close to being singular (ill-conditioned matrix). The above essentially means that convergence of iterative methods is fast when $\kappa_2(\mathbf{A}) \approx 1$ and slow when $\kappa_2(\mathbf{A}) \gg 1$.

To improve the convergence speed, it is necessary to apply a preconditioning mechanism, which transforms the initial linear system into an equivalent one with more favorable spectral condition number. The so-called preconditioner is a matrix $\mathbf{M}$ that approximates $\mathbf{A}$ in some way, such that the transformed system $M^{-1}Ax = M^{-1}b$ (which obviously has the same solution as the initial $Ax = b$) exhibits condition number $\kappa_2(M^{-1}A) \approx \kappa_2(I) = 1$. In practice, it is not necessary to invert the preconditioner M and apply it directly at the system $Ax = b$. It can be shown that the same thing can be accomplished by introducing an extra computational step within the iterative method, which entails solving a system $Mz = r$ with known right-hand side (RHS) vector r and unknown vector z in every iteration [23]. The general structure of a preconditioned iterative method is shown in Figure 8.3.

From the above, it follows that a good preconditioner M must satisfy two key properties:

1. It approximates the system matrix A well
2. A system $Mz = r$ can be solved more efficiently than the system $Ax = b$

where "more efficiently" can mean with less asymptotic complexity—ideally, an optimal or near-optimal complexity of $O(N)$ or $O(N \log N)$—and/or significantly more parallelism in the solution procedure. If the preconditioner is faithful enough to reduce the iterations substantially, then the whole burden of the algorithm is transferred to the preconditioner-solve step $Mz = r$.

Preconditioners for general-purpose linear systems have been developed, and they fall under the broad categories of incomplete factorizations or sparse approximate inverses [24]. The former perform an incomplete LU (ILU) factorization by either discarding completely fill-ins, thus keeping in LU only elements at nonzero positions in A (ILU(0)), or less often keeping fill-ins with magnitude above a certain threshold. The latter attempt to construct a sparse approximation $M^{-1} \approx A^{-1}$ to directly precondition the linear system $Ax = b$. Both types of preconditioners have been successful for a wide range of matrices, but they evidently cannot be as effective as preconditioners that are specially tuned to specific problems and matrices. Section 8.4.1

Inputs: System matrix A, preconditioner M, and RHS vector b
Output: Solution x of $Ax = b$

1: Set $x = x^{(0)}$ (initial guess); $r = b - Ax^{(0)}$ (residual)
2: **while** not_converged
3: …
4: Solve $Mz = r$
5: …
6: Update x and r (using z)
7: **end**

FIGURE 8.3
General structure of preconditioned iterative methods.

will present such a specialized preconditioner tailored for a specific type of circuit simulation problem.

For iterative methods there are typically not any canned software codes that implement them on particular architectures, because their algorithmic flows are common and fairly simple (see e.g., References 22, 25), and instead the main sophistication falls on the efficient implementation of the matrix–vector product as well as the inner products and the "saxpy's" on the target architecture. These core numerical operations are usually available in optimized basic linear algebra subprograms (BLAS) kernels [26] for any modern architecture, including parallel platforms such as multicore processors and GPUs. To juxtapose iterative methods with direct methods in large-scale circuit simulation problems, we consider again the circuit benchmarks modeling on-chip power distribution networks. Figure 8.4 depicts memory footprint and runtime against circuit size from the execution of a custom implementation of an iterative method with incomplete factorization preconditioner and a state-of the-art sparse direct solver (UMFPACK). The result

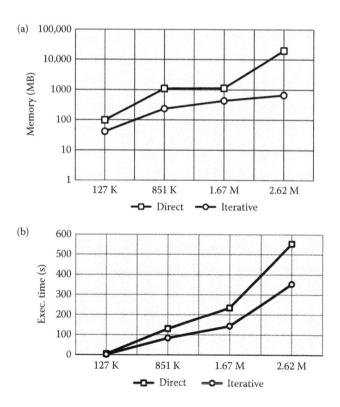

FIGURE 8.4

Comparison of direct and iterative methods: (a) memory footprint versus circuit size and (b) execution time versus circuit size.

that follows is that iterative methods become more efficient for circuits above 100 K nodes or so, both in memory requirements and execution time, and in fact direct methods fail completely for circuits larger than about a few million nodes because they run out of memory on standard machines.

Despite the above trends, direct methods are still the primary choice for industrial circuit simulators because of their overall robustness and faster execution in smaller scale problems (which cannot be overlooked in an industrial simulator). However, iterative methods are expected to gain ground in the future with the increasing sizes of circuits needed to be simulated in reasonable time, and in fact the industrial large-scale circuit simulator Xyce already incorporates an iterative GMRES solver with a circuit-specific preconditioning strategy for circuits above 100 K nodes [27].

8.4 Large-Scale Linear Circuit Simulation

Linear circuits composed of only linear RLC passive elements and active sources deserve a very different treatment in circuit simulation than nonlinear circuits (containing additionally nonlinear elements like semiconductor devices) for a variety of reasons. On one hand, linear circuits are simpler to simulate because the application of MNA directly gives a linear system of equations in both DC and transient analysis, and also because there is no stage of evaluation of device model equations at every time point. On the other hand, linear RLC circuits do not usually consist of deliberate circuit components, but are the outcome of modeling parasitic electrical effects in structures like power distribution networks, clock networks, multiconductor buses, or semiconductor substrate, and are often much larger than their nonlinear counterparts.

To derive the set of equations for the simulation of linear RLC circuits, it is advantageous to have all voltage sources transformed into Norton-equivalent current sources [28], thus rendering the b inductance currents the only current variables that are not eliminated in the MNA framework. The application of MNA for transient analysis then yields the following system of ordinary differential equations (ODEs):

$$\tilde{\mathbf{G}}\mathbf{x}(t) + \tilde{\mathbf{C}}\frac{d\mathbf{x}(t)}{dt} = \mathbf{b}(t) \tag{8.6}$$

where

$$\tilde{\mathbf{G}} = \begin{bmatrix} \mathbf{G} & \mathbf{A}_L \\ -\mathbf{A}_L^T & \mathbf{0} \end{bmatrix}, \quad \tilde{\mathbf{C}} = \begin{bmatrix} \mathbf{C} & \mathbf{0} \\ \mathbf{0} & \mathbf{L} \end{bmatrix}, \quad \mathbf{x}(t) = \begin{bmatrix} \mathbf{v}(t) \\ \mathbf{i}(t) \end{bmatrix}, \quad \mathbf{b}(t) = \begin{bmatrix} \mathbf{e}(t) \\ \mathbf{0} \end{bmatrix}$$

In the above system, $\mathbf{G}$ and $\mathbf{C}$ are the $n \times n$ node conductance and node capacitance matrices, $\mathbf{L}$ is the $b \times b$ branch inductance matrix, and $\mathbf{A}_L$ is the corresponding $n \times b$ node-to-branch incidence matrix (with elements $a_{ij} = \pm 1$ or $a_{ij} = 0$ depending on whether branch j leaves/enters or is not incident with node i). Also, $\mathbf{v}(t)$ and $\mathbf{i}(t)$ are the $n \times 1$ and $b \times 1$ vectors of node voltages and inductive branch currents, while $\mathbf{e}(t)$ is the $n \times 1$ vector of excitations from independent current sources at the nodes. Using the backward-Euler finite-difference approximation, we obtain the following linear system of equations for each time point t_i:

$$\left(\tilde{\mathbf{G}} + \frac{1}{h_i} \tilde{\mathbf{C}} \right) \mathbf{x}(t_i) = \mathbf{b}(t_i) + \frac{1}{h_i} \tilde{\mathbf{C}} \mathbf{x}(t_{i-1}) \tag{8.7}$$

where h_i is the time-step at t_i. The above linear system has dimension $N \times N$, but in the very common case where $\mathbf{L}$ is diagonal (i.e., when there are only branch self-inductances and not mutual inductances between branches) we can further reduce the system dimension to $n \times n$. Specifically, by inverting $\mathbf{L}$ and performing block-matrix operations in (8.7), we can obtain the following system of coupled recursive equations:

$$\left(\mathbf{G} + \frac{1}{h_i} \mathbf{C} + h_i \mathbf{A}_L \mathbf{L}^{-1} \mathbf{A}_L^T \right) \mathbf{v}(t_i) = \mathbf{e}(t_i) + \frac{1}{h_i} \mathbf{C} \mathbf{v}(t_{i-1}) - \mathbf{A}_L \mathbf{i}(t_{i-1}) \tag{8.8}$$

$$\mathbf{i}(t_i) = h_i \mathbf{L}^{-1} \mathbf{A}_L^T \mathbf{v}(t_i) + \mathbf{i}(t_{i-1}) \tag{8.9}$$

where at each time point t_i we must solve the $n \times n$ linear system (8.8) with system matrix $\mathbf{A} \equiv \mathbf{G} + \mathbf{C}/h_i + h_i \mathbf{A}_L \mathbf{L}^{-1} \mathbf{A}_L^T$ to find the vector of node voltages $\mathbf{v}(t_i)$, and subsequently compute the vector of branch currents $\mathbf{i}(t_i)$ from (8.9) (in DC analysis where there are neither LC elements nor time dependence, the above system reduces to the linear system $\mathbf{G}\mathbf{v} = \mathbf{e}$). Along with the reduced dimension, it can be shown that the system matrix of (8.8) is symmetric and positive definite (SPD), and also diagonally dominant whenever there are no mutual inductances and $\mathbf{L}$, $\mathbf{L}^{-1}$ are diagonal [28]. On the contrary, the Jacobian matrices arising from linearization of nonlinear circuit equations in (8.2) are in general nonsymmetric and indefinite. Thus, whenever the circuit to be simulated is linear, it can be extremely beneficial to exploit its special characteristics rather than resort to a standard SPICE implementation.

It must be noted that the case where $\mathbf{L}$ is dense and contains many mutual inductances rapidly gains importance in high-frequency applications, where the parasitic modeling of groups of long conductors requires them to be segmented lengthwise and/or subdivided into filaments across their cross-section, and mutual couplings to be specified between every conductor, segment, and filament. While $\mathbf{L}^{-1}$ is also dense in this case, it has been observed

[29] that it can be sparsified more stably and to a greater extent than $\mathbf{L}$ itself—and several techniques for suitable sparsification of $\mathbf{L}^{-1}$ have been proposed to this end [30]—thus rendering the system (8.8) and (8.9) still possible to use for efficient sparse simulation.

For SPD linear systems like (8.8), the Cholesky factorization is twice as efficient as the LU factorization and is the appropriate method to be used for direct system solution. The most well-known software package for sparse Cholesky factorization is CHOLMOD [31], for which there exist both serial and GPU implementations. Regarding iterative solvers, the appropriate Krylov-subspace method for SPD linear systems, which is also the first historically and the most well known, is the method of conjugate gradients (CG). The algorithmic loop of CG includes naturally one matrix–vector product alongside some inner products and "saxpy's" [22,25], whose execution can be carried out efficiently via optimized BLAS kernels on modern architectures. As discussed in Section 8.3.3, iterative methods turn out to be more efficient than direct methods on large-scale problems, but must be accompanied by an appropriate preconditioner (preconditioned conjugate gradients (PCG)) to compensate for their unpredictable convergence rate. General-purpose preconditioners such as the incomplete Cholesky (IC) factorization have been developed here as well, but again more efficient specialized preconditioners can be constructed for specific applications. The next section presents such a specialized preconditioner to demonstrate their worth, and motivate research for preconditioners in other circuit simulation problems.

8.4.1 Large-Scale Linear Circuits with Regular Spatial Geometry: The Case of On-Chip Power Distribution Networks

One of the most successful applications of specialized preconditioners is in the simulation of on-chip power distribution networks, so-called power grids. Contemporary power grids are truly gigantic, consisting of several millions or even billions of nodes, and are easily the largest circuits demanding thorough simulation to verify proper voltage supply to the active devices. The release of industrial benchmarks from IBM [32] with intention to motivate research in the area clearly demonstrates the importance of the problem. Unsurprisingly, the efficient simulation of power grids has seen an explosion of research works in the last decade [33–35]. The most successful recent efforts take advantage of the regular spatial geometry of power grids, and employ techniques from the field of partial differential equations such as multigrid and spatial fast Fourier transform (FFT). We will describe briefly the second technique that has reported the largest performance gains on modern massively parallel architectures like GPUs [36]. It also provides an excellent example of a specialized preconditioner that vastly outperforms general-purpose ones.

Practical power grids are created as orthogonal wire meshes with very regular geometries, with possibly some irregularities imposed by design constraints (e.g., some missing connections between adjacent nodes), and arranged in a few—typically 2–8—metal layers of alternating routing directions (horizontal and vertical). Due to the presence of vias between successive metal layers, the actual grid has the structure of a 3D mesh, with very few planes along the third dimension (Figure 8.5a). However, practical experience shows that the electrical resistances of vias are typically

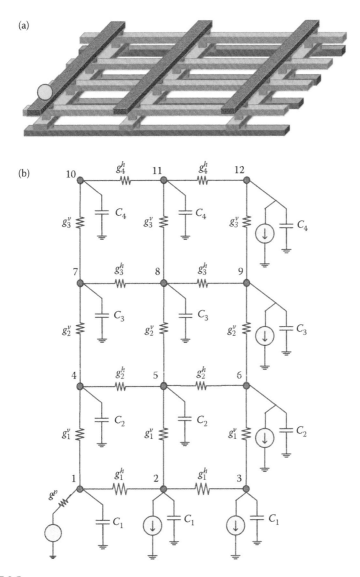

FIGURE 8.5
(a) Physical geometry of a 3D power grid with four metal layers and (b) electrical model of the regular 2D grid used to construct the preconditioner matrix **M**.

much smaller than wire resistances, leading to voltage drops much less than 1 mV [35]. Also, almost all horizontal and vertical resistances in each metal layer have the same values, with very few deviations due to grid irregularities [34].

Based on these observations—both of which are confirmed in the IBM benchmarks—the following regularization process of the given 3D grid to a regular 2D grid (Figure 8.5b) can be conceived to construct a specialized preconditioner for power grid simulation (parasitic inductances are ignored for simplicity, although they can be easily included):

1. Determine the distinct x- and y-coordinates of all nodes in the different layers of the 3D grid, and take their Cartesian product to specify the location of the nodes in the regular 2D grid.

2. By disregarding via resistances between layers, collapse the 3D grid onto the regular 2D grid by adding together all horizontal branch conductances g^h connected in parallel between adjacent nodes in the x-direction of the 2D grid, and all vertical branch conductances g^v connected in parallel between adjacent nodes in the y-direction of the 2D grid. If a conductance of the 3D grid occupies multiple nodes of the regular 2D grid, it is decomposed into a corresponding number of pieces. The node capacitances corresponding to the same regular grid nodes are also added together during the collapsing.

3. In the regular 2D grid, substitute horizontal branch conductances by their average value in each horizontal rail, and vertical branch conductances by their average value in each horizontal slice (enclosed between two adjacent horizontal rails). Substitute node capacitances in each horizontal rail by their average value as well.

4. In the regular 2D grid, amortize the total sum of supply pad conductances g^p of a specific horizontal rail to all nodes of this rail, that is, assume that all nodes of the jth horizontal rail have pad conductance

$$\bar{g}_j^p = \frac{\left(\sum g^p\right)_j}{m}, \text{ where } (\Sigma g^p)_j \text{ is the sum of the actual pad conductances}$$

attached to nodes of the jth horizontal rail and m is the total number of nodes in the rail.

As an example, the four-layer power grid shown in Figure 8.5a has $l = 4$ horizontal rails and $m = 3$ vertical rails in alternately routed layers, and will furnish the regular 2D grid depicted in Figure 8.5b after applying the first three steps of the previous process for constructing the preconditioner matrix **M**. If we use the indicated natural node numbering (proceeding horizontally, i.e., along the routing direction of the lowest level metal layer), the

MNA matrix $G + C/h_i$ that corresponds to the regular 2D grid will be the following block-tridiagonal matrix:

$$M = \begin{bmatrix} T_1 & -g_1^v I & & \\ -g_1^v I & T_2 & -g_2^v I & \\ & -g_2^v I & T_3 & -g_3^v I \\ & & -g_3^v I & T_4 \end{bmatrix}$$

where T_1, T_2, T_3, T_4 are 3×3 tridiagonal matrices (each one corresponding to a horizontal rail of the 2D grid) which have the following form:

$$T_1 = \begin{bmatrix} g_1^h + g_1^v + g^p + \dfrac{c_1}{h_i} & -g_1^h & \\ -g_1^h & 2g_1^h + g_1^v + \dfrac{c_1}{h_i} & -g_1^h \\ & -g_1^h & g_1^h + g_1^v + \dfrac{c_1}{h_i} \end{bmatrix}$$

$$T_2 = \begin{bmatrix} g_2^h + g_1^v + g_2^v + \dfrac{c_2}{h_i} & -g_2^h & \\ -g_2^h & 2g_2^h + g_1^v + g_2^v + \dfrac{c_2}{h_i} & -g_2^h \\ & -g_2^h & g_2^h + g_1^v + g_2^v + \dfrac{c_2}{h_i} \end{bmatrix}$$

$$T_3 = \begin{bmatrix} g_3^h + g_2^v + g_3^v + \dfrac{c_3}{h_i} & -g_3^h & \\ -g_3^h & 2g_3^h + g_2^v + g_3^v + \dfrac{c_3}{h_i} & -g_3^h \\ & -g_3^h & g_3^h + g_2^v + g_3^v + \dfrac{c_3}{h_i} \end{bmatrix}$$

$$T_4 = \begin{bmatrix} g_4^h + g_3^v + \dfrac{c_4}{h_i} & -g_4^h & \\ -g_4^h & 2g_4^h + g_3^v + \dfrac{c_4}{h_i} & -g_4^h \\ & -g_4^h & g_4^h + g_3^v + \dfrac{c_4}{h_i} \end{bmatrix}$$

In the above, g_j^h is the average horizontal conductance in the jth horizontal rail, g_j^v is the average vertical conductance in the jth horizontal slice, and c_j is the average node capacitance in the jth horizontal rail. The fourth step of the regularization process will change the block $\mathbf{T}_1$ of this particular example into

$$
\mathbf{T}_1 =
\begin{bmatrix}
g_1^h + g_1^v + \bar{g}_1^p + \dfrac{c_1}{h_i} & -g_1^h & \\
-g_1^h & 2g_1^h + g_1^v + \bar{g}_1^p + \dfrac{c_1}{h_i} & -g_1^h \\
& -g_1^h & g_1^h + g_1^v + \bar{g}_1^p + \dfrac{c_1}{h_i}
\end{bmatrix}
$$

where $\bar{g}_1^p = g^p/3$. It is not difficult to generalize the process to an arbitrary power grid with l horizontal and m vertical rails, and $n = l{\cdot}m$ nodes in the regular 2D counterpart. In that case, the $n \times n$ preconditioner matrix will comprise l diagonal blocks of size $m \times m$ and have the following form:

$$
\mathbf{M} =
\begin{bmatrix}
\mathbf{T}_1 & \gamma_1\mathbf{I} & & & \\
\gamma_1\mathbf{I} & \mathbf{T}_2 & \gamma_2\mathbf{I} & & \\
& \cdot & \cdot & \cdot & \\
& & \gamma_{l-2}\mathbf{I} & \mathbf{T}_{l-1} & \gamma_{l-1}\mathbf{I} \\
& & & \gamma_{l-1}\mathbf{I} & \mathbf{T}_l
\end{bmatrix}
\tag{8.10}
$$

where $\mathbf{I}$ is the $m \times m$ identity matrix and $\mathbf{T}_j$, $j = 1,...l$, are $m \times m$ tridiagonal matrices of the following form:

$$
\mathbf{T}_j =
\begin{bmatrix}
\alpha_j + \beta_j & -\alpha_j & & & \\
-\alpha_j & 2\alpha_j + \beta_j & -\alpha_j & & \\
& \cdot & \cdot & \cdot & \\
& & -\alpha_j & 2\alpha_j + \beta_j & -\alpha_j \\
& & & -\alpha_j & \alpha_j + \beta_j
\end{bmatrix}
$$

$$
= \alpha_j
\begin{bmatrix}
1 & -1 & & & \\
-1 & 2 & -1 & & \\
& \cdot & \cdot & \cdot & \\
& & -1 & 2 & -1 \\
& & & -1 & 1
\end{bmatrix}
+ \beta_j\mathbf{I}
$$

where $\alpha_j = g_j^h$, $\beta_j = g_j^v + g_{j-1}^v + \bar{g}_j^p + c_j/h_i$, $\gamma_j = -g_j^v$ (with $g_0^v = g_l^v = 0$).

The above preconditioner approximates the matrix of practical power grids well enough to reduce the iterations of PCG to nearly an order of magnitude in comparison to the IC preconditioner, as illustrated in Figure 8.6a and b (in fact, the larger benchmarks of Figure 8.6b are more regular and converge in fewer iterations).

But how can we solve efficiently a system $\mathbf{Mz} = \mathbf{r}$ in the preconditioner-solve step of the iterative algorithm? As it turns out, we can determine in advance the eigenvalues and eigenvectors of the tridiagonal matrices $\mathbf{T}_j$, and also the matrices of eigenvectors that diagonalize $\mathbf{T}_j$ are matrices corresponding to a spatial fast transform. More specifically, it can be shown [37] that each $\mathbf{T}_j$ has m distinct eigenvalues $\lambda_{j,i}$, $i = 1,...,m$, which are given by

$$\lambda_{j,i} = \beta_j + 4\alpha_j \sin^2 \frac{(i-1)\pi}{2m} = \beta_j + 2\alpha_j \left(1 - \cos \frac{(i-1)\pi}{m}\right) \qquad (8.11)$$

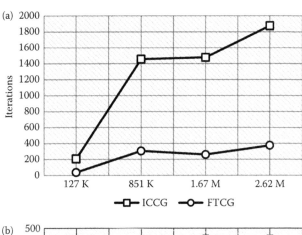

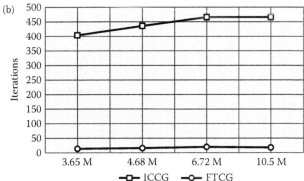

FIGURE 8.6
Comparison of convergence rate for incomplete Cholesky CG and fast transform-preconditioned CG: (a) circuits up to 3M nodes (quite irregular) and (b) circuits larger than 3M nodes (more regular).

as well as a set of m orthonormal eigenvectors $\mathbf{q}_i$, $i = 1,\ldots,m$, with elements

$$
q_{i,k} = \begin{cases} \sqrt{\dfrac{1}{m}}\cos\dfrac{(2k-1)(i-1)\pi}{2m}, & i = 1, & k = 1,\ldots,m \\[3mm] \sqrt{\dfrac{2}{m}}\cos\dfrac{(2k-1)(i-1)\pi}{2m}, & i = 2,\ldots,m, & k = 1,\ldots,m \end{cases} \tag{8.12}
$$

Note that the $\mathbf{q}_i$ do not depend on the values of α_j and β_j, and are the same for every matrix $\mathbf{T}_j$. If $\mathbf{Q} = [\mathbf{q}_1,\ldots,\mathbf{q}_m]$ denotes the matrix whose columns are the eigenvectors $\mathbf{q}_i$, then due to the eigen-decomposition of $\mathbf{T}_j$ we have $\mathbf{Q}^T\mathbf{T}_j\mathbf{Q} = \Lambda_j = \mathrm{diag}(\lambda_{j,1},\ldots,\lambda_{j,m})$. By exploiting this diagonalization of matrices $\mathbf{T}_j$, the system $\mathbf{M}\mathbf{z} = \mathbf{r}$ with $\mathbf{M}$ of the form of (8.10) is equivalent to the following system (due to $\mathbf{Q}^T\mathbf{Q} = \mathbf{I}$):

$$
\begin{bmatrix} \mathbf{Q}^T & & \\ & \ddots & \\ & & \mathbf{Q}^T \end{bmatrix} \mathbf{M} \begin{bmatrix} \mathbf{Q} & & \\ & \ddots & \\ & & \mathbf{Q} \end{bmatrix} \begin{bmatrix} \mathbf{Q}^T & & \\ & \ddots & \\ & & \mathbf{Q}^T \end{bmatrix} \mathbf{z} = \begin{bmatrix} \mathbf{Q}^T & & \\ & \ddots & \\ & & \mathbf{Q}^T \end{bmatrix} \mathbf{r}
$$

$$
\Leftrightarrow \begin{bmatrix} \Lambda_1 & \gamma_1\mathbf{I} & & & \\ \gamma_1\mathbf{I} & \Lambda_2 & \gamma_2\mathbf{I} & & \\ & \ddots & \ddots & \ddots & \\ & & \gamma_{l-2}\mathbf{I} & \Lambda_{l-1} & \gamma_{l-1}\mathbf{I} \\ & & & \gamma_{l-1}\mathbf{I} & \Lambda_l \end{bmatrix} \tilde{\mathbf{z}} = \tilde{\mathbf{r}} \tag{8.13}
$$

where

$$
\tilde{\mathbf{z}} = \begin{bmatrix} \mathbf{Q}^T & & \\ & \ddots & \\ & & \mathbf{Q}^T \end{bmatrix} \mathbf{z}, \quad \tilde{\mathbf{r}} = \begin{bmatrix} \mathbf{Q}^T & & \\ & \ddots & \\ & & \mathbf{Q}^T \end{bmatrix} \mathbf{r}
$$

If the $n \times 1$ vectors $\mathbf{r}$, $\mathbf{z}$, $\tilde{\mathbf{r}}$, $\tilde{\mathbf{z}}$ are themselves partitioned into l subvectors (blocks) of size $m \times 1$ each, that is,

$$
\mathbf{r} = \begin{bmatrix} \mathbf{r}_1 \\ \vdots \\ \mathbf{r}_l \end{bmatrix}, \quad \mathbf{z} = \begin{bmatrix} \mathbf{z}_1 \\ \vdots \\ \mathbf{z}_l \end{bmatrix}, \quad \tilde{\mathbf{r}} = \begin{bmatrix} \tilde{\mathbf{r}}_1 \\ \vdots \\ \tilde{\mathbf{r}}_l \end{bmatrix}, \quad \tilde{\mathbf{z}} = \begin{bmatrix} \tilde{\mathbf{z}}_1 \\ \vdots \\ \tilde{\mathbf{z}}_l \end{bmatrix}
$$

then we have $\tilde{\mathbf{r}}_j = \mathbf{Q}^T\mathbf{r}_j$ and $\tilde{\mathbf{z}}_j = \mathbf{Q}^T\mathbf{z}_j \Leftrightarrow \mathbf{z}_j = \mathbf{Q}\tilde{\mathbf{z}}_j$, $j = 1,\ldots,l$.

However, it can be shown that each product $\mathbf{Q}^T \mathbf{r}_j = \tilde{\mathbf{r}}_j$ corresponds to a discrete cosine transform of type-II (DCT-II) on $\mathbf{r}_j$, and each product $\mathbf{Q}\tilde{\mathbf{z}}_j = \mathbf{z}_j$ corresponds to an inverse discrete cosine transform of type-II (IDCT-II) on $\tilde{\mathbf{z}}_j$ [38]. This means that the computation of the whole vector $\tilde{\mathbf{r}}$ from $\mathbf{r}$ amounts to l independent DCT-II transforms of size m, and the computation of the whole vector $\mathbf{z}$ from $\tilde{\mathbf{z}}$ amounts to l independent IDCT-II transforms of size m. A modification of FFT can be employed for each of the l independent DCT-II/IDCT-II transforms [38], giving a total serial operation count of $O(lm \log m) = O(n \log m)$.

If now $\mathbf{P}$ is a $n \times n$ permutation matrix that reorders the elements of a vector or the rows of a matrix as $1, m + 1, 2m + 1,\ldots, (l - 1)m + 1, 2, m + 2, 2m + 2, \ldots, (l - 1)m + 2,\ldots, m, m + m, 2m + m,\ldots, (l - 1)m + m$, and $\mathbf{P}^T$ is the inverse permutation matrix, then the system (8.13) is further equivalent to

$$
\mathbf{P}\begin{bmatrix}
\Lambda_1 & \gamma_1\mathbf{I} & & & \\
\gamma_1\mathbf{I} & \Lambda_2 & \gamma_2\mathbf{I} & & \\
& \cdot & \cdot & \cdot & \\
& & \gamma_{l-2}\mathbf{I} & \Lambda_{l-1} & \gamma_{l-1}\mathbf{I} \\
& & & \gamma_{l-1}\mathbf{I} & \Lambda_l
\end{bmatrix} \mathbf{P}^T \mathbf{P}\tilde{\mathbf{z}} = \mathbf{P}\tilde{\mathbf{r}} \iff \begin{bmatrix}
\tilde{\mathbf{T}}_1 & & & \\
& \tilde{\mathbf{T}}_2 & & \\
& & \ddots & \\
& & & \tilde{\mathbf{T}}_m
\end{bmatrix} \tilde{\mathbf{z}}^P = \tilde{\mathbf{r}}^P
$$

$$(8.14)$$

where

$$
\tilde{\mathbf{T}}_i = \begin{bmatrix}
\lambda_{1,i} & \gamma_1 & & & \\
\gamma_1 & \lambda_{2,i} & \gamma_2 & & \\
& \cdot & \cdot & \cdot & \\
& & \gamma_{l-2} & \lambda_{l-1,i} & \gamma_{l-1} \\
& & & \gamma_{l-1} & \lambda_{l,i}
\end{bmatrix}
$$

$$(8.15)$$

and $\tilde{\mathbf{z}}^P = \mathbf{P}\tilde{\mathbf{z}}$, $\tilde{\mathbf{r}}^P = \mathbf{P}\tilde{\mathbf{r}}$. If the $n \times 1$ vectors $\tilde{\mathbf{z}}^P$, $\tilde{\mathbf{r}}^P$ are partitioned into m subvectors $\tilde{\mathbf{z}}_i^P$, $\tilde{\mathbf{r}}_i^P$ of size $l \times 1$ each, then the system (8.14) effectively represents m independent tridiagonal systems $\tilde{\mathbf{T}}_i\tilde{\mathbf{z}}_i^P = \tilde{\mathbf{r}}_i^P$ of size l which can be solved with respect to the blocks $\tilde{\mathbf{z}}_i^P$, $i = 1,\ldots,m$ (to produce the whole vector $\tilde{\mathbf{z}}^P$) in a total of $O(lm) = O(n)$ serial operations. For each such system the coefficient matrix (8.15) is known beforehand and is determined exclusively by the eigenvalues (8.11) and the values γ_j of matrix $\mathbf{M}$. The equivalence of the system $\mathbf{Mz} = \mathbf{r}$, with $\mathbf{M}$ as in (8.10), to the system (8.14) and (8.15) gives a procedure for fast (near-optimal) solution of $\mathbf{Mz} = \mathbf{r}$ which is shown in Figure 8.7.

Apart from the near-optimal serial complexity of $O(n \log m)$ operations, there is also a large amount of parallelism involved in the solution of $\mathbf{Mz} = \mathbf{r}$, both at the task level and the data level. At the task level, the l DCT-II/IDCT-II

Inputs: Eigenvalues (8.11) and off-diagonal entries γ_j of preconditioner **M**, right-hand side (RHS) vector **r**
Output: Solution **z** of **Mz = r**

1: Partition **r** into l subvectors $\mathbf{r}_j$ of size m, and perform DCT-II transform $(\mathbf{Q}^T \mathbf{r}_j)$ on each subvector to obtain transformed vector $\tilde{\mathbf{r}}$
2: Permute vector $\tilde{\mathbf{r}}$ by permutation **P**, which orders elements as 1, $m + 1$, $2m + 1,...,\ (l - 1)m + 1,\ 2,\ m + 2,\ 2m + 2,...,\ (l - 1)m + 2,...,\ m,\ m + m,$ $2m + m,...,\ (l - 1)m + m$, in order to obtain vector $\tilde{\mathbf{r}}^P$
3: Solve the m tridiagonal systems (8.14) with known coefficient matrices (8.15), in order to obtain vector $\tilde{\mathbf{z}}^P$
4: Apply inverse permutation $\mathbf{P}^T$ on vector $\tilde{\mathbf{z}}^P$ so as to obtain vector $\tilde{\mathbf{z}}$
5: Partition vector $\tilde{\mathbf{z}}$ into l subvectors $\tilde{\mathbf{z}}_j$ of size m, and perform IDCT-II transform $(\mathbf{Q}\tilde{\mathbf{z}}_j)$ on each subvector to obtain final solution vector **z**

FIGURE 8.7
Fast transform solution algorithm for the preconditioner-solve step **Mz = r**, with **M** of the form (8.10).

transforms and the m tridiagonal systems are completely independent to each other and can be solved in parallel. At the data level, the FFT and the tridiagonal system solution are highly parallel algorithms, allowing extremely efficient divide-and-conquer schemes, and off-the-shelf optimized codes are available for every modern parallel platform including multicore processors and GPUs. The above solution procedure was implemented on an NVIDIA Tesla C2075 GPU with only data-level parallelism exploited (i.e., optimized GPU routines for FFT and tridiagonal system solution were utilized, but all independent FFTs and tridiagonal solvers were executed sequentially), and was used to precondition the CG method. Figure 8.8 compares the runtime of the fast transform-preconditioned implementation of CG (FTCG) against the IC-preconditioned CG (ICCG) on the set of IBM power grid benchmarks. Even with no task-level parallelism, a nearly two orders of magnitude speed-up was observed on average, demonstrating clearly the benefits of specialized preconditioners that allow efficient (near-optimal and highly parallel) solution procedures.

A 3D extension of the preconditioner is possible in the case where vias between successive metal layers are significant and cannot be neglected [39]. Plenty of other efficient preconditioners for specific circuit simulation problems are still waiting to be discovered.

8.5 Conclusion

This chapter presented an overview of large-scale circuit simulation, both for nonlinear circuits (centered on MOS transistors) and linear circuits

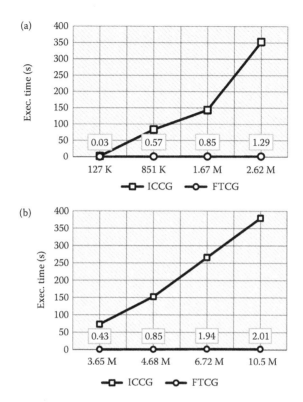

FIGURE 8.8
Comparison of execution time for incomplete Cholesky CG and fast transform-preconditioned CG: (a) circuits up to 3M nodes (quite irregular) and (b) circuits larger than 3M nodes (more regular).

(typically modeling large parasitic structures), with special focus on the solution of the relevant linear systems which dominates the large-scale computational cost. As the sizes of circuits to be simulated have long exceeded the million-node mark and the simulation needs are constantly escalating, we believe that they can only be met by wider adoption of iterative linear solution methods, as is already the norm in other fields like those governed by partial differential equations. The development of efficient preconditioners for specific circuit simulation problems is expected to play a central role here, like in the whole range of scientific computing. As Trefethen and Bau [40] put it, "With the subject of preconditioning, we find ourselves at the philosophical center of the scientific computing of the future. Nothing will be more central to computational science in the next century than the art of transforming a problem that appears intractable into another whose solution can be approximated rapidly. For Krylov subspace matrix iterations, this is preconditioning." Apart from the developments in the algorithmic front, the future will undoubtedly bring forth a broader exploitation of parallelism on modern mainstream machines that combine multicore processors and GPUs (i.e., blending MIMD and SIMD parallel paradigms) under the same hood. Especially for preconditioned iterative methods, the parallelization effort must go hand in hand with the development of the preconditioner, because their codesign for a specific application can lead to the largest boost of performance, as it has been eloquently demonstrated in the problem of simulating on-chip power distribution networks. We can certainly look forward to great advances for large-scale circuit simulation in the future.

Acknowledgments

The authors happily acknowledge the help and contribution of Konstantis Daloukas, who supported in the creation of the data for this chapter and whose former work is reflected in a substantial part of Section 8.4.2. Also, the General Secretariat of Research and Technology (GSRT) in Greece provided financial support under grant agreement 11SYN-5-719, which enabled writing of this chapter and part of the associated research.

References

1. Nagel, L., Pederson, D. Simulation program with integrated circuit emphasis. *IEEE Midwest Symposium On Circuit Theory*, 1973.
2. Roychowdhury, J., Mantooth, A. Simulation of analog and RF circuits and systems. In Scheffer, L., Lavagno, L., Martin, G. (eds.), *EDA for IC Implementation, Circuit Design and Process Technology*. CRC Press, 2006.
3. Schilders, W. Introduction to model order reduction. In Schilders, W., van der Vorst, H., Rommes, J. (eds.), *Model Order Reduction: Theory, Research Aspects and Applications*. Springer, 2008.
4. Rewienski, M. A perspective on fast-SPICE simulation technology. In Li, P., Silveira, L., Feldmann, P. (eds.), *Simulation and Verification of Electronic and Biological Systems*. Springer, 2011.
5. Najm, F. *Circuit Simulation*. Wiley, 2010.
6. Ascher, U., Petzold, L. *Computer Methods for Ordinary Differential Equations and Differential-Algebraic Equations*. SIAM, 1998.
7. Liu, W., Hu, C. *BSIM4 and MOSFET Modeling for IC Simulation*. World Scientific, 2011.
8. Sadayappan, P. Visvanathan, V. Circuit simulation on shared-memory multiprocessors. *IEEE Trans. Comput.*, 37(12), 1634–1642, 1988.
9. Gulati, K., Croix, J., Khatri, S., Shastry, R. Fast circuit simulation on graphics processing units. *IEEE/ACM Asia and South Pacific Design Automation Conference (ASP-DAC)*, 2009.

10. Kelley, C. *Iterative Methods for Linear and Nonlinear Equations*. SIAM, 1995.
11. Kelley, C. *Solving Nonlinear Equations with Newton's Method*. SIAM, 2003.
12. Yannakakis, M. Computing the minimum fill-in is NP-complete. *SIAM J. Algebraic Discrete Methods*, 2(1) 77–79, 1981.
13. Amestoy, P., Davis, T., Duff, I. Algorithm 837: AMD, an approximate minimum degree ordering algorithm. *ACM Trans. Math. Software*, 30(3), 381–388, 2004.
14. Gilbert, J., Tarjan, R. The analysis of a nested dissection algorithm. *Numerische Mathematik*, 50(4), 377–404, 1987.
15. Karypis, G., Kumar, V. A fast and high quality multilevel scheme for partitioning irregular graphs. *SIAM J. Scientific Computing*, 20(1), 359–392, 1998.
16. Li, X., Shao, M., Yamazaki, I., Ng, E. Factorization-based sparse solvers and preconditioners. *J. Physics: Conf. Series*, 180, 012015, 2009.
17. Davis, T. Algorithm 832: UMFPACK V4.3, an unsymmeric-pattern multifrontal method. *ACM Trans. Math. Software*, 30(2), 196–199, 2004.
18. Schenk, O., Gartner, K. Solving unsymmetric sparse systems of linear equations with PARDISO. *Future Generation Computer Systems*, 20(3), 475–487, 2004.
19. Demmel, J., Li, X. SuperLU_DIST: A scalable distributed-memory sparse direct solver for unsymmetric linear systems. *ACM Trans. Math. Software*, 29(2), 110–140, 2003.
20. Davis, T., Natarajan, E. Algorithm 907: KLU, a direct sparse solver for circuit simulation problems. *ACM Trans. Math. Software*, 37(3), 36:1–36:17, 2010.
21. Pothen, A., Fan, C. Computing the block triangular form of a matrix. *ACM Trans. Math. Software*, 16(4), 303–324, 1990.
22. Saad, Y. *Iterative Methods for Sparse Linear Systems*, 2nd edition. SIAM, 2003.
23. Axelsson, O., Barker, A. *Finite Element Solution of Boundary Value Problems*. SIAM, 2001.
24. Benzi, M. Peconditioning techniques for large linear systems: A survey. *J. Computational Physics*, 182(2), 418–477, 2002.
25. Barrett, R., Berry, M., Chan, T. et al. *Templates for the Solution of Linear Systems: Building Blocks for Iterative Methods*, 2nd edition. SIAM, 1992.
26. Blackford, L., Demmel, J., Dongarra, J. et al. An updated set of Basic Linear Algebra Subprograms (BLAS). *ACM Trans. Math. Software*, 28(2), 135–151, 2002.
27. Thornquist, H., Keiter, E., Hoekstra, R., Day, D., Boman, E. A parallel preconditioning strategy for efficient transistor-level circuit simulation. *IEEE/ACM International Conference on Computer-Aided Design* (ICCAD), 2009.
28. Chen, T., Luk, C., Chen, C. INDUCTWISE: Inductance-wise interconnect simulator and extractor. *IEEE Trans. Computer-Aided Design*, 22(7), 884–894, 2003.
29. Devgan, A., Ji, H., Dai, W. How to efficiently capture on-chip inductance effects: Introducing a new circuit element K. *IEEE/ACM International Conference on Computer-Aided Design* (ICCAD), 2000.
30. Apostolopoulou, I., Daloukas, K., Evmorfopoulos, N., Stamoulis, G. Selective inversion of inductance matrix for large-scale sparse RLC simulation. *IEEE/ACM Design Automation Conference* (DAC), 2014.
31. Chen, Y., Davis, T., Hager, W., Rajamanickam, S. Algorithm 887: CHOLMOD, supernodal sparse Cholesky factorization and update/downdate. *ACM Trans. Math. Software*, 35(3), 22:1–22:14, 2008.
32. Nassif, S. Power grid analysis benchmarks. *IEEE/ACM Asia and South Pacific Design Automation Conference* (ASP-DAC), 2008.

33. Kozhaya, J., Nassif, S., Najm, F. A multigrid-like technique for power grid analysis. *IEEE Trans. Computer-Aided Design*, 21(10), 1148–1160, 2002.
34. Shi, J., Cai, Y., Tan, S., Fan, J., Hong, X. Pattern-based iterative method for extreme large power/ground analysis. *IEEE Trans. Computer-Aided Design*, 26(4), 680–692, 2007.
35. Feng, Z., Zeng, Z., Li, P. Parallel on-chip power distribution network analysis on multi-core multi-GPU platforms. *IEEE Trans. VLSI Systems*, 19(10), 1823–1836, 2011.
36. Daloukas, K., Evmorfopoulos, N., Drasidis, G. et al. Fast transform-based preconditioners for large-scale power grid analysis on massively parallel architectures. *IEEE/ACM International Conference Computer-Aided Design (ICCAD)*, 2012.
37. Christara, C. Quadratic spline collocation methods for elliptic partial differential equations. *BIT Numerical Mathematics*, 34(1), 33–61, 1994.
38. Van Loan, C. *Computational Frameworks for the Fast Fourier Transform*. SIAM, 1992.
39. Daloukas, K., Evmorfopoulos, N., Tsompanopoulou, P., Stamoulis, G. A 3-D fast transform-based preconditioner for large-scale power grid analysis on massively parallel architectures. *IEEE/ACM Int. Symp. Quality Electronic Design (ISQED)*, 2014.
40. Trefethen, L., Bau, D. *Numerical Linear Algebra*. SIAM, 1997.

9

Mixed-Signal IC Design Addressed to Substrate Noise Immunity in Bulk Silicon toward 3D Circuits

Olivier Valorge, Fengyuan Sun , Jean-Etienne Lorival,
Francis Calmon, and Christian Gontrand

CONTENTS

9.1 Introduction

Design of mixed signal ICs supposes the resolution of coupling problems between analog and digital blocks placed on the same die. The switching events of digital gates involve variations of supply voltage function of digital signal transitions that are not compensated by decoupling capacitors [1]. Isolation between analog and digital parts with buried layers, guard rings, or separation of the supply nets are methods to reduce coupling mechanisms [2]. However, substrate parasitic capacitors and resistors allow the noise signal generated by the digital part to pass through the substrate and to reach the analog and radiofrequency (RF) blocks [3]. The literature is quite poor concerning the impact on analog and RF blocks, many papers being focused on noise generation and propagation into digital complementary metal-oxide semiconductor (CMOS) circuits [1,4]. This chapter deals with the impact of low-frequency substrate noise onto a fully integrated voltage-controlled oscillator (VCO). We have investigated the way the substrate noise (digital noise or injected noise by means of substrate taps inside VCO core) is converted close to the carrier frequency and impacts VCO spectral purity. Simple calculations and measurements without phase noise formalism are possible as we quantify the coupling mechanism by evaluating the side-bands power relative to the carrier (in dBc). Compared to previous work [5], this chapter aims to locally analyze coupling mechanisms between substrate noise and the VCO spectrum as two injected points are placed inside the VCO core. We do not use a global approach by considering only power supply bounces due to coupling between substrate and power supply metal rail. We analyze coupling mechanism in order to determine which devices are sensitive to bulk noise in our specific VCO test-chip including integrated injection taps. Advancements in the field of very-large-scale integration (VLSI) has led to more compact integrated circuits (ICs) having higher clock frequencies and lower power consumptions— see Systems-on-Chip (SoC) and Systems-in-Package (SiP). Nowadays, these conventional 2D planar technologies face several challenges from technological and financial points of view: physical limits, processing complexity, fabrication costs, etc. Consequently, technological approaches other than scaling are now investigated to continue following or get over Moore's law. Novel 3D integration technologies appear to be promising candidates; they are obtained by stacking vertically 2D integrated circuits, or layers in the same die. The electrical connections are ensured by new pre or postprocessed metallic structures: redistribution layers (RDL), which distribute power and high-speed signals on die top or backside surface, copper pillar, and mostly through silicon via (TSV) which is a key enabling technology for 3D integration, propagating signals through the whole silicon bulk. Many challenges are encountered with the 3D, notably in properly characterizing and electrically modeling the 3D interconnects; few CAD tools are available. Finally, the 3D interconnect global environment necessitates the modeling of the substrate couplings whose effects at high frequencies can no longer be neglected [6]. Section 9.2 reviews the context: silicon technology based on low- or high-resistivity substrate, the substrate noise generation and propagation. Noise generation will be simplified by digital power supply ringing transmitted to P-substrate or N-wells. In Section 9.3, we present the methodology based on the IC Emission Model—ICEM standard model [7,8], with an extension to substrate coupling. Associated tools are presented to quantify the voltage supply ringing, the substrate equivalent resistance and the analog ground fluctuations. Section 9.4 is an illustration of the methodology applied to a mixed-signal IC integrating different digital blocks and a voltage-controlled oscillator in a BiCMOS SiGe process; its core concerns an analytical approach of coupling mechanism by means of the impulse sensitivity function (ISF) formalism [9]. This way, we show that, in our specific test chip, coupling phenomena depend on the injection node locations inside the VCO core and on biasing conditions. At the end of our chapter—Section 9.5 presents compact models of some 3D interconnects systems, notably TSVs dedicated to low- and medium-frequencies (up to 10 GHz). They are derived from transmission line method (TLM), 3D numerical electromagnetic simulations, and parametrical extractions performed on test structures. The efficient and simple modeling approach that we propose includes the consideration of the global electrical context and aims at providing rapid information relative to signal integrity problems.

9.2 Substrate Noise in Mixed-Signal ICs in a Bulk Silicon Process

9.2.1 Ground and Substrate Noise Mechanisms

One of the main phenomena that induce substrate noise in a mixed-signal circuit is the power and ground supply voltage fluctuations that are

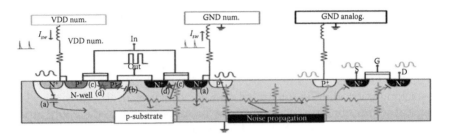

FIGURE 9.1
Substrate noise injection mechanisms in a digital inverter.

transmitted into the substrate through substrate biasing contacts [1,10,11]. The power and ground supply lines are not perfect and introduce several parasitic elements: resistances, inductances, and capacitances. In a large digital circuit, high peaks and fast slew rate on supply current create power-supply noise in the supply network due to *RLC*-like network formed by all these parasitic elements at different levels of a SoC: PCB, package, and the circuit itself. Other mechanisms create substrate disturbances in mixed-signal devices (see Figure 9.1).

Substrate coupling is essentially due to the power supply noise (Figure 9.1a). The signal transitions can also be coupled to the substrate through different physical structures: the input MOS capacitance (Figure 9.1c) of all the transistors, the output drain (Figure 9.1b) to substrate capacitances and the metal interconnection to substrate capacitances. Other phenomena can induce parasitic substrate currents like impact ionization (Figure 9.1d), photon-induced current and diode leakage current.

9.2.2 Substrate Noise Propagation in Low- or High-Resistivity Silicon Substrate

Propagation, then, has to be quantified through the substrate. Silicon material is characterized by a cutoff frequency $f_c = \sigma/2\pi\varepsilon$ where σ is the conductivity and ε the permittivity. With high resistivity substrate ($>10\ \Omega$ cm), silicon can be considered as purely ohmic for signal frequency below 10 GHz. Then, the silicon equivalent transfer function can be defined as an attenuation factor that can be calculated with point-to-point equivalent resistance. At this step, one has to consider the nature of the substrate: low- or high-resistivity silicon substrate.

Considering low resistivity substrate, the model depends on the comparison between the epitaxial layer thickness (W_{epi}) and the distance D between substrate taps. As shown in Figure 9.2a and b (case b); when the distance is up to $4 \times W_{epi}$, the entire current flowlines pass through the low-resistivity substrate that can be considered as a unique point. The resistance variation function of the distance D is represented in Figure 9.2c (substrate tap

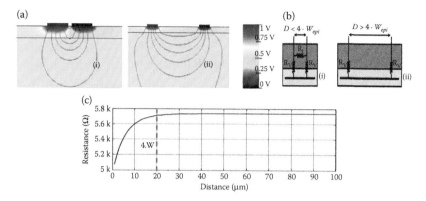

FIGURE 9.2
(a) Current flowlines for low resistivity substrate (single node) epitaxial layer (D is the distance inter-contact (pitch)) $D < W_{epi}/4 \Rightarrow$ (i) current flowlines go through the substrate and through the epi layer. $D > W_{epi}/4 \Rightarrow$ (ii) NO current flowlines go through the epi layer. (b) Equivalent model for low resistivity epi—(W_{epi} thickness) (i) $D < 4 \times W_{epi}$, (ii) $D > 4 \times W_{epi}$. (c) Equivalent resistance function of the distance for low resistivity substrate with epitaxial layer (W_{epi} thickness).

area is $20 \times 20 \ \mu m^2$, $W_{epi} = 5 \ \mu m$ at $10 \ \Omega \, cm$, substrate thickness is $300 \ \mu m$ at $0.05 \ \Omega \, cm$.

For high resistivity substrate, the die backside connection is first of interest; if it is floating, the resistance between two taps increases with the distance (Figure 9.3c). If the die backside is correctly connected to the ground, the resistance depends on the distance between the taps compared to the substrate thickness $W_{substrate}$ as illustrated in Figure 9.3a–c (substrate tap area is $20 \times 20 \ \mu m^2$, $W_{substrate} = 50 \ \mu m$ at $6.7 \ \Omega \, cm$).

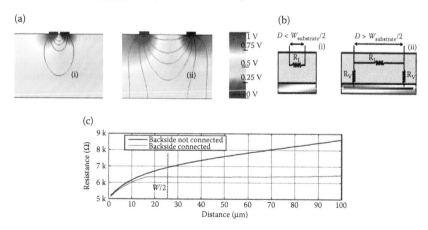

FIGURE 9.3
(a) Current flowlines for high resistivity substrate ($W_{substrate}$ thickness) (i) small taps distance (ii) large taps distance. (b) Equivalent model for high resitivity substrate—with grounded backside (i) $D < W_{substrate}/2$, (ii) $D > W_{substrate}/2$. (c) Equivalent resistance function of the distance for high resistivity substrate ($W_{substrate}$ thickness) with floating or grounded backside.

9.3 Modeling Methodology

9.3.1 Ground and Substrate Noise Modeling

We consider the power-supply noise as the unique substrate perturbation source. This method drastically simplifies the problem and gives quite good results. In order to model power and ground voltage oscillations, we have chosen to use the integrated circuit emission model (ICEM) approach [8]. The classical ICEM has been extended with a substrate network submodel [9] (see Figure 9.4).

The passive distribution network submodel is supposed to model the parasitic elements of power supply or signal lines. The internal activity submodel characterizes the dynamic current consumption of the device. The substrate-network submodel is expected to model the substrate propagation of parasitic signals from aggressor parts to victim parts.

9.3.2 Developed Tools

Basically, the conceptual extended-ICEM methodology can be described in its simplest form by the electrical diagram presented in Figure 9.5 (one digital block with one *Vdd-gnd* power supply pair). Nevertheless, some parameters have to be determined for each IC project. So, we use our two software applications developed in JAVA language. The first application is developed for ground and power supply bounce effects; the second application concerns the substrate extraction. These applications can be used at any step in the design phase: from the preliminary study to the final optimization (postlayout) as the parameters can be refined in time.

9.3.2.1 Application for Ground and Power Supply Bounce Effects

The first application deals with quantifying the ground and power supply fluctuations (bounces). First, the user describes the passive elements: R, L,

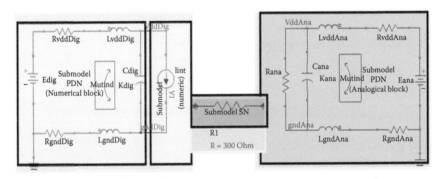

FIGURE 9.4
Schematic architecture of extended-ICEM model.

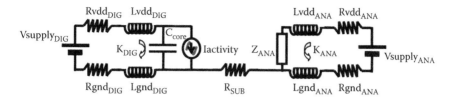

FIGURE 9.5
Basic architecture of extended-ICEM model.

C_{core}, and K. The parameters R and L are the parasitic resistance and inductance of the lines (PCB, socket, package, and on-chip metal rails). K is the coupling factor (mutual inductance) between the lines (mutual between bonding wires, typical value is around 0.5). In digital ICs, the power and ground parasitic elements are mainly due to the package effects [12], power-grid effects can be neglected in a first-order approach. Package lead and bonding wire lengths are often longer than on-silicon metal lines and induce mainly parasitic inductances. C_{core} is the equivalent digital capacitance; it depends on the technology (number of gates per mm²) and the digital area. Typical values are from 0.15 to 0.3 nF/mm² for 0.35 μm CMOS technology and from 0.8 to 1.6 nF/mm² for 90 nm. These passive parasitic elements can be estimated or determined from the Z_{11} S-parameter measurement (Figure 9.6).

In a second step, the internal activity of the digital block is described in a simple way. This current source is the superposition of periodical current pulses. The characteristics of each pulse (height, width, frequency, rising, and falling time) can be easily determined from general digital circuit

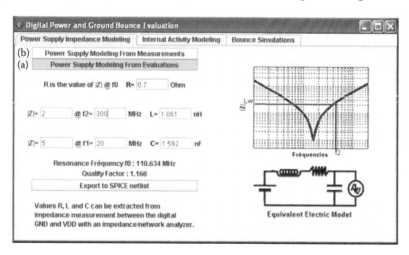

FIGURE 9.6
Dedicated application to quantify digital ground and power supply bounces (passive distribution network). (a) Passive parasitic elements can be estimated. (b) Passive parasitic elements can be determined from the Z_{11} S-parameter measurement.

parameters: its average consumption, its clock frequencies, its rising and falling transition times, its skews, and each clock *latency*. The last step concerns the ground and power supply bounce calculations. This is performed in the time domain (analytical resolution of *piece-wise* waveforms) or in the frequency domain (Fourier analysis).

9.3.2.2 Application Developed for Substrate Extraction

The second application concerns the substrate modeling by an equivalent {RC} netlist. The first step consists of depicting the resistivity and permittivity uniform thickness profile under P-taps.

We dedicate an application (Figure 9.7) to substrate extraction, with: geometry silicon: square $1000\ \mu m \times 1000\ \mu m$, four contacts ($100\ \mu m \times 100\ \mu m$, at the corner of the square) at $100\ \mu m$ from the silicon edges. Then, the substrate calculations are performed with a boundary element method based on TLM or Green functions [13].

9.3.3 Basic Rules to Reduce Digital Power Supply Network Ringing

With the presented tools, the user can estimate how the analog ground will fluctuate due to the activity of the digital(s) block(s), although the supplies are completely separated. Moreover, these tools can help one to understand malfunctions of digital blocks with an unstable digital power supply.

The passive distribution network presents a resonant frequency $f_{res} = \left(1/2\pi\sqrt{LC_{core}}\right)$ and a quality factor $Q = L\omega_{res}/R$. The passive distribution network will oscillate for $Q > 0.5$. To suppress or reduce power supply

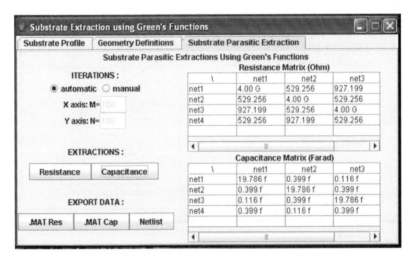

FIGURE 9.7
Dedicated application to substrate extraction ({RC} substrate matrix).

ringing, the designer can control the following parameters: R (line resistance from board or inside chip), L (line inductance), and C_{core} (core capacitance and additionally on-chip decoupling capacitance). Hereafter, we summarize the influence of these parameters on the digital power supply ringing

- Increasing R leads to a decrease in oscillation amplitude without any influence on the oscillation frequency, increasing L leads to a decrease in the oscillation frequency without any influence on the oscillation amplitude.
- Increasing C_{core} leads to a simultaneous decrease in oscillation amplitude and in oscillation frequency.

An intuitive way to decrease power and ground bounces is to add a huge decoupling capacitance on the power network of the PCB. In fact, this external decoupling capacitor just filters the power network parasitic voltage, and is not enough to decrease significantly the interferences .The main disturbing frequency is due to the ringing of the RLC path composed of the bonding wire inductance and resistance and by the core capacitance of the digital part of the circuit. Increasing as much as possible the off-chip capacitance will not avoid the on-chip natural ringing. Another simple solution, energy consuming and possibly inducing mistakes, is to add a serial resistor to the power network (on the PCB or inside the chip). The added resistor decreases the quality factor of the RLC network and, in the same time, damps the induced oscillations. The obtained low-pass filter gives some good results. These are the reasons why this kind of noise reducing technique is not recommended for a large digital circuit, but can be applied to a small digital application such as a frequency synthesizer.

The internal activity (digital equivalent current source) has also a direct link with the power and ground oscillation amplitude. For a clock frequency below resonant frequency, the power supply ringing will be enhanced when the resonant frequency is harmonic to the clock frequency (the worst case is for $f_{res} = 2f_{clock}$). For a resonant frequency below clock frequency, the passive distribution network acts as a low-pass filter. An increase in the average consumed digital current leads to an increase in the oscillation amplitude. The decrease in the transition times leads to a decrease in the oscillation amplitude.

9.4 Application to a Mixed-Signal IC

9.4.1 Test Chip Presentation

To validate the proposed methodology and to investigate the substrate isolation technique, a VCO is fully integrated in a standard 0.35 µm SiGe BiCMOS

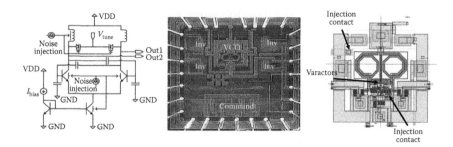

FIGURE 9.8
Simplified VCO core with injection contacts, microphotograph of the evaluation chip VCO layout.

process with high-resistive substrate. Figure 9.8 shows the simplified VCO schematic view. The main part of the VCO is the LC-tank. The negative resistance is obtained by a cross-coupled differential pair of hetero-junction bipolar transistors (HBT) biased by I_{bias} current.

To avoid frequency variations due to supply voltage noise, many on-chip decoupling capacitors are placed in the VCO (see layout in Figure 9.8). For the same purpose, substrate contacts and guard rings surrounding the devices presenting a large surface in contact with the bulk have been added.

The circuit supply voltage is 3.3 V. The tuning voltage can vary from 0 to 4 V to obtain an oscillation frequency in the range (4.25–4.60 GHz). The output signals pass through buffers in a common collector circuit configuration. The maximum carrier power with a 50 Ω load is closed to 0 dBm. The phase noise measured at 100 kHz from carrier is varying from −90 up to −100 dBc/Hz (depending on-chip dispersion and biasing).

The chip contains a VCO, digital blocks and many substrate taps. Especially two substrate taps have been placed inside the VCO core in order to inject a parasitic signal, directly into the VCO substrate, with an external source: one tap in the vicinity of the transistors of the cross-coupled pair and the current mirror (I_{bias} source), and the other one close to one of the inductors Note that the substrate tap close to the inductor is not far from the varactor, as the inductor does not have any substrate pattern shielding. (Figure 9.8). Chips are mounted in a RF package (VFQFPN/Very thin Fine pitch Quad Flat Pack No lead Packages) and a dedicated RF board has been designed.

The different parts that have to be modeled for building the extracted ICEM model are depicted in Figure 9.9.

9.4.2 Substrate Noise Reduction: Low Noise Version

The challenge is now to develop a low noise version of the chip. The first effort concerns the noise generator itself. As already mentioned, dynamic

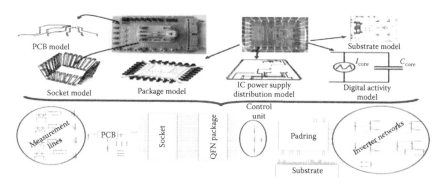

FIGURE 9.9
Building the complete model (based on ICEM approach with substrate extension).

signals involve dynamic current consumption and, so, high current spikes. By reducing the average activity of the digital device, we can decrease the global noise generation. In our case, we have chosen to decrease the dynamic activity of the digital part by enlarging the digital switching activity timing window. This window is called the skew. It is the time between the first logical gate commutation and the last logical gate commutation during a digital clock period. Increasing this skew value allows the digital circuit to absorb its necessary energy within a longer period of time. Thus, the global consumption current is less dynamic and involves less interference in the power supply network. To apply this method to our circuit, we have changed the commutation delay of each inverter network, in such a way that they do not switch at the same time (synchronism).

Second, we investigate the use of isolation techniques such as P guardrings. In both cases, when the guard ring is around the digital or the analog part, the simulated substrate parasitic levels are as high as the nonisolation configuration. The reason is that to be effective, the P+ guard ring has to be connected to the ground, as cleanly as possible. In our circuit, the ground line parasitic elements do not allow to correctly evacuate the parasitic substrate currents.

Finally, the passive distribution network is optimized. The on-silicon capacitance has been increased, to filter the on-silicon natural power ringing. A power supply resistance has also been added, in order to absorb all the oscillations.

We present in Figure 9.10, the new version measurements, with on tap closed inside the N well, the second tap being in the same substrate, near the perturbating digital blocks.

As expected, the perturbation frequencies are reduced due to the new value of the on-silicon capacitance. The *RLC* oscillations are also dampened by the power supply serial resistance.

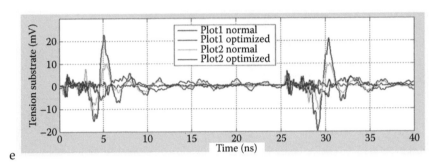

e

FIGURE 9.10
Different localizations of test pads: plot 1 ⇒ 42 mV → 11 mV RMS voltage values in the substrate of the test circuit; plot 2 ⇒ 22 mV → 6 mV RMS voltage values in an insulation N_{well}.

9.4.3 Voltage-Controlled Oscillator Spectrum in Normal and Low Noise Versions

Close to the carrier, the phase noise can amplify when the digital part is active. Due to frequency conversion, the clock frequency and its odd harmonics are observed in the VCO phase noise. Spur attenuation, between 5 and 6 dB, is measured for the low noise version, confirming the digital noise reduction (not presented here).

9.4.3.1 VCO Characterization

9.4.3.1.1 Static Sensitivity Functions: VCO Carrier Frequency Variations, Function of Bias Current, and Tuning Voltage

VCO carrier frequency depends on inductance and varactor choice, but also on bias current or tuning voltage. These variations of f_c, function of the bias current or tuning voltage, can be described by some VCO sensitivity functions respectively called K_{bias} (MHz/mA) and K_{tune} (VCO gain in MHz/V), that is, $K_{bias} = \partial f_c / \partial I_{bias}$ and $K_{tune} = \partial f_c / \partial V_{tune}$. We extract the VCO sensitivity functions by using derivative of the frequency-dependant curve relative to the tuning voltage or the bias current. We observe (not presented here) that the variations of the bias current results in a small change of the oscillator frequency. The voltage disturbance applied to varactors generates a sensitivity level much more important than the bias current sensitivity. Roughly, the sensitivity function magnitude for bias current K_{bias} is 10 times lower compared to tuning voltage sensitivity K_{tune}. This method allows us to establish the VCO sensitivities as functions of circuit bias conditions by measurements or simulations.

9.4.3.1.2 VCO Spurious Side-Bands Involved by Bias Harmonic Perturbations

VCO carrier depends on bias current or tuning voltage. The variations of VCO carrier function of the bias current or tuning voltage can be described by some sensitivity functions, respectively, called K_{bias} and K_{tune} (see below). A change in the I_{bias} current can modulate the carrier frequency in the same manner that a modification of V_{tune} voltage can change simultaneously the varactor capacitance and the output frequency. In this framework, considering a sine wave perturbation superimposed to the bias current or the tuning voltage

$$v_m(t) = A_m \cos(\omega_m t)$$ (9.1)

The carrier frequency changes according to the expression

$$f = f_c + K \cdot v_m(t)$$ (9.2)

where f_c is the VCO carrier frequency and K is the sensitivity function.

After frequency integration to determine the phase, the output signal of the oscillator has been determined by the following equation:

$$s(t) = A \cos\left(\omega_c t + \frac{A_m K}{f_m} \sin(\omega_m t) \right)$$ (9.3)

The modulation index ($A_m K/f_m$) depends on sensitivity function. For a supply voltage perturbation, the modulation index has a low value, so we can assume in our calculations that the modulation band is placed in a narrow range. We refer to the carrier in order to compute the perturbation level in the spectral domain. With these assumptions, the spectral power relative to the carrier in dBc units is given by

$$P_{sbc} = 20 \cdot \log\left[\frac{K \cdot A_m}{2 f_m} \right]$$ (9.4)

It measures the gap between the spurious side-band power and the carrier magnitude and the parasitic side-band power (Figure 9.11). A perturbation superimposed to the bias current or tuning voltage generates side-bands that have the power magnitude directly proportional to K sensitivity (Equation 9.4). In the same time, the K sensitivity is function of the bias current and of the tuning voltage. Thus, the noise power below carrier varies when biasing or tuning are modified.

9.4.3.1.3 VCO Side-Bands Involved by Substrate Harmonic Perturbations

An experimental approach demands to design many dedicated test-chips to analyze coupling phenomena. Simulations are also possible, with at first, the quantification and location of the perturbation signal (noise sources).

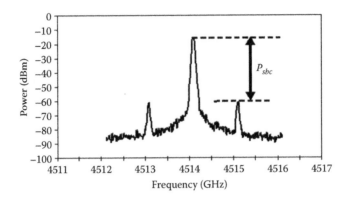

FIGURE 9.11
Measured P_{sbc} for a small sine-wave applied to the substrate contact near the inductor (50 mV-peak at 1 MHz).

Different studies focus on modeling injected noise by CMOS digital part in ICs [2,14,15]. Once equivalent noise sources are determined, these late ones are associated in a global scheme describing the board, the IC socket and package, the bondings, the pad ring, the power supply rails, and the coupling to substrate (resistive and capacitive) [16–19]. The next step concerns the propagation modeling. Designers can use substrate extractor (post-layout extraction) to obtain a substrate RC-network compatible with some SPICE-like simulator. The last point is the study of the sensitive bloc (e.g., the VCO in our case). Experimentally, when sinewave voltage is applied to substrate taps (inside the VCO core), we measure VCO side-bands (spurs) closed to the carrier frequency f_c (Figure 9.11). For a harmonic noise at f_m frequency, we observe side-bands at $f_c \pm f_m$ (the FM conversion mechanism is verified as spur magnitude is proportional to noise frequency). The VCO sensitivity, due to the injected perturbation (level, location, frequency, etc.), is quantified by P_{sbc} values. Figure 9.11 gives an example of P_{sbc} measurement for a small sinewave applied to the contact near inductors (50 mV-peak at 150 kHz). Measured P_{sbc} values, for a sinewave perturbation injected closed to the inductor/varactors and the bipolar transistors of the differential pair (150 mV-peak at 150 kHz), are done; from thes ones, we "observe" first that side-band power values (P_{sbc}) are larger when the injection is located closed to bipolar transistors of the cross-coupled differential structure and the current mirror; however, the decrease of P_{sbc} values with tuning voltage is more significant when injection occurs near inductors. Using a global approach, we can express a VCO carrier frequency change as

$$\Delta f_c = K_{bias}\Delta I_{bias} + K_{tune}\Delta V_{tune} + K_{VDD}\Delta V_{VDD} + K_{GND}\Delta V_{GND} + K_{sub}\Delta V_{sub} \quad (9.5)$$

where K_{sub} and ΔV_{sub} take into account the VCO carrier frequency sensitivity to substrate voltage.

When substrate noise is present, we can consider first, a propagation from substrate to on-chip ground through ground metal rail (K_{GND}, ΔV_{GND}) and second, a direct substrate path (K_{sub}, ΔV_{sub}). The other terms in Equation 9.5 can be neglected, due to the weak capacitive coupling at low substrate noise frequencies.

9.4.4 Analysis of Substrate Coupling Mechanisms via the ISF Approach

9.4.4.1 ISF Principle: Briefs Recalls

In this section, we use the linear time-variant model described in Reference 20 to analyze the VCO sensitivity to substrate perturbations. The method is based on the so-called ISF, which represents the excess phase after applying some perturbation impulse to an oscillator circuit. Varying the perturbation impulse event time (τ) during an oscillator period, we can build the ISF function $\Gamma(\tau)$. The perturbation impulse can be a current spike into a capacitive node (injected charge), or a voltage spike onto an inductive node. $\Gamma(\tau)$ is dimensionless and has the same period as the oscillator one (i.e., $\Gamma(\omega_c\tau)$ has a 2π-period) [9,20]. For a better clarity in this chapter, we will consider the pseudo-ISF function $\Gamma^\phi(\tau)$ in the following lines of this section. $\Gamma^\phi(\tau)$ has the dimension of radian/As or radian/Vs depending on the type of perturbation (current or voltage) [20]. Finally, we can write the excess phase for an impulse response as

$$h_\phi(t, \tau) = \Gamma^\phi(\omega_c\tau) \cdot u(t - \tau) \tag{9.6}$$

where $u(t)$ is the unit step.

In other words, $\Gamma^\phi(\tau)$ function is a direct representation of the excess phase (phase shift), normalized by injected charge (for current impulse); it can be extended in a Fourier series as follows:

$$\Gamma^\phi(\omega_c\tau) = \frac{c_0}{2} + \sum_{n=1}^{\infty} c_n \cdot \cos(n\omega_c\tau + \theta_n) \tag{9.7}$$

Considering, now, the harmonic perturbation (current or voltage) defined by its magnitude A and its angular frequency ω_m (frequency: f_m), that is,

$$p(t) = A \cdot \cos(\omega_m t) \cdot u(t - t_0) \tag{9.8}$$

Due to the linear time-variant system, the phase shift is straightly obtained according to the expression

$$\phi(t) = \int_{-\infty}^{t} \Gamma^\phi(\omega_c\tau) \cdot p(\tau) \cdot d\tau = \int_{t_0}^{t} \Gamma^\phi(\omega_c\tau) \cdot A \cdot \cos(\omega_m\tau) \cdot d\tau \tag{9.9}$$

Then, the phase shift can be written as follows:

$$\phi(t) = A \cdot \frac{c_0 \sin(\omega_m t)}{2\omega_m} + A \cdot \sum_1^\infty \frac{c_n \sin[(n\omega_c \pm \omega_m)t + \theta_n]}{2(n\omega_c \pm \omega_m)} + \phi_0(t_0) \quad (9.10)$$

In Equation 9.10, the term with $\omega_m = n\omega_c + \Delta\omega$ (with $\Delta\omega \ll \omega_c$ and $n = 0,1,2,...$) is preponderant; $s(t) = S \cdot \cos(\omega_c t + \phi(t))$. Finally, only low-frequency noise ($n = 0$) and noise disturbance around harmonics ($n = 1,2,...$) impact the phase. The c_n coefficient of the ISF function is used to calculate the phase shift introduced by the noise at pulsation $n\omega_c + \Delta\omega$. Then, the oscillator output can be written as

$$s(t) \approx S \cdot \left[\cos(\omega_c t) - \phi(t)\sin(\omega_c t)\right] = S \cdot \left[\cos(\omega_c t) - \frac{c_n A}{2\Delta\omega}\sin(\Delta\omega t)\sin(\omega_c t)\right]$$

$$(9.11)$$

According to a narrow side-band condition, a noise at $n\omega_c + \Delta\omega$ results in a pair of equal side-band at $\omega_c \pm \Delta\omega$ with a side-band power relative to the carrier, equals to

$$P_{sbc}(\Delta\omega) = 10 \cdot \log\left[\left(\frac{c_n A}{4\Delta\omega}\right)^2\right] \quad (9.12)$$

9.4.4.2 ISF Applied to an Injected Perturbation on Varactor Substrate

The ISF function $\Gamma^\circ(\tau)$ is calculated with a current perturbation directly applied to the substrate pins of the two varactors. The simulations are performed with a SPICE-like simulator (time-domain simulations) with a sufficient duration to be sure to obtain a stable phase shift. The current pulse (1 mA magnitude and 10 ps duration) is simultaneously applied to the substrate pins of the two varactors. We have also verified the linearity, not illustrated in this chapter (excess phase proportional to injected charge). We note that the ISF varies with twice the oscillator frequency (already reported in Reference 9).

To validate the ISF approach, we have compared the P_{sbc} values calculated with Equation 9.12 (using the c_n coefficients) with the values obtained directly by means of time-domain simulation coupled with discrete Fourier transform (DFT) postprocessing analysis. These comparisons were done for a 200 μA harmonic current noise at $nf_c + \Delta f$ with n: nth harmonic and $\Delta f = 50$ MHz.

Simulated waveform for P_{sbc} is performed for a current harmonic injected at 50 MHz with 200 μA magnitude. As already mentioned, the frequency domain curve is obtained with a DFT postprocessing analysis, with a very long transient time analysis to be sure to obtain the periodic steady state. Then we directly observe ($f_c \approx 4.4$ GHz) that P_{sbc} equals to –25 dBc for $n = 0$ (Figure 9.11) for the two side-bands at $f_c + 50$ MHz and $f_c - 50$ MHz. This value can be compared to the one calculated by ISF approach. This is derived by calculating the c_n Fourier coefficients of the ISF function with Equation 9.5; we extract $c_0 = 0.0048$ (we remind that the pulse characteristics for this ISF calculation were 1 mA, 10 ps). Using Equation 9.5, we obtain for the current harmonic 200 µA at 50 MHz, a P_{sbc} equals to –22 dBc.

9.4.4.3 Device Impulse Sensitivity Study

In this subsection, we are looking for an understanding on what are the devices the most affected by some substrate noise disturbances leading to observe side-bands in VCO spectrum [21]. From measurements, we note that measured P_{sbc} values depend on the VCO tuning voltage and the injected noise location inside the VCO (location close to one of the inductors or in the vicinity of the HBTs of the cross-coupled pair and the current mirror). Substrate noise induces GND power supply bounces (V_{tune} and VDD power supply bounces are negligible due to the weak capacitance coupling). In our specific test-chip, considering a high-resistive substrate (noise attenuation with distance: the substrate cannot be considered as a single node), numerous on-chip decoupling capacitors, injection nodes inside VCO: on the one hand, we can consider a propagation from injection substrate tap to device substrate through the ground metal rails and, on the other hand, a direct path between injected nodes and passive or active device bulk (varactor, inductor, HBTs, etc.). Hereafter we summarize the main points.

- Due to the differential structure, pulse injection in a single bulk node is most significant (compared to simultaneous injection in symmetrical device bulks).
- Mainly varactors and HBTs are concerned by substrate voltage disturbances Inductors, MIM capacitors, and polysilicon resistors seem not to be influenced by low frequency substrate noise.

In this section, we simulate the perturbation directly applied to the device bulk pin. The very reason is, that using a complete substrate RC extracted netlist, we obtain a false carrier frequency due to the limited frequency validity range after extraction, that is, substrate reduced netlist is not further valid up to a few GHz. Figure 9.12 presents the ISF curves for the highest sensitive devices; ISF are calculated when a voltage impulse (1 V during 20 ps) is applied to one varactor bulk or one HBT bulk (both calculated for $V_{tune} = 0$ V and 3 V). The ISF curve for HBT has the same period as the oscillator one, but it presents some dissymmetry leading to a significant $c_0/2$ term in Equation 9.7. For a 3 V tuning voltage, the impact in the low-frequency noise range is more important for the HBT than the varactor. This conclusion is drastically

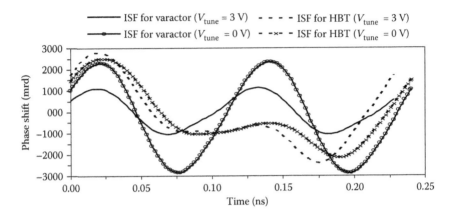

FIGURE 9.12
ISF function calculated for a 1 V/20 ps voltage impulse on one varactor bulk or one HBT bulk ($V_{tune} = 0$ and 3 V).

different for high-frequency noise, around carrier frequency or twice carrier frequency, as the Fourier coefficients differ (not discussed in this chapter). We also observe that the ISF curve for the varactor, at $V_{tune} = 0$ V, is higher and dissymmetrical, leading to an increase of the $c_0/2$ mean value. This is correlated to other results where P_{sbc} values are higher for 0 V tuning voltage. P_{sbc} values for injection close to HBT do not depend on the tuning voltage. For injection close to the inductor, the coupling behavior is driven by the varactor sensitivity and we observe P_{sbc} decreasing with the tuning voltage. P_{sbc} magnitudes are higher for injections close to the HBTs than for injection in the vicinity of the inductor, because the attenuation through substrate decreases the noise magnitude at the varactor bulk. In a future work, we will study the same structure with a shielded inductor (substrate pattern) in order to modify the substrate direct path between injection location and varactor.

In the last section, we do some insight in the 3D circuits case, which can encounter the drawbacks cited above, or worst.

9.5 A First Glance in the 3D Interconnect Modeling

A 3D structure is shown in Figure 9.13. We use the simulator Sentaurus [22]; which is a software package based on the finite element method (FEM) for the device study and the SPICE based models for connectics to the bulk regions. The TSV has a cylindrical shape filled with copper and with oxide thickness $T_{OXTSV} = 0.5$ μm. In order to simplify the study, only the coupling between the TSV and the nMOS is supposed. The nMOS transistor is placed very near the TSV.

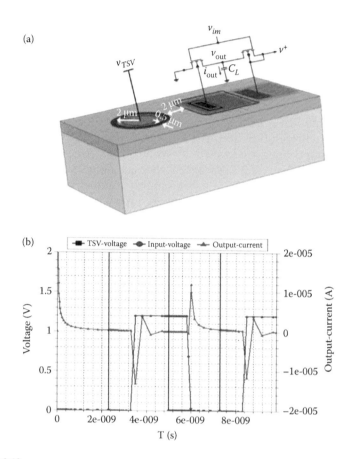

FIGURE 9.13
(a) 3D cross section of TSV-CMOS mixed mode coupling; (b) The voltages and output-current waveforms at $v_{TSV} = 0.0/1.2$ V and $v_{TSV} = 0.0/42.0$ V.

9.5.1 Steps of the Modeling Approach

3D interconnects necessitate the consideration of the global electrical context to correctly evaluate system performances. The compact models must hence be derived from semiconductor and electromagnetic theory instead of being developed and extracted outside of a realistic 3D IC system environment. The *RLCG* parametrical extractions are performed through S-parameters measurements with a vector network analyzer (VNA) for a frequency sweep from 70 kHz to 40 GHz. The Thru-Reflect-Line (TRL) de-embedding technique is applied to remove, from the S-parameters measurements, the contact and access lines effects, and thus obtain the test structures' actual RF behaviors. The modeling approach then relies on analyzing the test structures by means of a 3D full-wave electromagnetic tool, using FEM [23], to grasp the physical and electrical phenomena occurring in 3D interconnects during signal propagation, and finally on the use of TLM.

9.5.2 Presentation of a Test Structure

We will denote it hereafter as TSV chain (or nxTSV chain, consists in n medium-density TSVs) connected at the backside of the chip with a Back RDL (BRDL) redistribution line. The TSVs are connected to the BEOL level, where the ground lines are located, by means of square contact pads. The physical view of a chain of 2 TSVs (2xTSV chain) is represented in Figure 9.14.

The test structures are realized using a highly conductive silicon substrate having a thickness of 120 µm, a resistivity $\rho = 10$ mΩ cm, and a relative permittivity $\varepsilon_r = 11.7$. On the substrate, a thin epitaxial layer is deposited (thickness: 4.5 µm, resistivity: 10Ω cm, permittivity: 11.7). The lines at the BEOL level, for TSV chains, are isolated from the substrate by an oxide layer (thickness: 0.8 µm, permittivity: 6.2). Regarding the TSV chains, an oxide layer (thickness: 1 µm, permittivity: 6.5) is added at the backside of the structures to isolate the BRDL line from the substrate.

Figure 9.14 shows the top view of the 2xTSV chain. The medium-density TSV used is a large copper tubular conic structure, isolated from the silicon substrate with silicon oxide ($\varepsilon_{ox} = 6.5$) on the TSV walls and filled with polymer material. The SEM cross-section of such a TSV is observable in Figure 9.14.

Many analytical formulas are associated with vias [24], as

$$Rot = Rmt + Tot; \quad Rod = Rmd + Tod;$$

$$R_{TSV} = \frac{H}{\pi\sigma} \frac{1}{2(Rmt \cdot Tmd - Rmd \cdot Tmt)} \ln\left(\frac{Tmd}{2Rmd - Tmd}\right) \cdot \frac{2Rmt - Tmt}{Tmt};$$

$$C_{ox(TSV)} = \frac{\varepsilon_0 \varepsilon_{ox} \cdot 2\pi \cdot H}{Tot - Tod}\left[Rot - Rod + \frac{Rot \cdot Tod - Rod \cdot Tot}{Tot - Tod}\ln\left(\frac{Tod}{Tot}\right)\right]$$

and all others connecting systems.

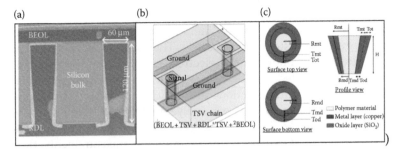

FIGURE 9.14
Physical views of some test structures (a) SEM cross-section of the medium-density TSV used in the TSV chains (W1 configuration); (b) a chain of 2 TSVs (2 × TSV chain); (c) a via model.

9.5.3 Interconnect Modeling with Global Electrical Context Consideration

9.5.3.1 Compact Models of the Medium TSV

In planar technologies, 3D interconnects can be built as *RLCG* equivalent electrical models described as Π or T networks, when the interconnect length is smaller than the tenth of the propagated signal wavelength. Otherwise, the two networks do not have the same RF behavior and they must be distributed in a certain number of elementary cells to give equivalent responses. Each of the 3D interconnect *RLCG* networks is modeled with serial elements (resistances or partial inductances) to model the signal propagation and parallel elements (capacitances or conductances) to model the interconnect environment. The proximity (coupling) effects are also included in the compact model description. The geometrical parameters describing the analytical expressions are all expressed in meters. The resistances are calculated for a DC value (the skin effect, or depletion layer neighboring the side of a cylindrical TSV, could be considered) and the inductances are calculated analytically depending on the partial inductance analytical expressions [24–26].

The interline capacitance C_{inter} (or coupling capacitance) is calculated according to the set of equations defined in Reference 25 from the concerned line surfaces, and the fringe capacitances taken at the extremities (Cf) and the middle (Cf') of the coupled-line system. The electrical parameter Cp corresponds to the coupling capacitance between the ground plane and the line surface concerned; it is also referred to in the literature as the line's self-capacitance. The polymer material filling the TSV does not significantly affect its RF behavior. Moreover, the variations of the TSV radius and the copper layer thickness are very low. Consequently, the analytical expressions of a filled cylindrical conductor are still used in order to calculate the self and mutual inductances.

9.5.3.2 Global Electrical Context Modeling

Since the substrate is highly conductive for the investigated coplanar waveguides, it can be modelled as a simple node. For each coplanar line, the vertical current path in the epitaxial layer is modelled by a resistance (R_{epi}) in parallel with a capacitance (C_{epi}).

The electrical parameter values must be adapted if the compact model is to depict a lumped model or an elementary cell of a distributed model. In the TSV chain, the currents are flowing vertically through the epitaxial and the oxide layers to reach the ground planes corresponding to the BEOL level ground lines (Figure 9.15). The coupling between the substrate and the TSVs is taken into account with the geometrical capacitance $C_{ox(TSV)}$ (here the silicon depletion region capacitance is neglected as it introduces minor effect due to the highly doped substrate). Only the TSV/TSV couplings have been taken into account. The TSV chain electrical context modeling involves

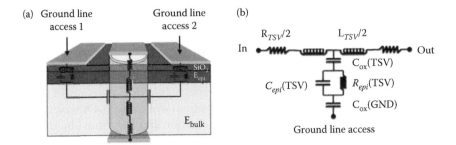

FIGURE 9.15
(a) Medium-density TSV; (b) *RLCG* compact modeling of the medium-density TSV including the current path.

dividing the BEOL ground lines into serial blocks, by dissociating the parts reached on the overall surface by the current coming from the TSVs.

Our efficient and simple modeling approach is based on subdividing the whole 3D interconnect path in different blocks for those simple analytical formulae are used to build a global model.

9.5.3.3 Substrate Modeling Approach

The modeling of a conductive substrate as a simple node is only viable for low- and medium-frequency domains. At high frequency, the substrate coupling effects must be added to the system electrical modeling by representing the substrate as a *RLCG* network to more accurately evaluate the IC's losses. Moreover, the substrate can be nonuniform with different doping values, that is, having different resistivities and permittivities throughout the considered volume. The substrate is therefore modeled as a stack of parallel heterogeneous dielectric layers [13,27].

We propose a 3D impedance extraction method [28] relying on the TLM or Green Kernels applied to a multilayered substrate and the electrostatic Green functions. Basically, a unitary current is applied on a contact pad which can be located atop or inside the substrate. A voltage is measured at the level of another contact, enabling us to deduce a square "transfer impedance" matrix $[Z_T]$ whose dimensions correspond to the number of contacts. The substrate coupling between two elements is generally modeled by a resistance in parallel with a capacitance.

9.5.4 Interconnects Extractor

9.5.4.1 Transmission Line Extractor 3D-TLE

We develop, using MATLAB® [29], a 3D extraction tool, 3D Transmission Line Extractor (3D-TLE), based on our modeling approach and which integrates the substrate extraction method algorithm (Figure 9.16). Through a

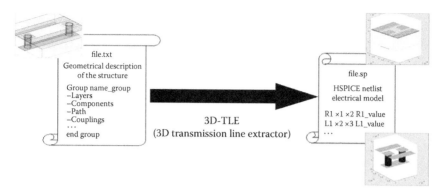

FIGURE 9.16
In-house 3D extraction tool 3D transmission line extractor (3D-TLE) environment.

hierarchical syntax and specific statements, the user depicts in a text file the geometrical and technological data of his/her system by defining: its layers; the type of components comprised in the system; the connectivity between chip elements, which are instantiated components; and the element interactions (couplings, interactions with layers for current path modeling).

The extraction tool generates from this text file a SPICE netlist, containing the system $RLCG$ electrical description, exportable toward CAD tools such as Agilent Technologies ADS® [30,31]. The viability of 3D-TLE is checked through S-parameter comparisons for given systems between their 3D-TLE SPICE netlists and their schematics designed under ADS®.

9.5.4.2 Modeling Approach Validation and Test Structures RF Behaviors

The modeling approach is validated, specifically in the frequency domain, via S-parameter comparisons between the measurements performed on the test structures and the simulation results from their equivalent electrical models. Some of these comparison results are presented in Figure 9.17. The equivalent $RLCG$ compact models proposed for the TSV chains demonstrate a very good accuracy for frequencies up to 10 GHz, respectively. Some spikes, due possibly to the coupling between the test structure substrate and its measurement environment, can be observed in measurements, but not in simulations at low-frequency (below 1 GHz). The spikes can be suppressed by isolating the test structure from its environment, implying the substrate has to be strongly grounded or biased.

9.6 Conclusion

In this chapter, first of all we resume some basics of the substrate noise generation and propagation in mixed-signal ICs. We also present the standard

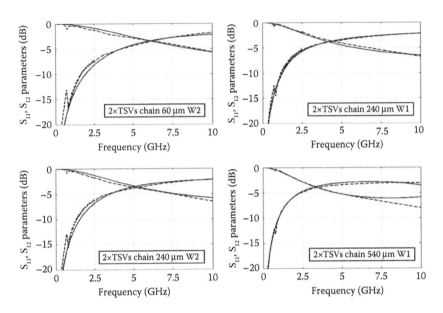

FIGURE 9.17
S-parameter comparisons, up to 10 GHz, between the measurements performed on 2 × TSVs chain test structures and the simulations of their electrical equivalent models for two configurations and different coplanar and back redistribution line lengths.

ICEM model and its extension allowing substrate noise simulations. This model considers only the noise generated in the substrate by the power supply and ground fluctuations of the digital part. Although this approach seems very simple, it gives good results regarding test-chip measurements. To reduce substrate noise, the designers can control some parameters concerning, at first, the noise generator (digital block). The skew and the transition time can be optimized in order to reduce the average current and therefore power supply ringing. The passive power supply distribution network, through its parasitic elements *RLC*, has also a key role. Adding decoupling capacitance (PCB and/or on-chip), resistance (if possible) may reduce oscillation frequency and magnitude. Finally, some substrate isolation techniques can be implemented (beginning with the substrate resistivity choice and the die back-side connectivity). Classical guard-rings and triple-wells are often solutions, but their efficiency depends on the biasing quality. Furthermore, triple-well will lose its isolation property when its area increases.

Nevertheless, some works remains to do, for example concerning the way to obtain the transient consumption current of an entire large digital part or an efficient method to simplify a layout and to extract an accurate substrate *RC* network. Moreover, the signal integrity issue in Systems-on-Chip—SoC or Systems-in-Package—SiP does not only concern conducted noise but also radiated propagation with electromagnetic effects: the ICEM standard model can also respond to this need.

To go further in noise study, we study the impact of low-frequency substrate noise on an analog circuit: a fully integrated 5 GHz voltage-controlled oscillator. The analytical approach for harmonic perturbations is resumed to the classical frequency modulation with observed spurious side-bands. We note that the side-band power below carrier is more important when the injection is located close to the bipolar transistors and does not depend on the tuning voltage. On the contrary, the side-band power below carrier decreases with the tuning voltage, when the injection is located close to the inductor and varactor. The next step is the study of the sensitive devices using the ISF. For an equivalent low-frequency substrate noise magnitude, the impact on the bipolar transistors is more important compared to the impact on the varactors. For high-frequency substrate noise, the behavior is drastically different and can be analyzed through the ISF Fourier series. In a realistic integrated circuit, noise source locations, layout implementation (power supply rails, shielding, guard ring, etc.) and substrate attenuation have to be taken into account to quantify the noise magnitude on each substrate device. In this chapter, we insist that a global approach by only considering resistive and/ or capacitive coupling between substrate and power supply rails is not sufficient for precise quantification of noise impact on analog device, such as RF VCOs.

Finally, a compact and pragmatic modeling approach for 3D interconnects is initiated, based on parametrical extractions performed on realistic test structures via some TLM. Complete equivalent electrical models are illustrated for TSV chains structures. This approach is validated through frequency analyses by comparing the S-parameters measured on the test structures with those obtained by running simulations of respective equivalent electrical models. At high frequency, the substrate effects must be integrated into the electrical model as a *RLCG* network; but the effects between all the elements sharing the substrate have also to be modelled. We highlight our 3D substrate extraction method, relying on the TLM over multilayered substrates and/or on Green kernels, to model these effects. Such simple approach presents the main advantage to be easily automatized; additionally it can be improved by including skin effect , eddy current, depleted layer at the via's flank level. The substrate extractor will be incorporated to electrically model a large panel of 3D structures, defining critical paths and to investigate signal integrity.

Acknowledgments

This work is supported by the EU Community FP7-2012-ITN project 317446 (INFIERI: INtelligent, Fast, Interconnected and Efficient devices for Frontier Exploitation in Research and Industry).

References

1. Hertzel, F., Razavi, B. A study of oscillators jitter due to supply and substrate noise. *IEEE Trans. Circuits Syst.—II: Analog Digit. Signal Process.*, 46(1), 56–62, 1999.
2. Blalack, T., Leclercq, Y., Yue, C.P. On-chip RF isolation techniques. *IEEE Bipolar/BiCMOS Circuits and Technology Meeting Conference*, October 2002, Monterey, CA.
3. Smedes, T., Van der Meijs, N.P., Van Genderen, A.J. Extraction of circuit models for substrate cross-talk. *Proceeding of IEEE International Conference of Computer-Aided Design ICCAD-95*, 5–9 November 1995, pp. 199–206.
4. Stanisic, B.R., Verghese, N.K., Rutenbar, R.A. et al. 1994. Addressing substrate coupling in mixed-mode IC's: Simulation and power distribution synthesis. *IEEE J. Solid-State Circuits*, 29(3), 226–238.
5. Nagata, M., Nigai, J., Hijikata, K. et al. 2001. Physical design guides for substrate noise reduction in CMOS digital circuits. *IEEE J. Solid-State Circuits*, 36(3), 539–549.
6. Pavlidis, V.F., Friedman, E.G. *Three-Dimensional Integrated Circuit Design*. New York: Elsevier, Inc., 2009.
7. International Electro-technica Commission, IEC 62014–3: Integrated Circuits Emission Model (ICEM). Draft technical report, IEC, 2004.
8. Levant, J.L., Ramdani, M., Preduriau, R. ICEM modeling of microcontroller current activity. *Proceedings of the EMC Compo 2002*, 88–91, 2002.
9. Hajimiri, A., Lee, T.H. The Design of Low Noise Oscillators. Boston: Kluwer Academic Publishers, 1999.
10. van Heijningen, M., Baradoglu, M., Donnay, S. et al. Substrate noise generation in complex digital systems: Efficient modeling and simulation methodology and experimental verification. *IEEE J. Solid-State Circuits*, 37(8), 1065–1072, 2002.
11. Valorge, O., Calmon, F., Andrei, C. et al. Mixed-signal IC design guide to enhance substrate noise immunity in bulk silicon technology. *Analog Integrated Circuits and Signal Processing*, 63(2), 185–196, 2010.
12. Yoshinaga, T., Nomura, M. Trends in R&D in TSV technology for 3D LSI packaging. *Science and Technology Trends*, quarterly review no. 37, October 2010.
13. Gharpurey, R. Modeling and analysis of substrate coupling in integrated circuits. Thesis in Engineering-Electrical Engineering and Computer Sciences, Berkeley: University of California, 1992.
14. Calmon, F., Andrei, C., Valorge, O. et al. 2006. Impact of low-frequency substrate disturbances on a 4.5 GHz VCO. *Elsevier Microelectron. J.*, 37(10), 1119–1127.
15. Soens, C., Van der Plas, G., Wambacq, P. et al. Performance degradation of an LC-tank VCO by impact of digital switching noise. *European Solid-State Circuits Conference*, Leuven, Belgium, 2004.
16. Nagata, M., Nigai, J., Morie, T. et al. Measurements and analyses of substrate noise waveform in mixed-signal IC environment. *IEEE Trans. Computer-Aided Design*, 19(6), 671–678, 2000.
17. Charbon, E., Ghapurey, R., Meyer, R.G. et al. Semi-analytical techniques for substrate characterization in the design of mixed-signal ICs. *IEEE ICAD*, 1996.

18. Baradoglu, M., van Heijningen, M., Gravot, V. et al. Methodology and experimental verification for substrate noise reduction in CMOS mixed-signal ICs with synchronous digital circuits. *IEEE J. Solid-State Circuits*, 37(11), 1383–1395, 2002.

19. Su, D.K., Loinaz, M.J., Masui, S. et al. 1993. Experimental results and modeling techniques for substrate noise in mixed-signal integrated circuits. *IEEE J. Solid-State Circuits*, 28(4), 420–430.

20. Hajimiri, A., Lee, T.H. A general theory of phase noise in electrical oscillators. *IEEE J. Solid-State Circuits*, 33(2), 179–194, 1998.

21. Liao, H., Rustagi, S.C., Shi, J. et al. Characterization and modeling of the substrate noise and its impact on the phase noise of VCO. *IEEE Radio Frequency Integrated Circuits Symposium*, 2003.

22. Sentaurus Device Electromagnetic Wave Solver User Guide (Version A-2008.09, September 2008 SYNOPSYS).

23. COMSOL Multiphysics. http://www.comsol.com/.

24. Xu, Z., Lu, J-Q. High-speed design and broadband modeling of through-strata-vias (TSVs) in 3D integration. *IEEE Trans. Compon. Packag. Manuf. Technol.*, 1(2), 154–162, 2011.

25. Liang, Y., Li, Y. Closed-form expressions for the resistance and the inductance of different profiles of through-silicon vias. *IEEE Electron Device Lett.*, 32(3), 393–395, 2011.

26. Kim, D.H., Mukhopadhyay, S., Lim, S.K. Fast and accurate modeling of through-silicon-via capacitive coupling. *IEEE Trans. Compon. Packag. Manuf. Technol.*, 1(2), 168–180, 2011.

27. Crovetti, P.S., Fiori, F.L. Efficient BEM-based substrate network extraction in silicon SoCs. *Elsevier Microelectronics Journal*, 39, 1774–1784, 2008.

28. Valorge, O., Sun, F., Lorival, J.-E. et al. Analytical and numerical model confrontation for transfer impedance extraction in three-dimensional radiofrequency circuits. *Circuits & Systems*, 3(2), 126–135, 2012, DOI: 10.4236/cs.2012.32017.

29. Matlab, http://mathworks.com.

30. Advanced Design System—ADS, http://www.home.agilent.com.

31. Xu, C., Kourkoulos, V., Suaya, R. et al. A fully analytical model for the series impedance of through-silicon vias with consideration of substrate effects and coupling with horizontal interconnects. *IEEE Transactions on Electron Devices*, 58(10), 3529–3540, 2011.

10

FIR Filtering Techniques for Clock and Frequency Generation

Ni Xu, Woogeun Rhee, and Zhihua Wang

CONTENTS

10.1 Introduction

The $\Delta\Sigma$ modulation has become an important technique not only in the area of data conversion but also in the area of clock and frequency generation these days. In wireless systems, frequency synthesizers based on the $\Delta\Sigma$ fractional-N phase-locked loop (PLL) greatly relax design requirements such as the in-band phase noise and the settling time [1–5]. Moreover, the $\Delta\Sigma$ fractional-N PLL enables direct digital frequency modulation for low-cost transmitter design [3,6–8], thus becoming a key building block in modern transceiver systems. Recent development of the digital-intensive PLL (DPLL) offers solid frequency generation over the gate leakage current and the loop parameter variation problems over process and temperature variations [9–11]. In the DPLL, the $\Delta\Sigma$ modulator is used to enhance the resolution of the digitally controlled oscillator (DCO) and to improve the performance of the time-to-digital converter (TDC) with dithering. In wireline systems, the ASIC-based system-on-chip (SoC) design with standard CMOS technology enforces the use of the DPLL for robust clock generation and good technology scalability.

The use of the $\Delta\Sigma$ modulator, however, often degrades the phase noise performance in high frequencies or the short-term jitter performance especially when wideband PLLs are employed. Several methods to mitigate the quantization noise problem of the $\Delta\Sigma$ modulator have been proposed in the literature [12–15], but most methods require good circuit matching and linearity or substantial hardware complexity. In this chapter, we present a semidigital (hybrid) finite-impulse response (FIR) filtering method that enhances the performance of the $\Delta\Sigma$ fractional-N PLL with good immunity against process and temperature variations. Various applications of the FIR filtering method in the area of clock and frequency generation are also introduced.

10.2 Overview of the $\Delta\Sigma$ Modulation Method

Figure 10.1 shows a simplified block diagram of the $\Delta\Sigma$ fractional-N PLL. Similar to the $\Delta\Sigma$ analog-to-digital converter (ADC), it utilizes an oversampled noise-shaping modulator to realize fractional frequency division with a coarse dual-modulus divider. As shown in Figure 10.1, the dual-modulus divider is analogous to the bi-level quantizer in the $\Delta\Sigma$ ADC. Since the frequency resolution of the $\Delta\Sigma$ fractional-N PLL is set by the number of bits of the digital modulator and the clock frequency, the resolution does not depend on the frequency of the voltage-controlled oscillator (VCO). Hence, a very fine frequency resolution can be obtained by simply increasing the number of bits of the $\Delta\Sigma$ modulator. The oversampling principle is similar to the random jittering method [16], but the $\Delta\Sigma$ modulation method also provides a noise-shaping property. Accordingly, it does not cause $1/f^2$

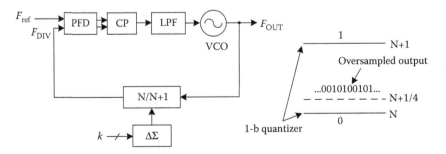

FIGURE 10.1
$\Delta\Sigma$ Fractional-N PLL.

phase noise at the VCO output, which is another important design aspect in the PLL design. A frequency-domain comparison of the random jittering and $\Delta\Sigma$ modulation methods is given in Figure 10.2. Since the second- or higher order $\Delta\Sigma$ modulators, in theory, do not generate fixed tones for DC inputs, they effectively shape the phase noise without causing any spur. The operation of the $\Delta\Sigma$ fractional-N PLL is based on the following key properties:

- Oversampled interpolation for a fine frequency resolution
- High-order modulation for spur-free randomization
- Noise shaping for low-frequency noise suppression
- Digital control for direct phase/frequency modulation capability

In the frequency synthesizer design, it is important to identify phase noise sources. The noise contribution from each source is shown in Figure 10.3 in which a type-II, fourth-order fractional-N PLL with a third-order $\Delta\Sigma$ modulator is assumed. The $\Delta\Sigma$ modulator can affect in-band noise or out-of-band noise, which depends on the open-loop gain of the PLL. The in-band noise may be limited by PLL nonlinearity. The out-of-band noise can be possibly determined by the residual quantization noise of the modulator rather than the VCO noise. Note that the quantization noise contribution does not depend on the division ratio, because it is generated by frequency modulation having the resolution of one VCO clock period. As depicted in Figure 10.3, the open loop gain of the PLL needs to be carefully designed to have

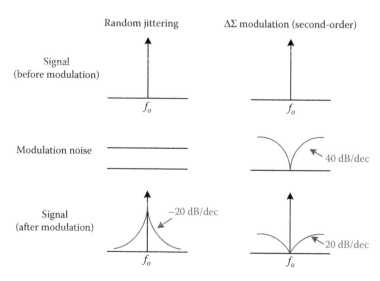

FIGURE 10.2
Random jittering methods vs. $\Delta\Sigma$ modulation method.

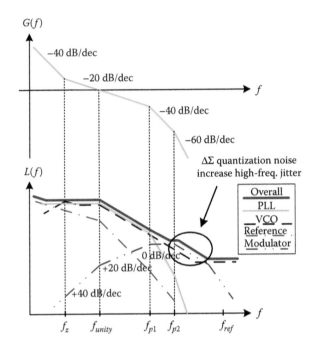

FIGURE 10.3
Phase noise sources and quantization noise effect in $\Delta\Sigma$ PLL.

the overall noise performance meet the system specification. Note that high-order poles are important not only for spur suppression but also for the out-of-band phase noise performance in the fractional-N PLL design.

    There are several techniques proposed for quantization noise reduction. The phase error cancellation method using a digital-to-analog converter (DAC) [13–15] achieves significant reduction of phase noise and spurs but the performance depends on the high-resolution DAC and analog matching, resulting in high design complexity.

10.3 Hybrid FIR Filtering Technique

The semidigital approach based on a FIR filtering method offers moderate quantization noise reduction without using the DAC [17–23]. With a wide bandwidth, out-of-band noise could be more problematic than in-band noise to meet the phase noise mask required by wireless standards or to achieve good short-term jitter performance in wireline applications. The hybrid FIR filtering method effectively reduces the out-of-band noise without affecting the PLL loop dynamics.

Figure 10.4 shows a conceptual diagram of how the FIR filter can be utilized to shape the periodic tone. For simplicity, a fractional-N PLL with 9-modulo fractional division is assumed. The 9-modulo fractional-N PLL with the reference frequency F_{ref} exhibits a periodic phase error with the frequency of $F_{ref}/9$. When a 3-tap FIR filter is applied, the fixed tone with the frequency of $F_{ref}/9$ is increased to higher frequency of $F_{ref}/3$ as depicted in Figure 10.4. As the fundamental frequency of the fixed tone increases, further spur reduction can be achieved by the PLL loop filter. If a 9-tap FIR filter is used, the fixed tone can be completely removed for the 9-modulor fractional division.

The FIR filtering method can be extended to the quantization noise reduction of the $\Delta\Sigma$ fractional-N PLL. A straightforward implementation and the equivalent discrete-time model are shown in Figure 10.5. To realize the hybrid FIR filtering method, multiple phase-frequency detectors (PFDs), a multi-input charge pump (CP), and multimodulus dividers (MMDs) are used. The number of PFDs and MMDs sets the number of the FIR filter taps k, as shown in the equivalent discrete-time model. Since the multi-input CP increases the phase detector gain by k times, the CP current for each input stage should be scaled down by k to maintain the same PLL open-loop gain. The output of the $\Delta\Sigma$ modulator is firstly loaded into the D-type flip flop (DFF) and then shifted by one clock cycle or several clock cycles per stage. Each MMD is sequentially controlled by the shifted output of each DFF. The multiple MMDs in parallel with sequential control bits perform

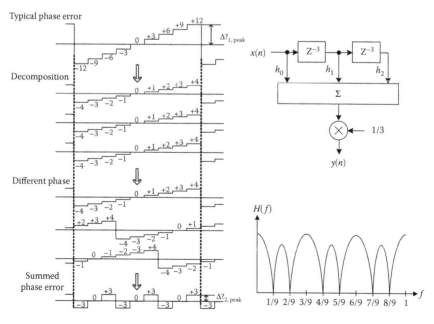

FIGURE 10.4
FIR filtering in time and frequency domains.

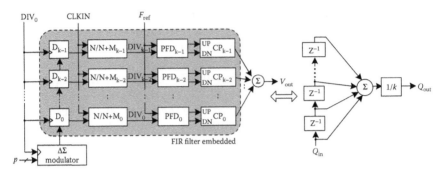

FIGURE 10.5
Hybrid FIR filtering for ΔΣ fractional-N PLL.

the FIR filtering with respect to the input control bits. Hence, the hybrid FIR filtering operation performs a discrete-time signal processing with respect to ΔΣ quantization noise. Since the input control bits are modulated by the ΔΣ modulator, the FIR filtering effectively reduces the modulator quantization noise. In Figure 10.5, the output of the modulator can be shifted by multiple clock cycles to realize different noise transfer function with the delay depth n in addition to the number of taps k. Figure 10.6a and b shows noise transfer functions of the 16-tap FIR filter ($k = 16$) with $n = 1$ and $n = 3$, respectively. With $n = 3$, the transfer function has the first zero frequency occur at the 1/48 of the clock frequency. Note that FIR filtering is effective only on the MMD control path, not on the PLL small signal path. Hence, the proposed method suppresses the modulator noise without affecting the PLL loop dynamics.

To verify the FIR filtering function, a PLL closed-loop transient simulation has been done with $k = 16$ and $n = 1$. In the simulation, the third-order ΔΣ modulator is used, and the PLL bandwidth is designed to be approximately 1/20 of the phase detector frequency. To clearly see the high-frequency quantization noise comparison, the high-order poles of the PLL are put at higher than

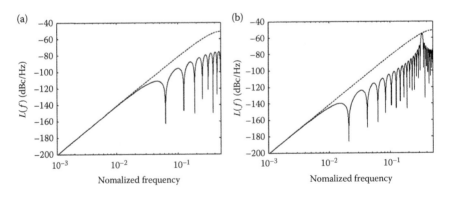

FIGURE 10.6
Quantization noise reduction with FIR filtering: (a) $k = 16$, $n = 1$ and (b) $k = 16$, $n = 3$.

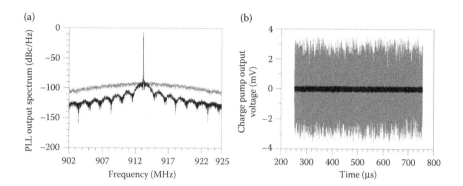

FIGURE 10.7
FIR filtering effects: (a) PLL output spectrum and (b) instantaneous phase error.

optimized frequencies. Figure 10.7a shows output spectra of two $\Delta\Sigma$ fractional-N PLLs: one with the conventional architecture (light color) and the other with the FIR-embedded architecture (dark color). To maintain the same bandwidth for the conventional fractional-N PLL case, all the 16 MMDs are controlled by the same output of the $\Delta\Sigma$ modulator. In this way, the loop parameter does not have to be modified. As seen clearly, the FIR-embedded fractional-N PLL has less phase noise contribution from the modulator in high frequencies. It also shows the zero of the FIR filter at 1/16 of the phase detector frequency as expected. Figure 10.7b shows the comparison of the instantaneous phase error between two PLLs where the peak-to-peak amplitude is significantly reduced with the hybrid FIR filter. This proves that high-frequency phase errors are well suppressed in the FIR-embedded fractional-N PLL after all 16 phase errors are summed at the output of the CP. It also shows that the linearity of the CP and the VCO could be improved with reduced voltage range over the loop filter.

Compared to the existing methods, the quantization noise suppression is more predictable with less dependency on PVT variations since noise reduction is done by the digital-domain filtering method. In addition, the power and area of the FIR filtering circuitry can be scaled with the advanced CMOS technology when the phase-interpolated MMD is employed [22,23], which is different from the DAC cancellation method.

10.4 Clock and Frequency Generation with FIR-Embedded $\Delta\Sigma$ Modulation

Since the hybrid FIR filtering method does not add latency and can be well predicted with the semidigital operation, it can be utilized in various applications in the area of clock and frequency generation. Several applications are introduced in the next subsections.

10.4.1 Digital Clock Generation with Wideband PLL

Design aspects of the $\Delta\Sigma$ fractional-N PLL for digital systems are somewhat different from those for RF systems. In many digital clocking systems, such as serial links and microprocessors, short-term jitter is as important as long-term jitter. Since short-term jitter is caused by out-of-band phase noise, high-frequency quantization noise not suppressed well by the PLL can easily degrade the overall system performance. Most PLLs in digital systems use low-cost ring VCOs. Accordingly, a wide PLL bandwidth is needed to suppress the VCO noise, but the low ratio of the phase detector frequency to the PLL bandwidth makes it difficult for the PLL to filter out quantization noise from the $\Delta\Sigma$ modulator.

By employing the hybrid FIR filtering method, the digital clock generator based on the fractional-N PLL can achieve reduced short-term jitter when the wide bandwidth is set to suppress the phase noise of the ring VCO. Figure 10.8 shows a simplified block diagram of the PLL with the hybrid FIR filter for wideband operation [17]. The multi-input CP only adds more switching devices but does not change the overall transistor size of the current mirrors since the same output current is used in total.

10.4.2 FIR-Embedded $\Delta\Sigma$ DLL with Fine Resolution

As data rate increases, clock skew becomes a significant portion of the overall timing margin and directly affects the bit-error-rate (BER) performance. A variable delay line (VCDL) or a delay-locked loop (DLL) is widely used for elastic timing control not only in source-synchronous serial links but also in

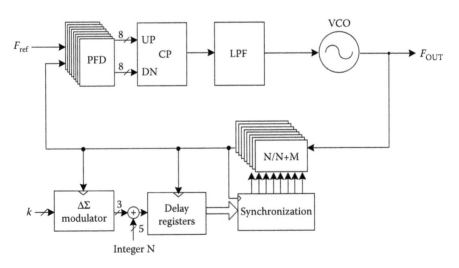

FIGURE 10.8
Digital clock generation with reduced short-term jitter.

clock-and-data recovery systems. Since the conventional analog delay line suffers from process and temperature variations, the semidigital DLL based on the phase rotator is considered for robust delay control [24–26]. However, the phase rotator utilizes discrete multiple phases, causing an algorithmic jitter in the semidigital DLL. Hence, achieving fine time resolution such as <1 ps is challenging in the digital-intensive DLL design.

Different from the fractional-N $\Delta\Sigma$ PLL whose frequency interpolation is achieved by the MMD modulation, the $\Delta\Sigma$ DLL achieves a fine phase resolution with an oversampled phase interpolator [27,28]. The quantized phases used for the oversampled interpolation can be put in the feedback loop. In theory, a phase resolution of as low as 1 ppm unit interval (UI) with good linearity can be easily achieved with the all-digital interpolation. However, the $\Delta\Sigma$ DLL has several design challenges. Different from the $\Delta\Sigma$ fractional-N PLL, the $\Delta\Sigma$ DLL does not have a frequency divider in the feedback loop. Therefore, the $\Delta\Sigma$ modulator needs to operate at the same frequency as the input clock frequency, resulting in high power consumption or use of a simple $\Delta\Sigma$ modulator. Also, the DLL, the first-order feedback system, provides worse noise filtering against $\Delta\Sigma$ quantization noise than the PLL does. To suppress out-of-band quantization noise, the hybrid FIR filtering method can be used as shown in Figure 10.9. By utilizing a divider circuit and a self-referenced multiphase generator, the $\Delta\Sigma$ modulator can operate at a much lower frequency without causing false lock or glitch problems [27]. As a result, the design complexity of the $\Delta\Sigma$ modulator and the CP can be reduced significantly.

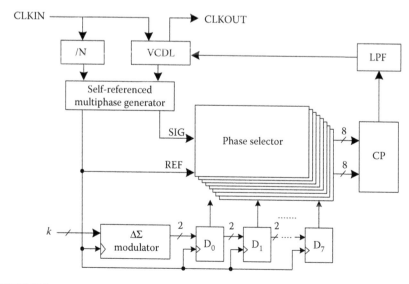

FIGURE 10.9
$\Delta\Sigma$ DLL with the hybrid FIR filter.

10.4.3 Fractional-N Frequency Synthesizer with Customized Noise Shaping

In most RF applications, frequency synthesizers need to meet both in-band and out-of-band phase noise requirements that are usually specified by a phase noise mask. Since the brick-wall phase noise mask gives a tight margin only at certain corner frequencies, the phase noise performances at those frequencies determine the overall noise performance of the frequency synthesizer. Since the noise transfer function of the FIR filter can be designed with proper filter coefficients, the customized noise shaping to meet the noise mask is possible as shown in Figure 10.10.

However, the use of the hybrid FIR filter requires multiple RF frequency dividers, which is not desirable for any low-power application. To reduce power and area, the multiple frequency dividers are designed based on a phase shifting method that can be designed to perform the hybrid FIR filtering at low frequency with standard CMOS logic. Figure 10.11 shows an example of the FIR-embedded fractional-N frequency division that requires only one RF frequency divider [18]. After the CML-based 1/4 frequency divider, the remaining multiple blocks designed with the CMOS logic do not add any substantial area or power when advanced CMOS technology is used. In other words, the hybrid FIR filtering block can be scaled with CMOS technology if the phase-selection-based MMD is designed.

10.4.4 Finite-Modulo Fractional-N PLL with Reduced Spur

Even though the $\Delta\Sigma$ fractional-N PLL offers superior performance to the integer-N PLL, it is more exposed to nonlinearity and coupling issues than other PLLs. Having moderate performance between the integer-N PLL and the $\Delta\Sigma$ fractional-N PLL, the traditional finite-modulo fractional-N PLL provides an alternative way of reducing the division ratio with negligible digital

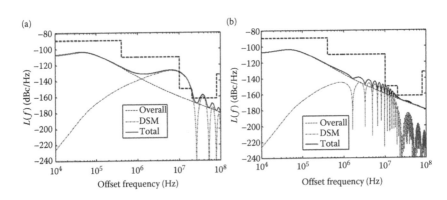

FIGURE 10.10
Digital clock generation with reduced short-term jitter. (a) Without FIR and (b) with 16-tap FIR phase noise of the synthesizer.

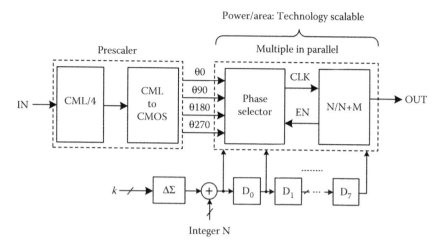

FIGURE 10.11
Multiple dividers with the phase shifting method for FIR filtering.

power and coupling. Similar to the $\Delta\Sigma$ fractional-N PLL, the FIR filtering method can also be employed for the finite-modulo fractional-N PLL [29,30]. Figure 10.12 shows an example of how the hybrid FIR filter is useful in the finite-modulo fractional-N PLL. In this example, the 8-modulo fractional-N PLL is designed with the 8-tap FIR filtering method [30]. A 3-bit accumulator combined with the 8-tap FIR filter performs periodic modulation without having the spur generation at the 1/8 of the clock frequency since the FIR filter generates a notch in the transfer function. If multiphase outputs are available from the VCO, further spur suppression can be achieved with finer frequency step, which is equivalent to have a fractional dual-modulus divider as illustrated in Figure 10.12.

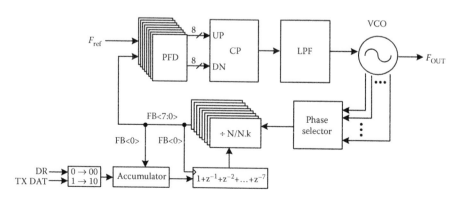

FIGURE 10.12
8-Modulo fractional-N PLL with the 8-tap hybrid FIR filter.

10.4.5 FIR-Embedded DCO Modulation

Different from the fractional-N divider, the quantization noise of the $\Delta\Sigma$ modulator for the DCO cannot be filtered by the loop filter of the PLL. In the digital control path of the DCO, the hybrid FIR filtering method can also be implemented for the DCO quantization noise reduction.

Figure 10.13 shows an example of the DCO block diagram with the FIR-embedded $\Delta\Sigma$ control. To perform 8-tap hybrid FIR filtering, eight DFFs are used. The gain of 1/8 at the output of the FIR filter is needed for the transfer function to have the unity gain at DC. The scaled gain of 1/8 can be realized in an analog way inside the DCO when the ring oscillator-based DCO is used [31]. After passing the FIR block, the dithered 2-bit input modulates the digital path with reduced quantization noise. Figure 10.14 shows the measured spectra at the DCO output with the FIR filter disabled and enabled. As clearly seen in Figure 10.14, the FIR-induced noise shaping is observed, showing that the out-of-band phase noise at 1–2 MHz offset frequencies is suppressed when the FIR filter is enabled.

It shows that the use of the FIR filter in the DCO can improve the short-term jitter performance for the given $\Delta\Sigma$ modulator and the same clock frequency. Similar to the previous MMD-based modulation case, the hybrid FIR filtering does not add any latency to the control path of the DCO, thus not affecting the loop dynamics of the digital PLL.

10.4.6 FIR-Embedded Two-Point Modulation for Transmitter

The $\Delta\Sigma$ PLL-based digital phase modulation significantly simplifies the overall transmitter architecture, which does not require DACs and mixers. To overcome the constraint between the PLL bandwidth and the required modulation symbol, a digital compensation method [32,33] or a two-point modulation method is employed [34–36]. In the digital compensation method, the transfer function of the digital compensation filter needs to be matched well with that of the PLL, which is highly sensitive to process and temperature

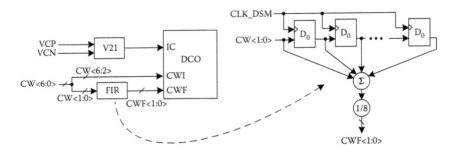

FIGURE 10.13
FIR-embedded $\Delta\Sigma$ modulation for DCO.

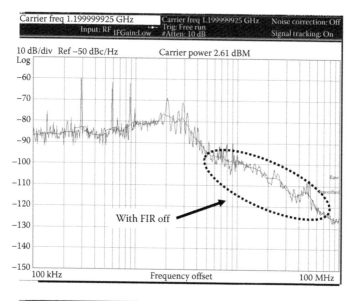

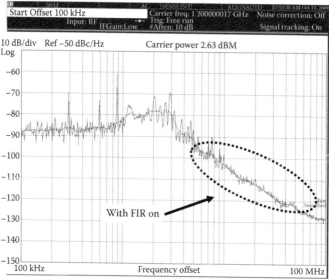

FIGURE 10.14
Measured FIR effect on DCO modulation.

variations. On the contrary, combining high-pass and low-pass modulation paths in the $\Delta\Sigma$ PLL, the two-point modulator overcomes the PLL bandwidth limitation and provides an all-pass modulation path regardless of the PLL bandwidth as illustrated in Figure 10.15. To achieve the all-pass transfer function, the transfer function of the high-pass modulation must be accurately scaled by the VCO gain [36]. In practice, the VCO gain is highly sensitive to

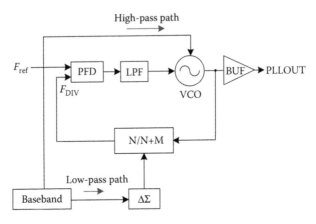

FIGURE 10.15
ΔΣ PLL-based two-point modulator.

PVT variations and is one of the most critical parameters to determine the overall performance of the two-point modulator. However, the all-digital PLL (ADPLL) design with DCO provides a robust way of calibrating the absolute DCO gain, but it is still challenging for the DCO nonlinearity calibration [37].

To overcome the VCO nonlinearity problem in the two-point modulator, a 1-bit modulation is considered for the high-pass modulation independent of the low-pass modulation with asymmetrically partitioned varactors in the DCO. For high data rate modulation, a wideband FM modulation is needed, requiring the DCO with a large single varactor for the 1-bit modulation. The DCO with the large varactor, however, generates substantial quantization noise and can be also very sensitive to coupling noise due to high DCO gain. To reduce the quantization noise and the sensitivity, the FIR filtering method can be considered similar to previous applications. As illustrated in Figure 10.16, the FIR filtering method not only reduces the high-frequency quantization noise caused by the 1-bit ΔΣ modulation but also realizes a time-interleaved modulation with uniformly-partitioned varactors in the high-pass modulation path, which significantly reduces the coupling sensitivity [38]. The averaging function of the FIR filtering method also reduces the mismatch effect of the partitioned varactors.

The ADPLL-based two-point modulation is considered a robust solution with the digital calibration of the gain and the delay mismatches. Figure 10.17 shows a block diagram of the digital modulator with 1-bit high-pass modulation based on the ADPLL architecture. The digital FIR filtering methods is employed for the high-pass and low-pass modulation paths to enhance linearity and reduce high-frequency quantization noise, respectively. The use of the dedicated 1-bit high-pass modulation path mitigates the nonlinearity problem of the DCO gain in the two-point modulator design, which can substantially simplify the two-point modulator architecture with good linearity.

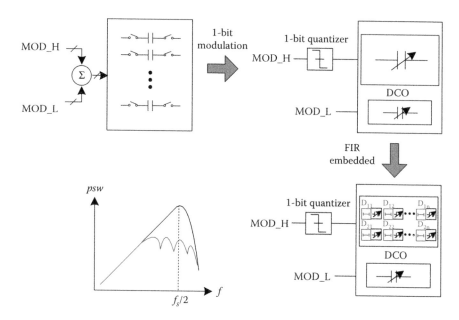

FIGURE 10.16
FIR-embedded 1-bit high-pass modulation.

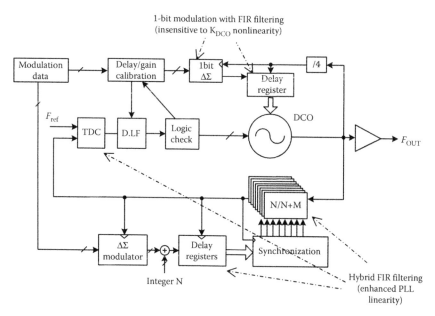

FIGURE 10.17
Two⁺-point modulator with FIR-embedded 1-bit high-pass modulation.

10.5 Conclusion

In this chapter, we introduce a hybrid FIR filtering method which can be useful in various clock and frequency generation applications. The hybrid filtering method with the digital-domain parallel operation and the analog-domain gain normalization provides a latency-free operation to the loop dynamics of the PLL and a robust transfer function over process and temperature variations. Several application examples of the hybrid FIR filter methods are presented: the clock generation PLL and the fine-resolution DLL for short-term jitter performance, the frequency synthesizer with customized noise shaping, the FIR-embedded DCO, and the low-noise two-point modulator designs. The FIR filtering technique not only enhances the performance of the $\Delta\Sigma$ fractional-N PLL with good immunity against process and temperature variations but also mitigates the quantization noise and the DCO noise coupling effects.

References

1. Miller, B., Conley, R. A multiple modulator fractional divider. In *Proceedings 44th Annual Frequency Control Symposium*, Baltimore, MD, May 1990, pp. 559–568.
2. Riley, T. A., Copeland, M., Kwasniewski, T. Delta-sigma modulation in fractional-N frequency synthesis. *IEEE J. Solid-State Circuits*, 28, 553–559, 1993.
3. Perrot, M. H., Tewksbury, T. T., Sodini, C. G. A 27-mW CMOS fractional-N synthesizer using digital compensation for 2.5-Mb/s GFSK modulation. *IEEE J. Solid-State Circuits*, 32(12), 2048–2060, 1997.
4. Rhee, W., Song, B., Ali, A. A 1.1-GHz CMOS fractional-N frequency synthesizer with a 3-b third-order Δ-Σ modulator. *IEEE J. Solid-State Circuits*, 35, 1453–1460, 2000.
5. Perrott, M. H., Trott, M. D., Sodini, C. G. A modeling approach for $\Sigma\Delta$ fractional-N frequency synthesizers allowing straightforward noise analysis. *IEEE J. Solid-State Circuits*, 37, 1028–1038, 2002.
6. Meyers, R. A., Waters, P. H. Synthesizer review for PAN-European digital cellular radio. In *Proceedings IEEE Colloquium on VLSI Implementations for 2nd Generation Digital Cordless and Mobile Telecommunication Systems*, London, March 1990, pp. 8/1–810.
7. Durdodt, C. The first very low-IF Rx, 2-point modulation TX CMOS system on chip Bluetooth solution. In *Proceedings IEEE RFIC Symposium*, Phoenix, AZ, June 2001, pp. 99–102.
8. Youssef, M., Zolfaghari, A., Mohammadi, B. et al. A low-power GSM/EDGE/WCDMA polar transmitter in 65-nm CMOS. *IEEE J. Solid-State Circuits*, 46(12), 3061–3074, 2011.
9. Staszewski, R. B., Wallberg, J. L., Rezeq, S. et al. All digital PLL and transmitters for mobile phones. *IEEE J. Solid-State Circuits*, 40(12), 2469–2482, 2005.

10. Straayer, M. Z., Perrott, M. H. A multi-path gated ring oscillator TDC with first-order noise shaping. *IEEE J. Solid-State Circuits*, 44(4), 1089–1098, 2009.
11. Nonis, R., Grollitsch, W., Santa, T. et al. digPLL-Lite: A low-complexity, low-jitter fractional-N digital PLL architecture. *IEEE J. Solid-State Circuits*, 48(12), 3134–3145, 2013.
12. Riley, T., Filiol, N. M., Du, Qinghong et al. Techniques for in-band phase noise reduction in $\Delta\Sigma$ synthesizers. *IEEE Trans. Circuits Syst. II*, 50, 794–803, 2003.
13. Gupta, M., Song, B. A 1.8 GHz spur cancelled fractional-N frequency synthesizer with LMS-based DAC gain calibration. *IEEE J. Solid-State Circuits*, 41(12), 2842–2851, 2006.
14. Swaminathan, A., Wang, K. J., Galton, I. A wide bandwidth 2.4 GHz ISM-band fractional-N PLL with adaptive phase noise cancellation. *IEEE J. Solid-State Circuits*, 42(12), 2639–2650, 2007.
15. Wu, Y.-D., Lai, C.-M., Lee, C.-C. et al. A quantization error minimization method using DDS-DAC for wideband fractional-N frequency synthesizer. *IEEE J. Solid-State Circuits*, 45(11), 2283–2291, 2010.
16. Reinhardt, V., Shahriary, I. Spurless fractional divider direct digital synthesizer and method. U.S. Patent 4,815,018, March 21, 1989.
17. Yu, X., Sun, Y., Rhee, W. et al. An FIR-embedded noise filtering method for $\Delta\Sigma$ fractional-N PLL clock generators. *IEEE J. Solid-State Circuits*, 44(9), 2426–2436, 2009.
18. Yu, X., Sun, Y., Rhee, W. et al. A $\Delta\Sigma$ fractional-N synthesizer with customized noise shaping for WCDMA/HSDPA applications. *IEEE J. Solid-State Circuits*, 44(8), 2193–2200, 2009.
19. Lee, I.-T., Lu, H.-Y., Liu, S.-I. A 6-GHz all-digital fractional-N frequency synthesizer using FIR-embedded noise filtering technique. *IEEE Trans. Circuits Syst. II Exp. Briefs*, 59(5), 267–271, 2012.
20. Yu, X., Sun, Y., Rhee, W. et al. A 65 nm CMOS 3.6 GHz fractional-N PLL with 5th-order $\Delta\Sigma$ modulation and weighted FIR filtering. *IEEE Asian Solid-State Circuits Conference*, Taipei, November 2009, pp. 77–80.
21. Kondou, M., Matsuda, A., Yamazaki, H. et al. A 0.3 mm^2 90-to-770 MHz fractional-N synthesizer for a digital TV tuner. *IEEE International Solid-State Circuits Conference (ISSCC) Digest of Technical Papers*, San Francisco, CA, February 2010, pp. 248–249.
22. Jee, D.-W., Suh, Y., Park, H.-J., Sim, J.-Y. A 0.1-fref BW 1 GHz fractional-N PLL with FIR-embedded phase-interpolator-based noise filtering. *IEEE International Solid-State Circuits Conference (ISSCC) Digest of Technical Papers*, San Francisco, CA, February 2011, pp. 94–95.
23. Jee, D.-W. Suh, Y., Kim, B. et al. A FIR-embedded phase interpolator based noise filtering for wide-bandwidth fractional-N PLL. *IEEE J. Solid State Circuits*, 48(11), 2795–2804, 2013.
24. Sidiropoulos, S., Horowitz, M. A semidigital dual delay-locked loop. *IEEE J. Solid-State Circuits*, 32(11), 1683–1692, 1997.
25. Bae, S. J., Chi, H. J., Sohn, Y. S., Park, H. J. A VCDL-based 60–760-MHz dual-loop DLL with infinite phase-shift capability and adaptive-bandwidth scheme. *IEEE J. Solid-State Circuits*, 40(5), 1119–1129, 2005.
26. Kim, B.-G., Kim, L.-S., Park, K.-I. et al. A DLL with jitter-reduction techniques for DRAM interfaces. *IEEE International Solid-State Circuits Conference (ISSCC) Digest of Technical Papers*, San Francisco, CA, February 2007, pp. 496–497.

27. Yu, X., Rhee, W., Wang, Z., Lee, J. B., Kim, C. A 0.4-to-1.6 GHz low OSR with self-referenced multiphase generation. *IEEE International Solid-State Circuits Conference (ISSCC) Digest of Technical Papers*, San Francisco, CA, February 2009, pp. 398–400.
28. Cheng, S.-J., Qiu, L., Zheng, Y., Heng, C.-H. 50–250 MHz ΔΣ DLL for clock synchronization. *IEEE J. Solid-State Circuits*, 45(11), 2445–2456, 2010.
29. Chen, F., Li, Y., Lin, D. et al. A 1.14 mW 750 kb/s FM-UWB transmitter with 8-FSK subcarrier modulation. *IEEE Custom Integrated Circuits Conference (CICC)*, San Joes, CA, September 2013, pp. 1–4.
30. Zhou, B., Qiao, J., He, R. et al. A gated FM-UWB system with data-driven front-end power control. *IEEE Trans. Circuits Syst. I*, 59(6), 1348–1358, 2012.
31. He, R., Liu, C., Yu, X. et al. A low-cost, leakage-insensitive semi-digital PLL with linear phase detection and FIR-embedded digital frequency acquisition. *IEEE Asian Solid-State Circuits Conference*, Beijing, November 2010, pp. 197–200.
32. Perrot, M. H., Tewksbury, T. T., Sodini, C. G. A 27-mW CMOS fractional-N synthesizer using digital compensation for 2.5-Mb/s GFSK modulation. *IEEE J. Solid- State Circuits*, 32(12), 2048–2060, 1997.
33. Shanan, H., Retz, G., Mulvaney, K. et al. A 2.4 GHz 2 Mb/s versatile PLL-based transmitter using digital pre-emphasis and auto calibration in 0.18 μm CMOS for WPAN. *IEEE International Solid-State Circuits Conference (ISSCC) Digest of Technical Papers*, San Francisco, CA, February 2009, pp. 420–421.
34. Durdodt, C., Friedrich, M., Grewing, C. et al. A low-IF RX two-point ΣΔ-modulation TX CMOS single-chip Bluetooth solution. *IEEE Trans. Microw. Theory Tech.*, 49(9), 1531–1537, 2001.
35. Lee, S., Lee, J., Park, H. et al. Self-calibrated two-point Delta-Sigma modulation technique for RF transmitters. *IEEE Trans. Microw. Theory Tech.*, 58(7), 1748–1757, 2010.
36. Youssef, M., Zolfaghari, A., Mohammadi, B. et al. A low-power GSM/EDGE/WCDMA polar transmitter in 65-nm CMOS. *IEEE J. Solid-State Circuits*, 46(12), 3061–3074, 2011.
37. Marzin, G., Levantino, S., Samori, C. et al. A 20 Mb/s phase modulator based on a 3.6 GHz digital PLL with −36 dB EVM at 5 mW power. *IEEE J. Solid-State Circuits*, 47(12), 2974–2988, 2012.
38. Xu, N., Rhee, W., Wang, Z. A hybrid loop two-point modulator without DCO nonlinearity calibration by utilizing 1-bit high-pass modulation. *IEEE J. Solid-State Circuits*, 2014.

11

Design-for-Test Methods for Mixed-Signal Systems

Mani Soma

CONTENTS

11.1 Introduction

Technology scaling in the last 15 years has led to a very high level of integration of numerous functions on a single chip: regular digital functions previously included in large-scale VLSI systems and mixed-signal functions recently incorporated to permit single-chip implementations increasingly required in computing and communication applications. The high level of integration leads directly to the issues in chip testing, especially given that the number of I/O has scaled very slowly while the number of devices on chip has increased much faster. Design-for-test (DFT) methods offer an alternative to reduce test cost, both at manufacturing time and also in the field. Analog/mixed-signal testing, as distinguished from digital system testing, focuses on functional and parametric tests since the lack of standard fault models and attendant measures (e.g., fault coverage) makes it impossible to perform convincing manufacturing tests as commonly accepted in digital system testing. Moreover, since device and circuit models for subnanometer analog designs are not well developed for advanced technologies, the assumption that the circuits work as designed is not well accepted, thus an important goal of analog/mixed-signal testing is to verify that the circuits work within specifications. In this sense, analog/mixed-signal testing should more properly be referred to as functional tests or functional verifications. Mixed-signal DFT thus should be more properly referred to as mixed-signal design-for-functional-test or design-for-measurement. However, given the popular DFT terminology, we will continue to use mixed-signal DFT in the descriptions of DFT methods in this chapter.

Digital DFT methods are relatively well understood, ranging from the popular scan techniques at chip and board levels to built-in self-test (BIST) techniques commonly employed in large memories. Mixed-signal DFT methods, on the other hand, are still under active development despite publications dating back to the late 1980s. This chapter reviews these existing mixed-signal DFT methods using several systematic frameworks and presents guidelines and practices to assist designers and test engineers in selecting the most appropriate methods for their systems.

11.2 Mixed-Signal DFT Classifications

Given a wide range of analog and mixed-signal functions and the essentially infinite analog voltage levels, it is difficult to create a generic DFT method, for example, similar to digital scan, that will work in all cases. It is possible, however, to consider frameworks that permit systematic reviews of analog and mixed-signal DFT methods.

Testing a circuit or system always involves some capabilities to examine the inputs and outputs of the circuit or system under test. Structurally, these I/Os can be accessed directly via test points or via propagating their waveforms to controllable or observable nodes. We will use "access-based DFT methods" to refer to all methods that place test points at circuit or system I/Os, and "reconfiguration-based DFT methods" to refer to all methods that seek to propagate these I/O waveforms, not just simply connecting them to buses, to controllable or observable nodes.

From the function implementation perspectives, a circuit could be all-analog, for example, an amplifier, or mixed-signal, for example, a phase-locked loop. An all-analog circuit is usually a small macrocircuit, which makes it more difficult to justify the incorporation of DFT methods without incurring a significant overhead. A mixed-signal circuit is usually more complex in terms of device counts, thus providing more flexibility in terms of possible DFT methods. DFT additions usually consist of both digital and analog circuits, thus any DFT-augmented analog circuit is mixed signal. From the function implementation perspectives, we will focus on circuit blocks with well-defined functions (e.g., oscillators). We intentionally exclude a major mixed-signal block, phase-locked loop (PLL), since PLL test methods deserve a chapter in themselves and due to its very high operating frequencies, very few DFT techniques have been developed and employed satisfactorily. Most of these techniques have more in common with digital test and DFT methods than with mixed-signal DFT.

The classification described above is intended to enhance clarity in DFT descriptions and does not imply that DFT methods in various classifications are mutually exclusive. An access-based DFT method could cover a function such as an analog filter, and a function-specific DFT method for an analog filter could well be implemented via I/O access.

Analog and mixed-signal designers, as distinguished from test researchers, also contribute significant advances in analog DFT methods, and in fact, their contributions are likely much more significant since they design the circuits and can easily incorporate their DFT methods, usually presented under various names such as design-for-calibration, parasitic-insensitive designs, etc. We will briefly present these methods at the end of this chapter. Note that the dichotomy between designer-created methods and test-researcher-created methods still exists and is now a focus of several recent conferences [1,2].

11.3 Mixed-Signal DFT Requirements and Metrics

While technology scaling and system complexity have led to developments in mixed-signal DFT methods to help reduce test cost and accelerate the product manufacturing schedule, any proposed method, be it from designers or test researchers, must meet many stringent requirements and metrics. To set the stage for the DFT descriptions in the next section, these requirements and metrics are discussed here, with emphasis only on the issues with major system impact since it is impossible to be exhaustive.

"Overhead" is a rather misleading term but since it is conventionally accepted, specific overhead considerations include the following:

1. *Space and layout.* This metric is probably the most understood and popular in all DFT methods, from digital to mixed signal. The metric

could actually be easily manipulated to show that the percent area required is small (e.g., smaller than 5%) depending on which system function area is used in the calculations. If the system function area is only the area of the block with DFT (e.g., a low-noise amplifier), the metric value could be high; if the system function area involves a larger block (e.g., the entire transceiver which includes the low-noise amplifier), and then the metric value would be much lower. The only guideline is "as low as possible." Options to use off-chip resources, for example, FPGA on load boards during the manufacturing test calibration, must be considered and implemented if possible.

2. *Additional parasitics.* This metric is particularly important for mixed-signal DFT methods. While adding a switch or a connection from the digital signal under test to a flipflop or a test bus does not really significantly impact the parasitics influencing this signal, the same DFT addition could have adverse effect on an analog signal. Leakage, noise, additional capacitive load, etc., are all important parasitics that must be considered when a DFT structure is connected to an analog signal. The fundamental difficulty involves the lack of understanding of parasitic models (e.g., noise models) especially in advanced technologies, thus compensation for these parasitics is impossible in many cases.

3. *Possible reduction or shift in signal frequency.* The additional capacitive and resistive load clearly would impact the frequency of the signal under test, and the question is whether this impact is significant enough to render the DFT method inapplicable.

4. *Device mismatches.* Analog and mixed-signal circuits are always carefully designed to match devices and layout components (e.g., wire lengths to devices in a differential pair). Thus any DFT method that, due to the additional circuit elements or wires, disturbs this matching would likely be rejected since it would introduce nonideal effects adversely influencing circuit performance.

5. *Power requirement.* With the popularity of mobile devices, system power requirement is becoming more stringent. Any mixed-signal DFT method that requires significant power would likely not be considered as a possibility. Off-chip resources as mentioned above would make a DFT method more amenable especially for manufacturing test. Note that the power metric could also be easily manipulated as described above for the space and layout area metric.

These metrics and requirements need to be kept in mind as background for the next presentations.

11.4 Mixed-Signal DFT Methods: Principles and Conceptual Designs

11.4.1 Structural DFT Methods

11.4.1.1 Access-Based Methods

Access-based methods seek to place test points to examine signals deemed critical to designers and test engineers in verifying system functions. Circuit I/Os are obvious test point candidates but in many cases (e.g., transceivers), it is desirable to observe circuit internal signals as well. The fundamental trade-offs in test point placement are as follows:

  a. *Loading of the circuit node.* Any test point, even a simple wire connecting the signal to a chip I/O pad, adds additional capacitive and resistive load to the circuit under test. This additional load and possible leakage current must be taken into account during the design phase, not added on as an afterthought.

  b. *Number of test points.* Once a test point is selected, the issue is how to bring its signal waveform to a chip I/O or another observable point on chip. This is a nontrivial problem and due to the limited I/O and observable points for test purposes, the number of test points must be limited as well.

A systematic method and software tool to select test points based on the classical controllability and observability measures [3] were developed and applied to an industry power electronics case study. While the definitions of analog controllability and observability are still not standardized, the test points selected by this tool agree with the test points selected by the designers, thus validating the concepts and the algorithms. Another test point selection method based on graph theoretic measures [4] was developed for analog fault diagnosis of unpowered circuit boards, with encouraging results: access to 50% of the nodes is sufficient for diagnosis using multitone test stimuli.

Once the test points are selected by whichever methods, the overhead in chip I/O pins to observe these test points could be reduced by multiplexing test points. The fundamental considerations in this case involve the additional loading on the nodes due to the multiplexer circuits. Moreover, analog multiplexers themselves also affect the waveforms to be transmitted from the selected test points to an observable node. In its simplest model, an analog multiplexer is an MOS transistor biased to behave as a resistor, thus the amplitude, frequency, and other signal characteristics of the waveform transmitted via this resistor are affected by the multiplexer resistance and the capacitive load of the observable node. Compensation and calibration must be incorporated, usually via software on a test system after the measurement of the waveform, to ensure that the observed waveform truly reflects the waveform at the test point. This requirement, coupled with many others, in turn, leads to design-for-calibration and design-for-self-healing methods that will be discussed at the end of this chapter.

Returning to the DFT methods based on test point access, the most well-known access-based methods are the IEEE Mixed-signal Test Bus Standard 1149.4 [5]. The standard is intended for board-level and system-level applications and has several essential components as follows:

a. An I/O pad design that incorporates two analog buses AB1 and AB2, facilities for connecting a chip pin to either logic 1 or 0 to be used for regular interconnect test, digitizer that acts as a programmable 1-bit digitizer, and a core-disconnect switch to activate the test bus structure during interconnect test or analog measurement (Figure 11.1). Note that the switches in Figure 11.1 do not imply simple MOS switches but could be tri-state buffers or other circuits that function like switches but have much smaller effects on the signal paths.

b. A test bus controller identical to the IEEE Std. 1149.1 TAP controller, with additional instructions for analog test purposes.

c. A set of additional instruction for analog test purposes.

d. A test bus structure on chips to reduce noise introduced by these buses.

e. A possible extension to use the board-level test bus to examine waveforms of signals within a chip.

Since IEEE Std. 1149.4 has been well described [5], we will only mention that its applications have been limited by several factors:

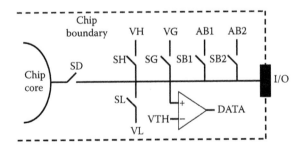

FIGURE 11.1
Input–output pad design based on IEEE Std. 1149.4. (From Osseiran, A. (ed.), *Analog and Mixed-Signal Boundary-Scan: A Guide to the IEEE 1149.4 Test Standard*, Kluwer Academic Publishers, 1999. With permission.)

a. *Loading issue on analog/mixed-signal nodes at each chip I/O.* This loading also could limit the frequency range where the standard is applicable.

b. *Lack of tools to automatically incorporate the standard into chip-level designs.* While these tools are available for IEEE Std. 1149.1, similar tools still do not exist for IEEE Std. 1149.4.

c. *Lack of tools to simulate the effects of the test buses on the signals observed.* The modeling of the test bus structures is not trivial, and the lack of these tools has impeded the application of the standard.

d. Note that the issue above regarding multiplexer resistance is exacerbated in the I/O pad design since there are MOS-based switches inserted in the signal path as part of the standard. There are other possible design methods that employ controllable tri-state buffers to avoid the use of MOS-based switches but this demands that designers have strong analog design background to design and layout these structures.

11.4.1.2 Current-Based Analog Scan

Scan methods form the cornerstone for many digital DFT techniques, so it is natural to enquire if a similar scan could be developed for analog and mixed-signal applications. Digital DFT relies on latches or flipflops in scanning signals, and these circuits are capable of restoring logic levels to 0 or 1 as appropriate in a long scan chain. There is no restoring circuit equivalent for the analog case. A unity-gain amplifier is too large to be used as an element for voltage scanning. Current-based analog scan [6,7] demonstrated the design and implementation of a scan path, whose fundamental element is analog current mirror (Figure 11.2a), which is much smaller in size and consumes much less power. A voltage-to-current converter, which may be implemented with seven MOS transistors, is required to sample each node voltage and convert it to a current to be scanned. The coupled current mirror performs a similar function to the master/slave latch structure of a digital flipflop, with the fundamental difference that the mirror gain, even after compensation due to mismatches and process variations, is never exactly equal to unity. Figure 11.2b [7] shows the performance of an experimental current mirror, with unity gain and error smaller than 0.1% over the entire current range. A scan path with 16-bit shift register (32 current mirrors) has been demonstrated for continuous-time analog design [7], which given the small size of many analog circuits to be tested, might be sufficient. While the results of analog scanning are encouraging in experimental case studies, the limitations in terms of space and power overhead and inherent mismatches are considerable and demand, just as in the IEEE 1149.4 case above, strong design skills in its implementation.

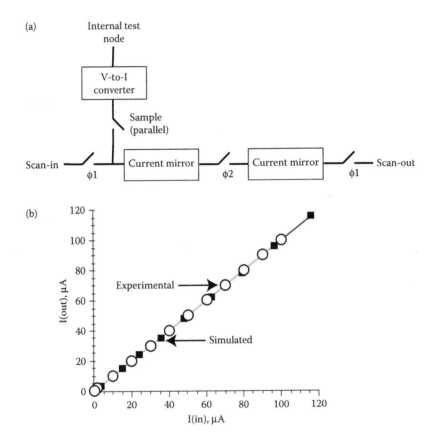

FIGURE 11.2
Current-based analog scan. (a) Scan cell design. (b) Current-mirror performance. (Soma, M. et al. *Proceedings of IEEE International Test Conference*, Washington, DC, November 3–6. © 1997. IEEE.)

Note that for sampled-data analog and mixed-signal circuits, there are much better methods for scanning signals than the current-based analog scan. These methods are presented in the next section, employing reconfiguration of existing circuits to reduce overhead and test pin count.

11.4.1.3 Voltage-Based Analog Scan

Voltage-based analog scan is a direct mapping of digital scan methods using analog primitives to implement the scan cell and scan chain. A structure for analog circuit fault diagnosis published as a BIST method [8] is an example of such implementation. Figure 11.3 shows the topology of the basic design: sampling switches connected to the test points within a circuit under test, a unity-gain buffer connecting this test point analog voltage into the scan chain, the basic master–slave analog scan cell built with unity-gain amplifier.

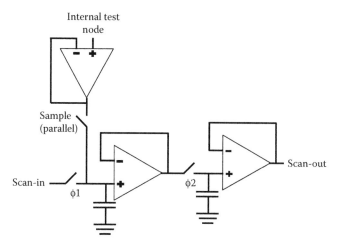

FIGURE 11.3
Voltage-based analog scan: Scan cell design. (Wey, C. L. *IEEE Trans. Instr. Meas.*, 39(3), 517–521.
© 1990. IEEE.)

The structure was demonstrated experimentally for a six-stage scan path. The method shares the same advantages and disadvantages as the current-based analog scan method described earlier, notably the nonideal properties of the analog scan cell, the additional power required, the layout size especially involving capacitance values to hold analog signals, etc.

11.4.1.4 Reconfiguration-Based Methods

To a large extent, all reconfiguration-based methods involve the modifications of the original circuits to make them more testable; thus these methods are more invasive although they do not require test point placement as in the access-based methods. For active analog filters, with the assumption that the operational amplifiers are fault free (which can be verified during the test procedures implemented by this method), the passive networks at the input, output, and feedback loop of each filter stage could be systematically transformed by the addition of digital switches [9] to "scan" signals through the filter structure with little impact on signal frequency. Figure 11.4 shows an example of a filter design both in the normal operating mode and in the reconfiguration test mode, with the three additional switches labeled as Tn. This method has been extended to switched-capacitor filters [10] where no switches need to be added, thus lowering the overhead and performance impact.

Another filter DFT method employing the same paradigm of reconfiguration for scanning exploits the fact that many analog active filters are designed using the same fundamental biquadratic section (biquad). Cascading these universal biquad structures (UBSs) permits the realization of many analog

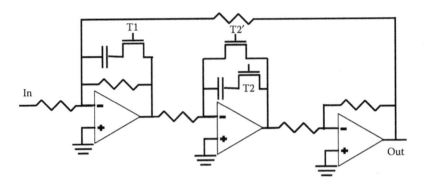

FIGURE 11.4
Reconfiguration-based DFT method for analog active filters. (Soma, M. *Proceedings of IEEE International Test Conference*, Washington, DC, September 10–14. © 1990. IEEE.)

filtering functions. During testing, as controlled by the test mode signal T (Figure 11.5), each UBS or a series of UBS may be reconfigured into simple gain stages via programming of the biquad coefficients [11]. Fault detection within one single UBS or a series of UBSs may be accomplished by placing a monitoring circuit to compare the output of the reconfigured filter under test and a programmable gain (K) circuit. The theoretical foundation of this method is the Laplace transfer function for continuous-time filters and the z-transform for discrete-time filters. To alter a UBS from a filter stage to a gain stage requires pole-zero cancellation, which is never exact; thus the tolerance must be built in as shown in the error generation circuit at the output of the filter in Figure 11.5. Note that this inexact cancellation may also be used for fault detection as well [11]. The method has been experimentally demonstrated using a second-order analog filter. As in many DFT methods described in this chapter, the overhead involves the programming circuits to alter the capacitor values, the capacitors themselves with sufficient built-in programmable value range, T-monitor circuit, power, layout area for additional wirings, etc.

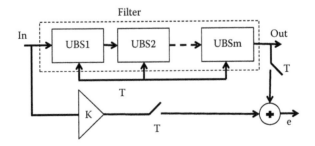

FIGURE 11.5
DFT method for universal-biquad active analog filters. (Vazquez, D., Rueda, A., Huertas, J. L. *Proceedings of IEEE VLSI Test Symposium*, Cherry Hill, NJ, April 25–28. © 1994. IEEE.)

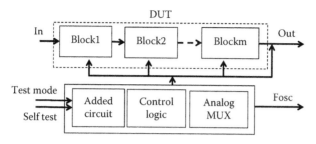

FIGURE 11.6
Simplified test structure of oscillation-test strategy. (Arabi, K., Kaminska, B. *Proceedings of IEEE VLSI Test Symposium*, Princeton, NJ, April 28–May 1. © 1996. IEEE.)

The filter applications so far employ reconfiguration to scan signal waveforms through the filter stages. Reconfiguration has been used in BIST methods, of which oscillation-based test [12] is an example. In this instance, a normal analog circuit is reconfigured into an oscillator in the test mode (Figure 11.6) and the frequency of oscillation is an indicator of the original circuit performance and fault susceptibility. This method avoids the fundamental difficulty of measuring an analog signal amplitude and relies on frequency measurement, which is easier and more robust since there have been many published works on zero-crossing measurements approximating frequency measurements. The additional overhead depends on the circuit under test and could be optimized to test operational amplifiers, data converters, etc.

An extension of oscillation-based test combines the method with the universal-biquad method above to provide a practical scheme to test integrated filters [13]. The fundamental block whose transfer function is $H(z)$ is modified (Figure 11.7), both internally and via the feedback path, to introduce oscillation during the test mode. Implementations of this method in a switched-capacitor biquad filter and a dual-tone multifrequency (DTMF) receiver show that the method could help detect both manufacturing faults and large parametric deviations of the original circuits under test when both frequency and amplitude of the oscillation are measured.

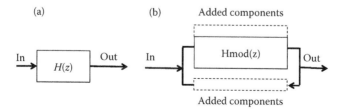

FIGURE 11.7
Oscillation-based test applied to biquad filter structures. (a) Basic stage. (b) Modified stage for oscillation. (Huertas, G. et al. *IEEE Design and Test of Computers*, pp. 64–82. © 2002. IEEE.)

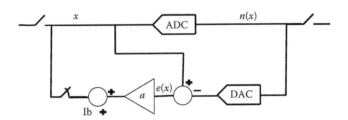

FIGURE 11.8
Complex chaotic oscillation DFT method applied to data converters. (Callegari, S. *Proceedings of IEEE International Symposium Circuits Systems,* Seattle, WA, May 18–21. © 2008. IEEE.)

An interesting variation of this method is to induce complex chaotic oscillation [14,15] where only time-domain measurements are needed to characterize the performance of the original analog circuits and the possible faults that could affect this performance. Complex chaotic oscillation method may be applied to data converters [14] and filters [15]. Figure 11.8 illustrates the modification of an original analog-to-digital converter (ADC) to introduce complex chaotic oscillation. The feedback digital-to-analog converter (DAC) together with the linear feedback function, $f(x) = (x - b)/a$ where x is the ADC input, b is the feedback offset, and a is the feedback gain, creates a one-dimensional dynamical system. With proper values of a and b to ensure that there is no equilibrium points or stable limit cycles, the entire circuit in Figure 11.8 will become a chaotic oscillator. Parameters such as gain, integral, and differential nonlinearities may be determined based on the measurements of this oscillation. A similar application to second-order bandpass filters [15] introduces jumps in the state variable outputs of the two-stage filters, thus creating a phase diagram where the chaotic oscillation path may be controlled. Parametric measurements of the original filter quality factor and center frequency could be deduced from the oscillation waveform and parameters.

11.4.2 Function-Oriented DFT Methods

Another possible taxonomy of DFT methods for mixed-signal circuits focuses on circuit functions. These methods tend to add on circuits at the I/O of the functional blocks, with a few exceptions in inserting test devices within the circuit design, and sometimes extend to BIST methods with incorporation of schemes to activate the tests, measure, and decide on Pass or Fail. We will cover mostly DFT methods and intentionally exclude BIST-intensive methods since they deserve a separate and complete exploration.

11.4.2.1 Operational Amplifiers and Analog Filters

Very few circuits remain all-analog: basic operational amplifiers, continuous-time active filters (e.g., see Section 3.1.4), and basic amplifiers. These circuits

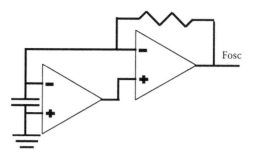

FIGURE 11.9
Operational amplifier DFT for two amplifiers in series. (Arabi K., Kaminska, B. *IEEE J. Solid-State Circuits*, 33(4), 573–581. © 1998. IEEE.)

remain small in terms of device counts and sizes, making it less amenable to DFT incorporation since the overhead tends to be higher. However, operational amplifier is a fundamental functional block of analog and mixed-signal circuits, thus becomes a focus of many DFT methods. A representative method [16] converts an operational amplifier or a series of operational amplifiers into an oscillator (Figure 11.9) and extends the oscillation-based test method (see Section 4.1.4), where the operational amplifier's parameters can be inferred from the measurements of the oscillator characteristics.

Note that this vector-less DFT method may also be classified as BIST under an appropriate controller.

Analog filters, due to their popularity in signal processors, also receive intense attention for DFT incorporation. Besides the switched-capacitor methods in Section 11.4.1.4, another method seeks to measure capacitor ratio [17], which is a fundamental parameter in any switched-capacitor implementation. An accurate capacitor ratio implies that the filter meets performance specifications without detailed and costly functional measurements. Figure 11.10 shows the added ADC to convert the capacitor ratio C_a/C_b to a digital value easily scanned out to determine whether the ratio meets the original design specifications. The accuracy of this scheme could be extended by longer test time, and a 12–15-bit resolution of the measured ratio is possible. While the overhead seems high due to the additional converter, the same converter may be used to measure all capacitor ratios in a switched-capacitor filter structure, thus making it more acceptable as an implementation method.

11.4.2.2 Fully Differential Functions

Analog circuits, by nature, tend to be differential to counteract the errors due to process variations, device imperfections, and other manufacturing uncertainties. A general DFT method for fully differential circuits, using switched-capacitor filters as a case study, places checkers at the output and several other locations (e.g., filter inputs) within the filter structure [18]. The layout and placement of the checkers, as in all cases of analog DFT and

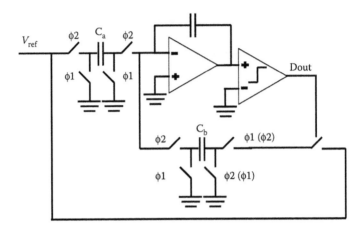

FIGURE 11.10
Analog-to-digital converter to measure C_a/C_b in a switched-capacitor filter. (Vinnakota, B., Harjani, R. *IEEE Trans. CAD Integrated Circuits Syst.*, 19(7), 789–798. © 2000. IEEE.)

especially in fully differential circuits, are critical so as not to introduce additional measurement errors due to the DFT structures. For a fabricated fifth-order Butterworth filter, an input checker (Figure 11.11) monitors the input common-mode voltage and generates a local measurement of each operational amplifier within the filter structure. The experimental design shows

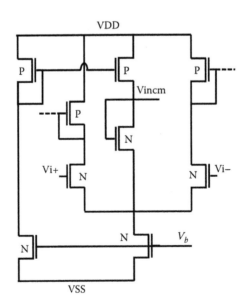

FIGURE 11.11
Input checker design for fully differential analog circuit DFT. (Stessman, N. J., Vinnakota, B., Harjani, R. *IEEE J. Solid-State Circuits*, 31(10), 1526–1534. © 1996. IEEE.)

that the method performs quite well in analog fault coverage and parametric fault detection, meeting the input-checker performance requirement to be used as part of the DFT structure.

11.4.2.3 Automatic Gain Control and Related Functions

A comprehensive DFT method [19] has been presented for automatic gain control (AGC) circuits, widely used in many signal processing and system applications. The method and its inherent design focusing on controllability and observability are extendable to other classes of analog and mixed-signal circuits since the DFT structures are added at circuit I/Os, without a requirement for detailed knowledge of the internal design. The method also permits a wide range of functional tests, for example, AC performance in all AGC gain parameters, DC offsets, transient response to load, and transient response to input signals at various frequencies. The method adds a sampling circuit with a 6-bit ADC at the AGC output and an augmented controller to conduct various test functions. Figure 11.12 shows a possible implementation with an ADC and controller. The experimental implementation shows good performance in AGC gain step size tests, AGC ramp test, and parametric fault coverage figures. Note that while the added structures seem quite significant in terms of devices, most can be on the test system and the loadboard, not on the AGC chip itself.

11.4.2.4 Oscillator and Related Functions

Oscillators represent a large class of analog and mixed-signal circuits fundamentally used in many system applications to generate clocks and

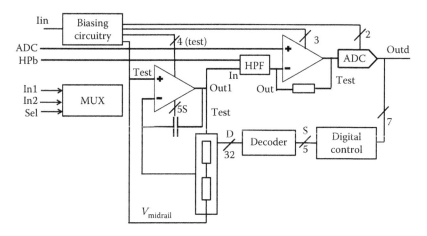

FIGURE 11.12
AGC test evaluation DFT structure. (Lechner, A. et al. *Proceedings of IEEE International Mixed-Signal Test Workshop*, Den Haag, The Netherlands, June 8–11. © 1998. IEEE.)

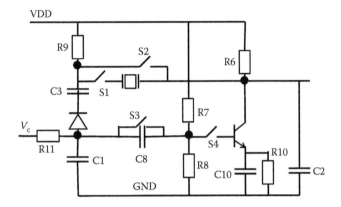

FIGURE 11.13
DFT switches for a Pierce crystal oscillator. (From Santo-Zarnik, M., Novak, F., Macek, S. *J. Electronic Testing: Theory Appl.*, 11, 109–117, 1997. With permission.)

control waveforms. A DFT method that adds 4 switches into a Pierce crystal oscillator circuit (Figure 11.13) has been implemented [20] to permit the observation of various waveforms internal to the oscillator and to infer if the oscillator will work as per the design. These switches are used to open or short circuit paths during the normal mode and two possible test modes to measure resistances, resistance ratios, capacitances, and capacitance ratios. The sizes of the DFT switches were analyzed and appropriately chosen to reduce their impact on the normal mode of the oscillator. The DFT design has been demonstrated on a 16.384 MHz crystal oscillator, with full temperature testing (even though temperature compensation was not included in the DFT method), fault-oriented tests, and regular performance tests and measurements.

11.4.3 Designer-Created DFT Methods

The many methods reviewed in the earlier sections represent contributions from researchers with predominantly test perspectives. While these methods are DFT with the emphasis of course on design and have been proven in test chip implementations, their incorporation into real analog and mixed-signal systems remains a challenge. Analog and mixed-signal tests, reflecting similar design methodologies in the early 1970s for digital systems, usually are not considered until the designs and layouts have been optimized for performance. Any DFT incorporation would disturb this optimization and make it less appealing to designers. In fact, traditional analog and mixed-signal designers have not been idle at all with respect to test issues of their circuits. Many DFT methods actually implemented on real products come from analog and mixed-signal designers, not from test researchers and designers. These design-oriented methods are usually referred to as designs-for-calibration, parasitic-aware designs, etc. We will review two examples of these methods in this section.

The predominance of high-performance ADC leads directly to many design-for-calibration methods to extend the number of bits in the presence of process variations and mismatches. A 10-bit 1-MHz ADC design [21] shows how calibration is used to remove capacitance mismatch effects and correct for gain errors. Figure 11.14 shows the conceptual calibration scheme implemented at the ADC output, with backend calibration logic. Note the essential feature that the calibration logic is implemented mostly off chip, thus only adds minimal overhead in terms of layout, power, additional noise, etc. The fabricated chip, tested with the calibration logic implemented mostly on an FPGA, meets all performance specifications after the calibration algorithm is executed.

A digitally reconfigurable with auto-amplitude calibration of a wide-turning range VCO has been designed with reduced phase noise [22], again with calibration methods built in by designers, not by test engineers. This circuit, with operating frequencies between 3 and 6 GHz, is also insensitive to process variations, temperature, and operating power supply. The auto-amplitude calibration circuit in Figure 11.15 uses digital circuits (including a finite-state machine) and very few analog components. Noise injection into the analog circuits is well controlled and the calibration is accomplished in 3–16 cycles.

To a large extent, these methods should be referred to as design-for-calibration. In fact, this phrase "design-for-calibration" could even be considered as redundant since designers view calibration circuit and methods as part of the design process, not as part of the test process.

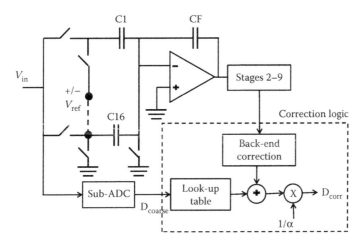

FIGURE 11.14
Conceptual illustration of digital correction at the ADC output. (Sahoo, B. D., Razavi, B. *IEEE J. Solid-State Circuits*, 48(6), 1442–1452. © 2013. IEEE.)

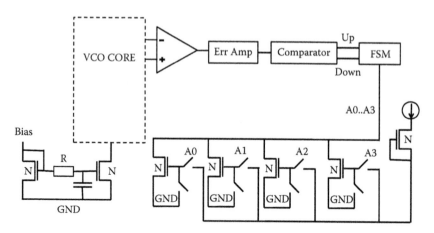

FIGURE 11.15
Automatic amplitude calibration loop of a VCO. (Zhang, Z. et al. *Proceedings of 10th IEEE International Conference of Solid-State and Integrated Circuit Technology,* Shanghai. © 2010, pp. 542–544. IEEE.)

11.5 Guidelines and Practices

While the description of various mixed-signal DFT methods was presented earlier in three general frameworks, their implementation given a specific system function is not as clear-cut. The access method, for example, must take into account whether the signal being accessed has only low-frequency contents or sufficient high-frequency components that should not be ignored in testing. Most designs and existing products incorporating analog and mixed-signal subsystems use functional tests to verify that the mixed-signal subsystems work properly, and designers do not view mixed-signal DFT as a requirement. Note the distinction between design-for-calibration, which is almost always incorporated as part of the design process for analog and mixed-signal subsystems, and DFT, which in the sense of the many methods reviewed above, is not frequently implemented due to the various issues and system performance impact as noted in Section 11.3. Defect-oriented analog and mixed-signal test and DFT methods are not well developed for many reasons, one of which is that defect-oriented analog fault models have not been well accepted. This fault model topic is rather outside the scope of this chapter so will not be reviewed in depth herein.

The lack of tools to implement, simulate, and verify the performance of the mixed-signal DFT circuits makes it costly and time-consuming for designers to incorporate these methods. Tool development has proceeded very slowly: after almost 30 years, analog and mixed-signal DFT tools are still not available or proven from CAD tool vendors. Designers' experience, just as in normal analog and mixed-signal designs, becomes the paramount factor in determining which calibration or functional test enhancements to be used for a specific mixed-signal function, thus minimizing the roles of mixed-signal test researchers and engineers.

The most appropriate guidelines for selecting and implementing mixed-signal DFT methods need to consider trade-offs between system performance specifications (including power, noise, frequency, etc.), cost versus benefits of mixed-signal DFT, whether functional tests at the system level are adequate to ensure mixed-signal performance without DFT. Note that commercial products using state-of-the-art technologies, especially mobile systems, have short lifetime (a few years at most), and thus the long-term benefits for mixed-signal DFTs could not be justified based on field service, repair, and maintenance. These products are often discarded by users even when they are still perfectly functioning. Thus while there are theoretical drivers to create new mixed-signal DFT methods, economic drivers are frequently much weaker.

11.6 Conclusion

Mixed-signal DFT methods, under intense development in the mid 1980s and early 1990s, have settled down to a more steady pace, consisting of both well-structured techniques and ad hoc techniques as reviewed in this chapter. Designers' contributions, in the sense of design-for-calibration, have dominated the realistic incorporation of digitally assisted structures to ensure that the circuits meet specifications. System functional tests in many cases are sufficient to guarantee mixed-signal circuit performance without a significant investment in DFT techniques. While new developments are always sought after and standardized structured methods, equivalent to digital scan, are still being sought, test researchers and designers need to be aware of these constraints to ensure that their creations do satisfy the currently unmet requirements for mixed-signal DFT.

References

1. Soma, M., Fiez, T., and Li, M. P. Designers' and test researchers' roles in analog DFT. *Proceedings of 32nd IEEE VLSI Test Symposium*, Napa, CA, April 14–16, 2014, pp. 196.
2. Fiez, T., Kumar, S., Paramesh, J. and Sunter, S. Analog-design-for-test: What's the real story? *Proceedings of IEEE International Test Conference*, Seattle, WA, October 21–23, 2014.
3. Huynh, S., Kim, S., Soma, M., Zhang, J. Automatic analog test signal generation using multi-frequency analysis. *IEEE Trans. CAS-II*, 46(5), 565–577, 1999.

4. Huang J. L., Cheng, K. T. Test point selection for analog fault diagnosis of unpowered circuit boards. *IEEE Trans. CAS-II*, 47(10), 977–987, 2000.

5. Osseiran, A. (ed.), *Analog and Mixed-Signal Boundary-Scan: A Guide to the IEEE 1149.4 Test Standard*. Kluwer Academic Publishers, 1999.

6. Soma, M. Structure and concepts for current-based analog scan. *Proceedings of IEEE Custom Integrated Circuits Conference*, Santa Clara, CA, May 1–4, 1995, pp. 517–520.

7. Soma, M., Bocek, T. M., Vu, T. D., Moffatt, J. D. Experimental results for current-based analog scan. *Proceedings of IEEE International Test Conference*, Washington, DC, November 3–6, 1997, pp. 768–775.

8. Wey, C. L. Built-in Self-Test (BIST) structure for analog circuit fault diagnosis. *IEEE Trans. Instr. Meas.*, 39(3), 517–521, 1990.

9. Soma, M. A design-for-test methodology for active analog filters. *Proceedings of IEEE International Test Conference*, Washington, DC, September 10–14, 1990, pp. 183–192.

10. Soma, M., Kolarik, V. A design-for-test technique for switched-capacitor filters. *Proceedings of IEEE VLSI Test Symposium*, Cherry Hill, NJ, April 25–28, 1994, pp. 42–47.

11. Vazquez, D., Rueda, A., Huertas, J. L. A new strategy for testing analog filters. *Proceedings of IEEE VLSI Test Symposium*, Cherry Hill, NJ, April 25–28, 1994, pp. 36-41.

12. Arabi, K., Kaminska, B. Oscillation-test strategy for analog and mixed-signal integrated circuits. *Proceedings of IEEE VLSI Test Symposium*, Princeton, NK, April 28–May 1, 1996, pp. 476–482.

13. Huertas, G., Vazquez, D., Peralias, E. J., Rueda, A., Huertas, J. L. Practical oscillation-based test of integrated filters. *IEEE Design Test Comp.*, November–December 2002, pp. 64–82.

14. Callegari, S. Introducing complex oscillation based test: An application example targeting analog to digital converters. *Proceedings of IEEE International Symposium Circuits Systems*, Seattle, WA, May 18–21, 2008, pp. 320–323.

15. Callegari, S., Pareschi, F., Setti, G., Soma, M. Complex oscillation based test and its application to analog filters. *IEEE Trans-CAS I*, 57(5) 956–969, 2010.

16. Arabi, K., Kaminska, B. Design for testability of embedded integrated operational amplifiers. *IEEE J. Solid-State Circuits*, 33(4), 573–581, 1998.

17. Vinnakota, B., Harjani, R. DFT for digital detection of analog parametric faults in SC filters. *IEEE Trans. CAD Integr. Circuits Syst.*, 19(7), 789–798, 2000.

18. Stessman, N. J., Vinnakota, B., Harjani, R. System-level design for test of fully differential analog circuits. *IEEE J. Solid-State Circuits*, 31(10), 1526–1534, 1996.

19. Lechner, A., Richardson, A., Hermes, B., Perkin, A., Zwolinski, M. Design for testability strategies for a high performance automatic gain control circuit. *Proceedings of IEEE Int. Mixed-signal Test Workshop*, Den Haag, The Netherlands, June 8–11, 1998, pp. 376–385.

20. Santo-Zarnik, M., Novak, F., Macek, S. Design for test of crystal oscillators: A case study. *J. Electronic Testing: Theory Applications*, 11(2), 109–117, 1997.

21. Sahoo B. D., Razavi, B. A 10-b 1-GHz 30-mW CMOS ADC. *IEEE J. Solid-State Circuits*, 48(6), 1442–1452, 2013.

22. Zhang, Z., Li, J., Sun, Y., Rhee, W., Wang, Z. A digitally reconfigurable auto amplitude calibration method for wide tuning range VCO design. *Proceedings of 10th IEEE International Conference on Solid-State Integrated Circuit Technology*, Shanghai, 2010, pp. 542–544.

12

Built-In Testing and Tuning of Mixed-Signal/RF Systems: Exploiting the Alternative Testing Paradigm

Abhijit Chatterjee and Jacob Abraham

CONTENTS

12.1 Introduction

The use of scaled complementary metal-oxide semiconductor (CMOS) technologies for high speed 10–100 GHz+ communication devices is posing daunting technological challenges due to the effects of manufacturing process variations on the performance of these devices [1]. Traditionally, negative feedback has been used in the design of analog circuits to provide resilience to process variability effects [2]. However, continuous feedback-based stabilization techniques result in design tradeoffs against available device bandwidth and are not adequate in the face of extreme process variations. In such situations, the use of system-level performance testing and tuning mechanisms is necessary for minimizing manufacturing yield loss.

In the past, designers have invented self-testing mechanisms that are tailored toward *specific* (critical) mixed-signal/radio frequency (RF) perfor-mance metrics of *specific* circuit architectures (transmitter, receiver, etc.). What is desired, however, is the ability to *test for multiple performance metrics concurrently* and trade them off against one another in an optimal manner, through performance tuning mechanisms, to satisfy system-level quality of service (QoS) guarantees. A key goal is to design the built-in-testing and tun-ing mechanisms in such a way that they can be exercised without the need to access external test instrumentation. Further, the methods employed must be scalable across different device types/circuit architectures and supported by CAD tools that enable *automation* of self-tuning design procedures (see Figure 12.1).

In this chapter, recent research advances are presented that allow low cost and rapid self-tuning of complex mixed-signal/RF systems and enable the development of design automation tools to support such activity across diverse performance metrics and circuit types. In the following, we first give an overview of standard mixed-signal/RF manufacturing test practice in industry and the difficulties involved in applying some of the key underly-ing principles to the problem of built-in test and tuning. Then, we discuss the alternative testing paradigm which forms the backbone of the built-in test and tuning mechanisms developed in this research. This is followed by a more detailed description of the core infrastructure and algorithms needed to perform built-in test and tuning, followed by test cases and a description of experimental results.

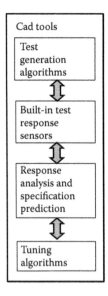

FIGURE 12.1
CAD support for test and tuning infrastructure.

12.2 Standard Testing Practice

To understand the alternative testing paradigm, we need to first revisit how mixed-signal/RF integrated circuits (ICs) are tested as per current industry practice. If a device under test (DUT) has N performance specifications, N tests are applied to the device sequentially. Each test is designed to extract a corresponding performance metric and involves instrument setup, stimulus application, a wait time to allow the device to respond and analysis/measurement of a performance specification (see Figure 12.2). While the actual test time per test may be small (few ms depending on the specifications of the DUT), the tester loadboard relays need to be reconfigured in between tests to allow instrument setup and test reconfiguration (typically tens of ms). So, for example, if 10 tests are conducted sequentially, the total test time can easily exceed 250 ms even when the test time per device is very small.

There are other issues also with the standard testing procedure of Figure 12.2 that make it difficult to directly apply such practice to built-in test and tuning of mixed-signal/RF devices. First, existing built-in test and tuning techniques are generally geared toward very specific performance metrics of the DUTs concerned. In the recent past, *digitally assisted* test and tuning techniques have been applied to a variety of mixed-signal/RF circuits. In this

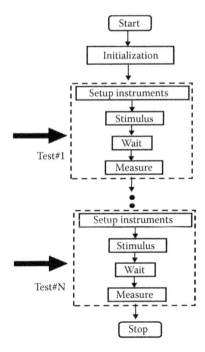

FIGURE 12.2
Standard testing procedure.

design paradigm, on-chip digital logic is used to intelligently compensate for loss of mixed-signal/RF performance due to process variations. In this context, there has been significant work in the area of digitally assisted tuning of analog circuits including PLLs, frequency synthesizers, and digital radio [3–6]. In Reference 7, a digitally assisted algorithm for testing the IIR spec of a Weaver image-rejection receiver is presented. In Reference 8, algorithms are used to perform digitally assisted effective number of bits (ENOB) test of pipelined data converters. In Reference 9, digitally assisted data converter architectures are discussed. As evident from the above discussion, most of this research focuses primarily on testing and tuning of *specific* (or few) analog/RF/analog to digital converter (ADC) specs. This requires the design of on-chip circuitry that is *specific to the targeted performance metrics* of the DUT. In general, the larger the number of specifications to be measured on-chip, the larger the amount of on-chip circuitry needed to support the associated test measurements. Solutions need to be found that are *scalable across a multitude of (complex) specifications* without incurring additional hardware cost.

A second major limiting factor in directly applying standard test methods to the problem of built-in test and tuning is that standard testing of high-speed devices requires the use of sophisticated test instruments for acquiring high-speed signals in the frequency and time domain. Such instruments (e.g., spectrum analyzers) are too complex to be ported to circuitry "on-chip" for testing and tuning purposes. Clearly, low-overhead alternatives to standard specification test practice are needed to enable the same.

12.3 The Alternate Testing Paradigm

Consider variations in the process parameter space P of Figure 12.3 that affect the specifications S of the DUT as well as a set of carefully design

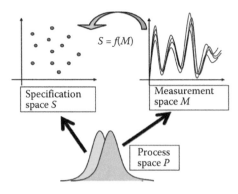

FIGURE 12.3
Basis of the alternate testing paradigm.

alternative measurements M made on the DUT as well. For low-cost/low-overhead test, this *alternate set of measurements* is determined such that the *test measurements are strongly correlated with variations in the test specification values* of the DUT under observed process variations. The resulting set of measurements defines the *measurement space M* of Figure 12.3. As an example, for RF devices, the measurements are made on the output of a sensor (e.g., envelope detector) connected to the RF device or a mixer-down converted version of the RF signal. Any deviation in the observed measurements from the expected, implies a corresponding deviation of the measured RF specifications of the DUT from the expected in the *specification space S*, due to the strong correlation between the measurements M and the specifications S of the DUT under perturbations in the process space P.

A key aspect of the alternative test method lies in the design of the test measurements M. Accordingly, the alternate test stimulus to be applied to the DUT is carefully optimized using stochastic or directed search methods to *maximize the correlation* between the specifications S and the alternate measurements M across a selected set of "critical" devices under given process variation statistics. Conceptually, such "critical" devices exhibit maximally uncorrelated responses to random stimulus for the stated process statistics and describe the different ways in which the DUT response to the applied alternate test can be affected by process variations. It is seen that unless the alternate test stimulus is optimized carefully based on the above criteria, the statistical correlation between the specifications S and measurements M of test devices cannot be guaranteed. Hence, it is very important to optimize the stimulus carefully to achieve good alternative test performance (ability to avoid false positives and false negatives with respect to standard specification test).

Once such an alternative test [11–20] is obtained, a nonlinear regression mapping model $S = f(M)$ using MARS [21] is developed using measurements on a "training set" of devices (hardware) that maps observed the DUT responses to the applied alternate test to their measured performance specification values. Such training on hardware devices is necessary to offset the inaccuracies associated with simulation models and to calibrate for test instrumentation nonidealities and measurement noise. During production test, the standard (expensive) tests are not applied. Instead the alternate test is applied to every manufactured DUT and its performance specifications are predicted using the regression model $S = f(M)$. Pass/fail decisions are subsequently made on the predicted specification values resulting in significant reductions in test time (the alternative test is typically less than a few millisecond and a *single* test is sufficient to predict *all* the test specifications of the DUT). The production test implementation of the above alternate test procedure is shown in Figure 12.4.

A key benefit of the alternate test approach above, is that *multiple DUT specifications* can be predicted from a *single data acquisition*. This leads to significantly reduced test time as compared to conventional specification-based testing methods. However, considerable effort is expended in developing the

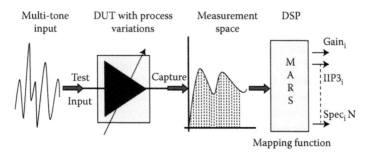

FIGURE 12.4
Alternate testing in production. (From Akbay, S., Chatterjee, A. Built-in test of RF components using feature extraction sensors. *Proceedings, VLSI Test Symposium*, CA: Palm Springs, pp. 243–248, May 2005. With permission.)

regression-based mapping $S = f(M)$. Specifically, the training set of devices for developing the mapping using supervised learning techniques must include devices across as many process corners as possible. In practice, a *defect filter* [22–24] is used to first determine if the DUT specifications can be predicted accurately from its response (the specifications of devices outside the performance domain of devices in the training set are not predicted accurately by the alternate test procedure). If not, then standard specification tests are applied to these "outlier" devices for pass/fail classification and the resulting data is used to "update" the supervised learner. In the presence of unanticipated process shifts, it is expected that the resulting outlier devices are "kicked-back" to standard testing procedures and the resulting information fed to the supervised learning algorithm. In such a scenario, the system "learns" about process shifts and adjusts appropriately over time. However, the learning process does not eliminate the use of standard test mechanisms.

12.3.1 Alternate Testing: Precision Op-Amp

The alternate test approach was first demonstrated via the FASTest test suite in References 25,26. This involved several steps:

Development of a fast transient stimulus: In the first step, a set of transient stimuli is developed that can be applied to several pins of the DUT simultaneously. The fast transient stimulus when applied to the DUT results in a response that has one or more components correlated to each of the DUT specifications.

Extraction of FASTest runtime transforms: The functional relationship between the DUT specification parameters and the DUT response to the fast transient stimulus is captured by a nonlinear transformation or mapping, referred to as the *runtime transform* given by $S = f(M)$ as described earlier. To build the transform, several devices are tested using the (current) standard test methods and using the FASTest method. The runtime transform coefficients are derived by using statistical regression techniques and noise reduction algorithms on the measured data.

Final production test: During final production testing, the fast transient test stimulus is applied to the DUT, and individual components corresponding to the DUT parameters are extracted from the captured DUT response and transformed into standard DUT parameters using the Runtime nonlinear transforms. The extracted DUT parameters are then used for pass/fail classification and for generating datalogs as in standard specification test.

The fast transient test configuration uses three stimulus inputs: VSET, VDD, and VSS. The stimulus for the three inputs is optimized to maximize the correlation between the response of the precision op-amp and its performance specifications of interest. Figure 12.5 shows the optimized alternate test stimulus used by FASTest and Figure 12.6 shows the response of the op-amp to the applied test. Figure 12.7 shows the predicted (vertical axis) versus measured (horizontal axis) offset voltage of the op-amp in mVolts across a large sample set of devices. Each dot in the graph corresponds to a unique device.

Table 12.1 shows test guardbands as a percentage of the test limits for standard as well as the alternate fast transient testing procedure for a range of test specifications of the op-amp. It is remarkable that the test guardbands for alternate test are in fact *superior* to that for standard test. This happens because the repeatability of the test measurements using alternate fast transient testing is superior to the repeatability of standard test measurements because of the fact that the performance metrics of the DUT are predicted from a large number of samples of the transient DUT response to the applied

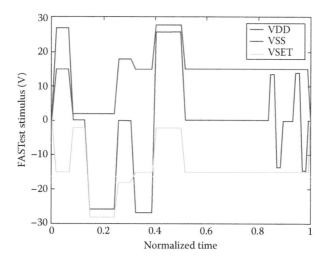

FIGURE 12.5
Optimized op-amp test stimulus. (From Voorakaranam, R. et al. *Proceedings, International Test Conference*, Charlotte, NC, pp. 1174–1181, September 2003. With permission.)

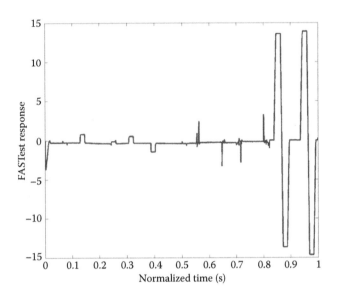

FIGURE 12.6
Op-amp response to fast alternate test. (From Voorakaranam, R. et al. *Proceedings, International Test Conference*, Charlotte, NC, pp. 1174–1181, September 2003. With permission.)

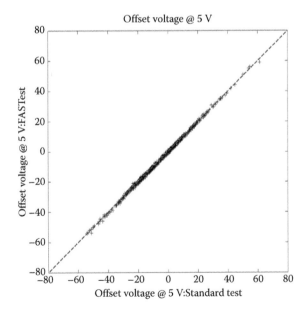

FIGURE 12.7
Predicted (*y*-axis) versus measured (*x*-axis) offset voltage. (From Voorakaranam, R. et al. *Proceedings, International Test Conference*, Charlotte, NC, pp. 1174–1181, September 2003. With permission.)

TABLE 12.1

Test Guardbands as a Percentage of the Test Limits for Standard and Alternate
Fast Transient Testing Procedure

Parameter	Standard Guardband (%)	FASTest Guardband (%)
Offset voltage at 5 V	12.6	11
Offset voltage at 15 V	14	10
PSRR	9	8
CMRR	10.6	11
Large signal gain	12.4	14
Positive swing at 2 K load	0.2	0.3
Negative swing at 2 K load	0.2	0.2
Positive bias current	2.7	7.6
Negative bias current	3.7	8.5

Source: From Voorakaranam, R. et al. Production deployment of a fast transient testing
methodology for analog circuits: Case study and results. *Proceedings, International
Test Conference*, Charlotte, NC, September 2003, pp. 1174–1181. With permission.

test stimulus of Figure 12.5 as opposed to a single DUT specification measurement (which can be prone to measurement errors).

12.3.2 Alternate Testing: Low Noise RF Amplifier

Figure 12.8 shows a second test case, that of an RF low noise amplifier (LNA)
[26]. The methodology uses modulated RF sources and baseband waveform
generators (AWGs) to stimulate the RF and baseband inputs to the DUT. The
transient response of the DUT to the stimuli is captured using RF receivers

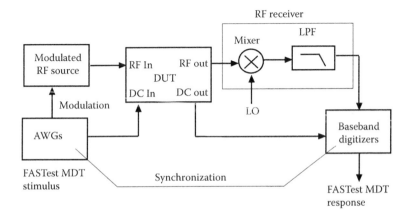

FIGURE 12.8
Test configuration for RF LNA. (From Cherubal, S. et al. *Proceedings, VLSI Design Conference*,
Bombay, India, pp. 1017–1022, January 2004. With permission.)

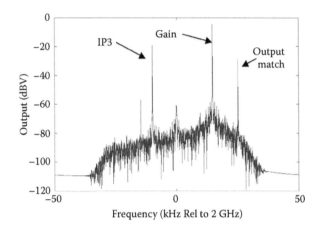

FIGURE 12.9
Response of LNA to applied stimulus. (From Cherubal, S. et al. *Proceedings, VLSI Design Conference*, Bombay, India, pp. 1017–1022, January 2004. With permission.)

and baseband digitizers. The stimuli required for alternate testing is carefully constructed such that the automatic test equipment (ATE) can accurately generate the ATE stimuli, and the resulting DUT response can easily be captured. Gain compression and intermodulation tests are replaced with a single test that consists of a three-tone modulated stimulus. The three tones are selected such that two tones of high amplitude are out of the bandwidth of the RF receiver, but the third order intermodulation product is within the RF receiver bandwidth. The frequency of the third tone is set to be within the RF receiver bandwidth.

The amplitudes of the three tones are set such that at the output of the DUT, the inter modulation product between the two high amplitude tones has approximately the same amplitude as the smaller amplitude tone within the RF receiver bandwidth. The response of the LNA to the applied stimulus in the frequency domain is shown in Figure 12.9. Table 12.2 shows the

TABLE 12.2

LNA Specifications Absolute Accuracy That Can Be Predicted Using the Alternate Test Methodology

Specification	Limit	Extraction Error (RMS)
Bias point	650 mV	3.0 mV
Noise figure	1.8 dB	0.029 dB
Gain	13.0 dB	0.033 dB
IIP3	10.0 dBm	0.097 dBm
Insertion loss	3.5 dB	0.014 dB

Source: From Cherubal, S. et al. Concurrent RF test using optimized modulated RF stimuli. *Proceedings, VLSI Design Conference*, Bombay, India, January 2004, pp. 1017–1022. With permission.

absolute accuracy with which each of several specifications of the LNA can be predicted using the alternate test methodology.

12.4 Alternate Built-In Test Using Sensors

Figure 12.10 shows a built-in test configuration for an orthogonal frequency-division multiplexing (OFDM) transmitter and receiver connected in loop-back mode. A carefully optimized test stimulus from the baseband is used to stimulate the I and Q inputs of the transmitter. An envelope detector connected to the output of the transmitter as a built-in test response "sensor" facilitates the built-in test procedure. The performances of the up-conversion mixer and power amplifier (PA) are extracted from analysis of the signal produced by the envelope detector. The ability to do this in the presence of cascaded device nonlinearities was first shown in References 28 through 30 and followed by further developments in References 27, 31 through 34. Cherubal and Chatterjee in References 31,32 showed that Spice model/process parameters could be diagnosed, using a nonlinear solver, from the response of analog devices to carefully optimized test stimulus. Erdogan and Ozev [33] used a nonlinear solver to compute behavioral model parameters of RF transceivers from the transient response of the RF system to stimulus applied from the baseband unit for testing purposes. The obtained behavioral model parameters were then used to determine the relevant specifications of the transmitter and the receiver. This was followed by the work of Banerjee [27,34] in which the stimulus applied from the baseband unit was carefully optimized

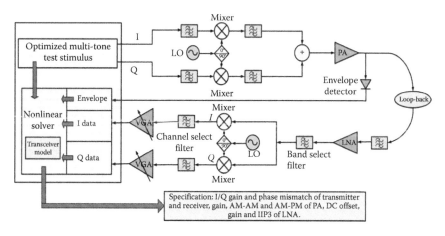

FIGURE 12.10
Built-in test configuration. (Banerjee, A. et al. *29th IEEE VLSI Test Symposium (VTS)*, pp. 58–63. © May 2011. IEEE.)

to enable efficient estimation of the transmitter, receiver and RF module performance specifications.

After the transmitter parameters are calculated from the envelope detector output, the receiver parameters are computed from the down converted I and Q data using the calculated transmitter parameters to calibrate for any transmitter performance loss. The mathematical models used are discussed in Reference 35. It is shown in References 27, 35 that when the test stimulus is optimized carefully to maximize the correlation between the DUT response to the applied stimulus and the DUT test specifications, the behavioral model parameters of the up and down conversion mixers as well as those of the RF PA and LNA in the loopback configuration of Figure 12.10 can be determined using (a) a regression mapping that maps the observed DUT response to the model parameters [36] or (b) by using a nonlinear solver [31–34] that solves for the model parameters in such a way that the response to the applied test produced by the model matches the observed DUT response (this is formulated as a cost metric for test stimulus optimization). The latter approach is shown in Figure 12.11. Figure 12.12 shows the optimization of the cost metric versus the number of GA iterations. The optimization is repeated for three

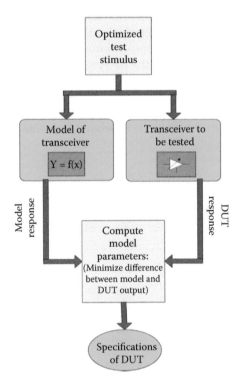

FIGURE 12.11

Model parameter computation from observed DUT response. (Banerjee, A. et al. *29th IEEE VLSI Test Symposium (VTS)*, pp. 58–63. © May 2011. IEEE.)

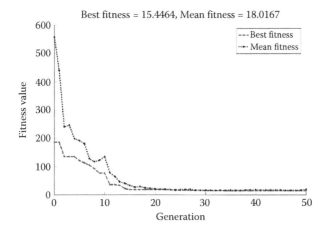

FIGURE 12.12
Convergence of genetic algorithm. (Banerjee, A. et al. *29th IEEE VLSI Test Symposium (VTS)*, pp. 58–63. © May 2011. IEEE.)

types of test stimuli for the I and Q inputs of the transmitter (result of optimization shown in Figure 12.13): orthogonal tones (top, Figure 12.13), nonorthogonal tones (middle, Figure 12.13), and time domain transient waveform (bottom, Figure 12.13). The left and right waveforms in each case represent the optimized I and Q inputs, respectively. It is seen that for both cases of multi-tone stimulus, the use of four tones is optimal and the cost function improves negligibly if more than four tones are used.

A MATLAB®-based nonlinear equation solver [37–39] is used for computation of model parameters using three types of test stimuli as shown in Figure 12.13. The computation is performed in two ways: (a) with envelope detector connected to the output of the transmitter and (b) loopback without the envelope detector. In both cases, the nonlinear solver was able to solve for the nonlinearity parameters of the transmitter and receiver. The parameters that are computed for cases (a) and (b) are I/Q gain and phase mismatch of transmitter and receiver, all coefficients of fifth-order nonlinearity of the transmitter, AM-PM distortion of the PA and coefficients of the third-order nonlinear model of the receiver. For case (a), in addition to the parameters above, the third-order nonlinearity of the up-conversion mixer and the third-order nonlinearity of the down conversion mixer are also determined. For both cases and with all three test stimuli, the average error in model parameter computation ranges from 0.5% to upto 10% (see Figure 12.14 for some RF parameters diagnosed from loopback including PA phase distortion).

Transceiver test case study: As a further test case, a 940 MHz RF front end test chip with on-chip detectors, including a 940 MHz LNA and a 940 MHz to 40 MHz down conversion mixer was implemented in 0.18 μm CMOS technology using UMC foundry models [41]. The LNA was designed using a

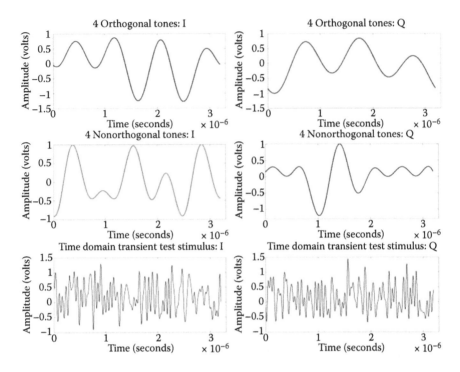

FIGURE 12.13
Different optimized test inputs. (Banerjee, A., Chatterjee, A. *IEEE Trans. VLSI*, 23(2), 342–255. © 2015. IEEE.)

differential structure with inductive source degeneration. The nominal gain was 7 dB with IIP3 specification of –1.5 dBm. The mixer design used a differential Gilbert cell with several improvements including current injection and inductive source degeneration. Nominal gain and IIP3 of the mixer was 4 dB and –1 dBmm, respectively. A transmitter path with an up-conversion mixer and a preamplifier was also implemented. The full die photo is shown in Figure 12.15. The upper half portion is the TX path. Since no on-chip inductor was used, the area was much smaller than that of the RX path, where on-chip inductors were used for linearity and noise performances.

The system under test is in a system-on-chip (SoC) environment, which includes TX and RX channels with a DSP core. In this setup, the TX output is connected to the RX input through a lossy link, which simulates the real space loss. Usually, there is an up-conversion mixer and a PA in the TX path; LNA and a down conversion mixer in the RX path, together with some preamplifiers, buffers and filters, depending on different applications. By inserting on-chip RF detectors along the RF transceiver path, all the components are monitored. Since the on-chip detector outputs are low frequency, multiple monitor points can be time-multiplexed to one on-chip ADC.

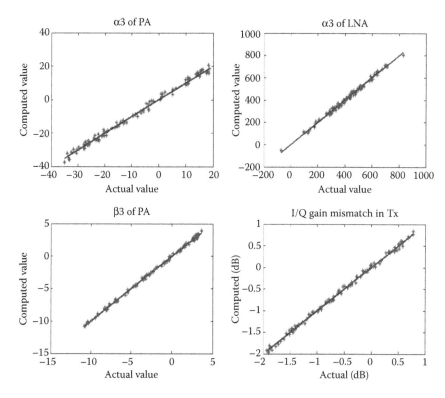

FIGURE 12.14
Accuracy of parameter computation. (Banerjee, A., Chatterjee, A. *IEEE Trans. VLSI*, 23(2), 342–255. © 2015. IEEE.)

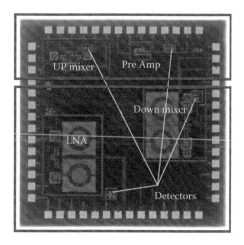

FIGURE 12.15
Die photo with components marked. (From Zhang, C., Gharpurey, R., Abraham, J. A. *J. Electron. Test. Theory Appl. (JETTA)*, 557–569, 2012. With permission.)

Two signal generators were used to generate input signal tones, and the Tektronix digital oscilloscope was used to capture detector outputs. As a measurement comparison to the on-chip detector result, the spectrum analyzer was used to measure chip specifications (gain and IIP3). The spectrum analyzer was controlled by a PC through local area network (LAN) connection. The digital oscilloscope was controlled by a laptop through LAN connections. Agilent VEE and Labview programs were developed to control the instruments and obtain the data automatically.

When measuring the detector outputs, a two-tone input signal with frequency components 40.1 and 39.9 MHz at power levels of –13 dBm was used. This is one advantage of loopback test: The stimulus tones are much lower frequency than RX only test (around 940 MHz). The LO signal was 900 MHz at a –2 dBm power level. The detectors were inserted after each of the components along the transceiver chain. Detector outputs were sampled at 10 MHz rate with the Tektronix DPO 7104 Digital Oscilloscope. Five microseconds of data for each of the detectors were logged. Then the digitized waveform was transformed with a fast Fourier transform (FFT). The total test time was only 20 μs for measurement plus the FFT computation time. The low frequency components were used as the input parameters for the nonlinear mapping function. One hundred and fifty instances were measured, among which 120 instances were randomly picked as training cases to generate the mapping function between detector outputs (DC and low frequency tones) and the circuit specifications (gain and IIP3) using the mathematical tool, MARS. Here, we did not attempt direct computation because of the more complicated nonideality of the loopback setup. The other 30 instances were used to verify the accuracy of the method.

Using the 30 instances, the measured specifications from high frequency RF instruments and the calculated ones from the detector outputs with the mapping function were combined in scatter plots. A 45° reference line was inserted in each plot. With distributed on-chip detectors, the specifications of the individual components in both of the transmitter and receiver sides are also calculated and the results are shown in Figures 12.16 and 12.17 [41]. The plots show close matching of the measured and calculated performances for the transceiver. Two main specifications, the gain performance and the linearity performance were illustrated to show the accuracy of the prediction. With our approach, the complexity of measurement and test time are greatly reduced, while maintaining high accuracy.

12.5 Built-In Test Driven Tuning Techniques

In order for postmanufacture tuning of mixed-signal SoCs to be economically feasible, it is necessary that *testing and tuning* of the same be accomplished in

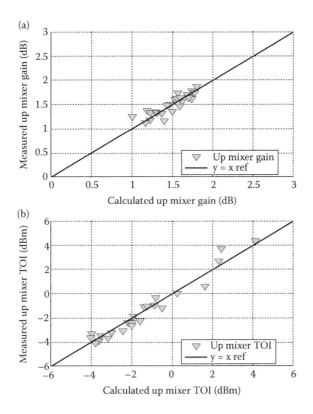

FIGURE 12.16
Calculated versus measured module performances (a) Up mixer gain, (b) Up mixer TOI. (From Zhang, C., Gharpurey, R., Abraham, J. A. *J. Electron. Test. Theory Appl. (JETTA)*, 557–569, 2012. With permission.)

an amount of time not much larger than the time taken to just *test* the system as per current practice. Figure 12.18 shows a testing and tuning architecture for an OFDM transceiver. The test architecture of Figure 12.18 consists of sensors designed into the RF signal path to monitor the quality of signals at different internal test points in response to test stimulus applied from the baseband DSP. The sensors used include envelope, peak or rms detectors and other special test response feature extractors. In test mode, either OFDM-modulated signals or carefully designed multitones are applied to the transmitter and the signals "looped back" to the input of the receiver to be analyzed by the baseband digital signal processor (DSP). Figure 12.18 does not show the loop-back path, but such loop-back may be performed via an external frequency translation circuit (mixer) located on the tester load-board or integrated into the device package. Algorithms running on the DSP are used to analyze the looped-back test response from the receiver as well as data from all the embedded sensors to tune the individual circuit components using an iterative test-tune-test procedure. Tuning consists of adjusting

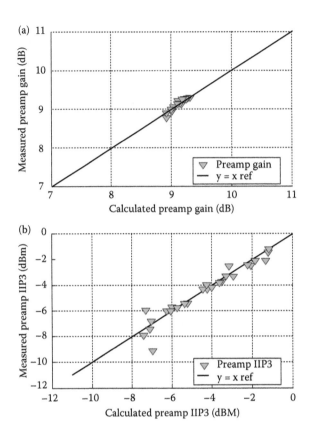

FIGURE 12.17
Calculated versus measured module performances (a) Preamp gain, (b) Preamp gain. (From Zhang, C., Gharpurey, R., Abraham, J. A. *J. Electron. Test. Theory Appl. (JETTA)*, 557–569, 2012. With permission.)

the bias currents/voltages of the mixers, PA and LNA among other tuning parameters so that transmitter and receiver noise and linearity specifications are met. In order to minimize test and tuning time, the following characteristics of the test and tuning infrastructure are desirable [42]:

a. The ability to measure *multiple RF specifications* of embedded devices using a *single* data acquisition (or few acquisitions).

b. The ability to perform test and tuning *autonomously* with little or no external RF test equipment support and *minimal on-chip hardware overhead* (and consequently *minimal impact on device performance*).

c. The ability to ensure near-optimal tuning avoiding local minima (with regard to a performance-based cost function) with the *least (negative) impact on device power consumption*.

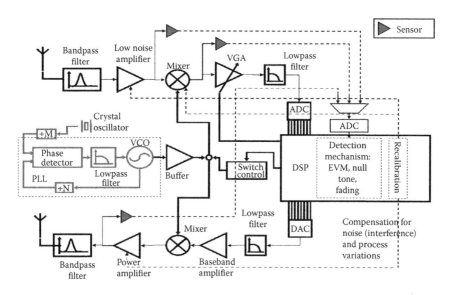

FIGURE 12.18
Test and tuning infrastructure.

With regard to (a), it is worthwhile to point out that existing tuning techniques generally support testing/tuning of *a design specification at a time*. For each such tested specification, dedicated on-chip hardware or external RF test equipment with appropriate test access is used. To complicate matters, tuning is usually accomplished using *iterative procedures* that need *repeated test measurement* of multiple specifications across as many as 50 tuning iterations. Through each iteration, *multiple design specifications must be monitored* as it is not possible, in general, to control multiple design specifications *independently* with a limited number of design tuning parameters. Hence, *tradeoffs between design specifications must be considered* during the tuning procedure. Note that the alternative built-in test approach is ideal for supporting the objectives (a) and (b), above. If power consumption is also included as a performance measure, then the alternative test-based tuning built-in test and tuning procedure is suitable for objective (c) also (Figure 12.19).

12.5.1 Learning Driven Tuning Algorithms

Consider process variations in the process parameter space P of Figure 12.3 that affect the specifications of the DUT. Consider a statistically significant number of samples of the DUT across the entire range of process variations that are tuned accurately in a power-conscious manner using a known tuning algorithm that avoids local minima. For each device in the process space P, there exists an alternate measurement M for the untuned DUT. If each such DUT is tuned using the above tuning algorithm, then a set of tuning

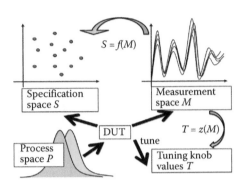

FIGURE 12.19
Predicting optimum tuning knob values directly from alternate test measurements M.

parameter values T is obtained corresponding to each alternate measurement M of the untuned device. Consequently, a regression mapping $z(M)$ is built using training algorithms that is used to directly predict the best tuning parameter values $T = z(M)$ for each DUT directly from the alternate test measurements M made on the untuned device. This is shown in Figure 12.20. The alternate test applied must satisfy certain diagnosability conditions [19,42] for the mapping $z = f(M)$ to be feasible. A detailed discussion of this is beyond the scope of this discussion but results are shown below for an LNA. Note that the tuning procedure is not iterative and can be performed via *single* test application [41].

Test Case: A CMOS LNA with folded PMOS IMD (intermodulation distortion) sinker [19] was used as a test vehicle and was designed in TSMC 0.25 µm CMOS technology. Its simplified schematic is shown in Figure 12.20. Two tuning knobs were employed to control the programmable bias of the transistor M_1 and M_p. A fabricated self-tuning chip (different design, TSMC 0.25 µm CMOS) is also shown in Figure 12.20.

A two-tone sinusoidal waveform was utilized as the test stimulus (–20 dBm at ±5 MHz offset from the center frequency 1.9 GHz). Hence, the fundamental frequency of the envelope response of the envelope detector sensor at the output of the LNA as shown in Figure 12.20 was placed at 10 MHz. The specifications of the LNA were extracted from the envelope detector output. Figure 12.21 shows the histograms for NF and Idd of the LNA before and after tuning. It is seen that across multiple specifications (NF, TOI, S21, and Idd), yield is improved from 41.8% to 93.3%.

12.5.2 Implicit Tuning Using Golden Signatures

In the implicit tuning methodology, a "golden" signature [43] is determined in such a way that the signature is maximally sensitive to *any perturbation in the process or the tuning knob values that also affect the DUT's specifications*. The signature above consists of the DUT response to the applied test stimulus

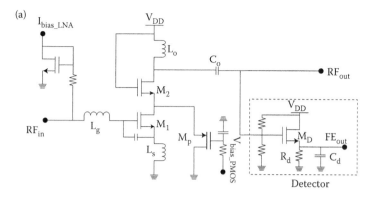

FIGURE 12.20
CMOS tunable LNA design and test chip. (From Han, D., Kim, B., Chatterjee, A. *TVLSI*, 18(2), 305–314, 2010. With permission.)

as well as the response obtained from any sensors inserted at internal test points of the DUT. Once such a signature is determined (via simulation and hardware calibration), *no training algorithms are needed*. Tuning is performed to *minimize the difference* between the golden and observed signatures and when this difference is minimized, the tuning procedure terminates.

12.5.3 Concurrent Tuning Using Specification Prediction

For a given RF system, let $K = [K_1, K_2, \ldots, K_N]$ be a vector describing the values of N control knobs. These control knobs may include *digital predistortion* parameters in the baseband also. For a process-perturbed device, the tuning algorithm first makes a *power-aware intelligent initial guess of the value of K* using the learning-based tuning technique described earlier. The complexity of the training

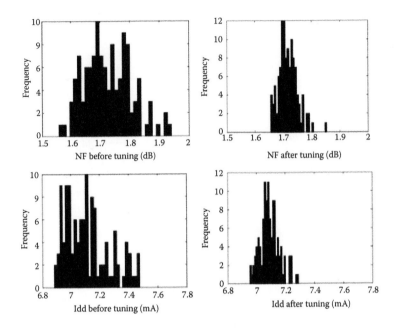

FIGURE 12.21
NF and Idd histograms before and after tuning. (From Han, D., Kim, B., Chatterjee, A. *TVLSI,* 18(2), 305–314, 2010. With permission.)

procedure here is traded off for simplicity allowing an approximate solution to be generated that is close to the global minimum. Then, starting from this initial guess, optimization algorithms [44,45] are used to minimize a cost function that represents the difference between the desired and the observed performance specifications of the RF system. This cost function requires the use of alternate test-based specification prediction using the mapping $S = f(M)$ of Figure 12.3. Training algorithms are needed to generate this mapping as discussed before. The search algorithm is used to determine the "next" tuning knob values at each iteration of the tuning procedure. For these tuning knob values, the alternative tests are rerun, the cost function recomputed, and the procedure repeated until convergence is achieved. Figure 12.22 shows the histogram plot of Gain, IIP2, and IIP3 specifications, for an RF transmitter before and after tuning (simulated). The data shows significant potential for postmanufacture yield improvement [46]. Data from "real" hardware measurements are presented in Reference 46 and further validate the tuning methodology.

12.6 Conclusions

A novel approach to built-in test and tuning of mixed-signal/RF devices has been presented. The approach holds the promise of (a) scalability across

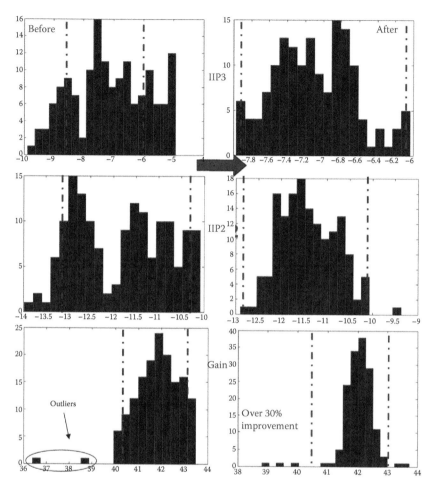

FIGURE 12.22
Yield results for tuning of transmitter using augmented LaGrange approach. (Banerjee, A. et al. *IEEE Design and Test of Computer* (to appear). © 2015. IEEE.)

different performance metrics, (b) suitability for built-in test, and (c) amenability to design automation via CAD tools. Future work will investigate scalability of the proposed method to tens of GHz frequencies and to complex systems such as MIMO transceivers and phased-array radar.

Acknowledgments

We would like to acknowledge the numerous graduate students and colleagues who have helped in this research directly or indirectly. Specifically, at

Georgia Tech, we would like to acknowledge P. Variyam, R. Voorakaranam, S. Cherubal, S. Chakrabarti, V. Gomes, J. Hou, A. Halder, S. Bhattacharya, G. Srinivasan, D. Han, S. Goyal, S.S. Akbay, V. Natarajan, R. Senguttuvan, S. Sen, A. Banerjee, and S. Devarakond who worked on different aspects of this project over the span of more than a decade. At UT Austin, we would like to thank C. Zhang and R. Gharpurey for their advice, inputs, and contributions to proving the ideas on a full transceiver chip. We would also like to thank D. Goodman and D. Majernik who played key roles in the first commercial validation of the alternate test methodology. Finally, we would like to thank the US National Science Foundation, MARCO-DARPA (Berkeley Gigascale Research Center), SRC, and National Semiconductor Corp. (acquired by Texas Instruments) for providing funding for this research.

References

1. Borkar, S., Karnik, T., Narendra, S. et al. Parameter variations and impact on circuits and micro-architecture. *Proceedings of DAC*, 2003, Anaheim, CA, pp. 338–342.
2. Sen, S., Chatterjee, A. Design of process variation tolerant radio frequency low noise amplifier. *IEEE International Symposium on Circuits and Systems*, Seattle, WA, 2008, pp. 392–395.
3. Perrot, M.H., Tewkbury, III T.L., Sodini, C.G. A 27-mW CMOS fraction-N synthesizer using digital compensation for 2.5-Mb/s GFSK modulation. *IEEE J. Solid-State Circuits (JSSC)*, 32(12), 2048–2060, 1997.
4. Staszewski, R.B. Wallberg, J.L., Rezeq, S. et al. All-digital PLL and transmitter for mobile phones. *IEEE J. Solid-State Circuits (JSSC)*, 40(12), 2469–2482, 2005.
5. Kim, S., Soma, M. An all-digital built-in self-test for high-speed phase-locked loops. *IEEE Trans Circuits Syst II, Analog Digit. Sig. Process.*, 48(2), 141–150, 2001.
6. Veillette, B.R., Roberts, G.W. On-chip measurement of jitter transfer function of charge pump phase-locked loops. *IEEE J Solid-State Circuits (JSSC)*, 33(3), 483–491, 1998.
7. Chang, H.-M., Lin, M.-S., Cheng, K.-T. Digitally assisted analog/RF testing for mixed-signal SoCs. *17th Asian Test Symposium*, Sapporo, November 2008, pp. 43–48.
8. Chang, H.-M., Lin, K.-Y., Chen, C.-H. et al. A built-in self-calibration scheme for pipelined ADCs. *Proceedings, International Symposium on Quality Electronic Design*, San Jose, CA, March 2009, pp. 266–271.
9. Murmann, B. A/D converter trends: Power dissipation, scaling and digitally-assisted architectures. *IEEE Custom Integrated Circuits Conference*, San Jose, CA, March 2008, pp. 105–112.
10. Akbay, S., Chatterjee, A. Built-in test of RF components using feature extraction sensors. *Proceedings of VLSI Test Symposium*, CA: Palm Springs, May 2005, pp. 243–248.
11. Variyam, P., Chatterjee, A. Specification driven test generation for analog circuits. *IEEE Trans. Computer-Aided Design Integr. Circuits Syst.*, 19(10), 1189–1201, 2000.

12. Variyam, P., Cherubal, S., Chatterjee, A. Prediction of analog performance parameters using fast transient testing. *IEEE Trans. Computer-Aided Design Integr. Circuits*, 21(2), 349–361, 2002.

13. Voorakaranam, R., Akbay, S.S., Bhattacharya, S. et al. Signature testing of analog and RF circuits: Algorithms and methodology. *IEEE Trans. Circuits Syst.*, 54(5), 1018–1031, 2007.

14. Natarajan, V., Sen, S., Banerjee, A. et al. Analog signature-driven post manufacture multidimensional tuning of RF systems. *IEEE Design Test Comput.*, 27(6), 6–17, 2010.

15. Natarajan, V., Sen, S., Devarakond, S. et al. A holistic approach to accurate tuning of RF systems for large and small multiparameter perturbations. *28th VLSI Test Symposium (VTS)*, Santa Cruz, CA, April 2010, pp. 331–336.

16. Devarakond, S., Sen, S., Natarajan, V. et al. Digitally assisted concurrent built-in tuning of RF systems using hamming distance proportional signatures. *19th IEEE Asian Test Symposium (ATS)*, Shanghai, December 2010, pp. 283–288.

17. Devarakond, S., Banerjee, D., Banerjee, A. et al. Efficient system-level testing and adaptive tuning of MIMO-OFDM wireless transmitters. *IEEE European Test Symposium*, Avignon, 2013, pp. 1–6.

18. Natarajan, V., Senguttuvan, R., Sen, S. et al. Built-in test enabled diagnosis and tuning of RF transmitter systems. *J. VLSI Design*, 2008, 1–10, 2008.

19. Han, D., Kim, B., Chatterjee, A. DSP driven self-tuning of RF circuits for process-induced performance variability. *TVLSI*, 18(2), 305–314, 2010.

20. Banerjee, A., Natarajan, V., Sen, S. et al. Optimized multitone test stimulus driven diagnosis of RF transceivers using model parameter estimation. *24th IEEE International Conference on VLSI Design*, Chennai, January 2011, pp. 274–279.

21. Friedman, J. H. Multivariate adaptive regression splines. *Ann. Stat.*, 19, 1–141, 1991.

22. Voorakaranam, R., Chatterjee, A. , Cherubal, S. et al. Method for using an alternate performance test to reduce test time and improve manufacturing yield. US Patent 2006/0106555A1, May 2006.

23. Akbay, S.S., Chatterjee, A. Fault based alternate test of RF components. *International Conference on Computer Design*, Lake Tahoe, CA, 2007, pp. 518–525.

24. Stratigopoulos, H., Mir, S., Acar, E. et al. Defect filter for alternate RF test. *Proceedings of European Test Symposium*, Seville, May 2009, pp. 101–106.

25. Voorakaranam, R., Newby, R., Cherubal, S. et al. Production deployment of a fast transient testing methodology for analog circuits: Case study and results. *Proceedings, International Test Conference*, Charlotte, NC, September 2003, pp. 1174–1181.

26. Cherubal, S., Voorakaranam, R., Chatterjee, A. et al. Concurrent RF test using optimized modulated RF stimuli. *Proceedings, VLSI Design Conference*, Bombay, India, January 2004, pp. 1017–1022.

27. Banerjee, A., Sen, S., Devarakond, S. et al. Automatic test stimulus generation for accurate diagnosis of RF systems using transient response signatures. *29th IEEE VLSI Test Symposium (VTS)*, Dana Point, CA, May 2011, pp. 58–63.

28. Halder, A., Bhattacharya, S., Srinivasan, G. et al. A system-level alternate test approach for specification test of RF transceivers in loopback mode. *Proceedings of VLSI Design Conference*, Kolkata, 2005, pp. 289–294 (best paper award).

29. http://smartech.gatech.edu/bitstream/1853/11456/1/halder_achintya_200608_phd.pdf.

30. Han, D., Goyal, S., Bhattacharya, S. et al. Low cost parametric failure diagnosis of RF transceivers. *11th IEEE European Test Symposium*, Southampton, May 2006, pp. 207–212.

31. Cherubal, S., Chatterjee, A. Parametric fault diagnosis for analog systems using functional mapping, *Design, Automation and Test in Europe*, 195–200, 1999.

32. Cherubal, S., Chatterjee, A. Test generation based diagnosis of device parameters for analog circuits. *Design, Automation and Test in Europe*, 596–602, 2001.

33. Erdogan, E.S., Ozev, S. Detailed characterization of transceiver parameters through loop-back-based BiST. *IEEE Trans. VLSI Syst.*, 18(6), 901–911, 2010.

34. Banerjee, A., Natarajan, V., Sen, S. et al. Bhattacharya, S. Optimized multitone test stimulus driven diagnosis of RF transceivers using model parameter estimation. *24th IEEE International Conference on VLSI Design*, Chennai, January 2011, pp. 274–279.

35. Banerjee, A. Design of digitally assisted adaptive analog and RF circuits and systems. PhD Thesis, School of ECE, Georgia Tech, December 2013.

36. Han, D., Bhattacharya, S., Chatterjee, A. Low-cost parametric test and diagnosis of RF systems using multi-tone response envelope detection. *IET Computers & Digital Techniques*, 1(3), 170–179, 2007.

37. Mathworks documentation, MATLAB optimization toolbox, available online: http://www.mathworks.com/help/toolbox/optim/ug/bqnk0r0.html.

38. Nocedal, J., Wright, S.J. *Numerical Optimization*, Springer, , New York, 2006.

39. Coleman, T.F., Li, Y. On the convergence of interior-reflective Newton methods for nonlinear minimization subject to bound. *Math. Program.*, 67(1–3), 189–224, 1994.

40. Banerjee, A., Chatterjee, A. Signature driven hierarchical post manufacture tuning of RF systems for performance and power. *IEEE Trans. VLSI*, 23(2), 342–255, 2015.

41. Zhang, C., Gharpurey, R., Abraham, J.A. Built in test of RF subsystems with integrated detectors. *JETTA*, 557–569, 2012.

42. Chatterjee, A., Han, D., Natarajan, V. et al. Iterative built-in testing and tuning of mixed-signal/RF systems. *Invited Paper, Proceedings of IEEE International Conference on Computer Design (Invited Paper)*, Lake Tahoe, CA, October 2009, pp. (s): 319–326.

43. Natarajan, V., Senguttuvan, R., Sen, S. et al. ACT: adaptive calibration test for performance enhancement and increased testability of wireless RF front-ends. *VLSI Test Symposium*, San Diego, CA, 2008, April 27–May 1 2008, pp. 215–220.

44. Shang, Y., Wah, B.W. A discrete lagrangian based global search problem for solving satisfactory problems. *J. Global Optim.*, 66–98, 1998.

45. Snyman, J.A. *Practical Mathematical Optimization*. Springer Science, Business Media, Inc. New York, 2005.

46. Natarajan, V., Banerjee, A., Sen, S. et al. Yield recovery of RF transceiver systems using iterative tuning driven power conscious performance optimization. *IEEE Design Test Comput.* 32(1), 61–69, 2015.

13

Spectrally Pure Clock versus Flexible Clock: Which One Is More Efficient in Driving Future Electronic Systems?

Liming Xiu

CONTENTS

13.1 Clock Generator Must Be Mixed-Signal Circuit

Since an electrical voltage/current level (which is proportional to the number of electrons) is the only physical property that can be sensed by any electronic device, the flow-of-time inside electrical world must be quantified through this medium. This is achieved by employing an electrical pulse train called a clock signal. The moments in the flow-of-time are determined from this clock pulse train's voltage level transitions: The moment of its voltage crosses a predefined threshold in the low-to-high (or high-to-low) transition. Therefore, any noise associated with the voltage level around the threshold area will be converted into timing noise [1,2]. This mechanism is illustrated in Figure 13.1. As shown, the voltage level uncertainty ΔV is reflected as timing uncertainty Δt when viewing from a timing perspective. For this reason, all circuits designed for generating a clock signal (i.e., the clock generator or

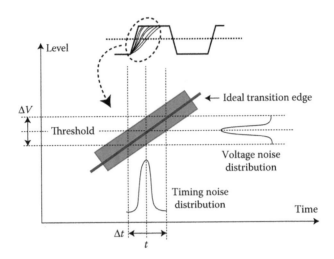

FIGURE 13.1
Voltage noise is converted into timing noise: Every voltage point matters.

timing circuit) must possess analog signal processing capability since every point in the neighborhood of the threshold of the clock pulse's voltage level matters.

In Figure 13.2, the general architecture of timing circuit is depicted. The key component in any timing circuit is the one that produces the electrical oscillation (and thus the final clock pulse train). There are two main techniques for serving this purpose: the approach of using an oscillation circuit and the method of employing an edge combination/selection circuit. In today's CMOS domain, the oscillation circuit can either be delay-cell-based ring or LC-tank-based oscillator [3–5]. Ring-based structure is very cost-effective and commonly used for driving digital application. LC structure has much better noise performance thanks to its higher Q factor. But its

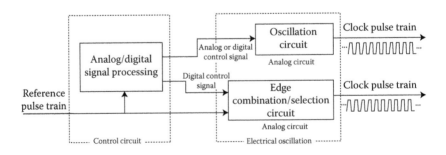

FIGURE 13.2
The general architecture of clock generator (timing circuit).

output frequency range is limited and the implementation cost is high. It is mainly used in RF application for generating carrier signal. In this approach of using oscillation circuit, the output from the free run oscillation is often not directly usable since its oscillating frequency is not precisely known. The oscillation must be controlled by referencing it to a known frequency source. Since the final clock pulse train is not directly associated with its reference pulse train but only connected through the control/processing circuit, this kind of circuit is often characterized as indirect approach. It is an autonomous system because the oscillation is generated internally inside the timing circuit. Its noise level is mainly determined by the quality of the oscillator.

The edge combination/selection method directly constructs the output clock pulse by selecting edges from a reference signal having a plurality of phases, or from a group of reference signals. It then combines them into final pulse train [6–8]. The key circuit component here is the multiplex used for edge selection. To ensure timing quality, the multiplex must be designed as an analog circuit (digital multiplex is not recommended). As a whole, there is no self-oscillation in this type of timing circuit. It is a driven system. The output's timing quality is heavily dependent on that of reference signal.

In both approaches, as shown in Figure 13.2, there is a need of a control/processing circuit. This part of circuit does not have direct connection with the final clock output. Based on the reference signal and the system configuration, its function is to generate control signal for directing the oscillation circuit or the edge combination/selection circuit. It shows its impact only implicitly through the following oscillation circuit. Since we care the noise associated with the voltage level at which the pulse makes transition, the oscillation circuit and the edge combination/selection circuit must be analog in nature. The control/processing circuit can be either analog or digital.

Modern designs of timing circuit have large amount of digital circuit in its system since advanced process nodes are more digital-oriented and digital implementation is very cost-effective. Those circuits are often denoted as digital-intensive frequency synthesis. However, the term "all digital clock circuit" (or "all digital frequency synthesis") is inherently incompatible with the philosophy of timing circuit since digital signal processing only cares about two levels: high and low. When the term "direct digital frequency synthesis" is used [9,10], digital-to-analog converter (DAC), signal filter, and frequency divider are usually used in its circuit implementation. DAC and filter are analog components. Frequency divider is made of flip–flops. As a base cell, high-quality flip–flop is designed as an analog circuit. Furthermore, "all digital frequency synthesis" requires a reference clock (usually at high frequency) to function as the originating source of electrical oscillation. This reference clock must come from somewhere. It must be generated in an analog way wherein every point in level matters. Therefore, *all clock generators (timing circuits) must be mixed-signal circuits.*

13.2 Phase-Locked Loop: Today's Mainstream Clock Generator and Its Limitation

Today's mainstream timing circuit is phase-locked loop (PLL). It is one of the foundational circuits in the arsenal of today's IC design. It is a beautiful blend of analog and digital circuit in one functional block. It can be found in almost every modern IC chip. In some complex chips, the number of on-chip PLLs could be in the order of tens. The PLL functional block diagram is shown in Figure 13.3. In the past several decades, PLL has been studied thoroughly [11–13]. Referring back to the generic timing circuit of Figure 13.2, the part producing the electrical oscillation in this case is the oscillator which could be a CMOS ring or LC-tank-based oscillator. Its output, sometime after a post divider, is the generated clock signal. The oscillator is controlled by a control signal outputted from a control/processing circuit. In PLL case, the control/ processing circuit is a feedback loop executing the action of "compare then correct." The frequency and phase detector, low-pass filter and frequency divider form the loop. The oscillator output and the reference pulse train are the two signals that are compared. From a fixed and known frequency f_1, the PLL can generate other frequencies $f_2 = N{\cdot}f_1$ where N is an integer (integer-N PLL) [14,15]. It is also possible for N to use different values at different times, bearing some predesigned patterns. In this case, the output and the reference frequencies will averagely maintain a noninteger ratio (fraction-N PLL), at the expense of degraded noise performance [16].

Although extremely popular, PLL has its drawbacks. This first one is that the available output frequencies from PLL are limited. In other words, its frequency granularity is the reference frequency. This is not always satisfactory for many applications. Fractional-N PLL can improve its granularity. But it achieves this goal at the expense of output signal's noise quality, output frequency range, and design complexity (thus the cost). The second problem with PLL is the response speed. Due to the feedback mechanism employed, any request of changing output frequency has to wait a long period of time (e.g., hundreds of cycles) for the PLL to finish its work and the new frequency to become effective. Due to these two issues, PLL has difficulty in

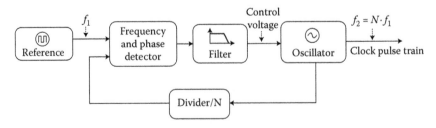

FIGURE 13.3
Block diagram of PLL.

satisfying the complicated requirements imposed by modern systems as will be explained in next section. In short summary, PLL-based frequency generator is considered as a rigorous clock source. In contrast, a flexible clock source is defined as the one that (1) has small frequency granularity (can generate many frequencies) and (2) can switch its output frequency from one to another in quick fashion.

13.3 Future Electronic System Demands More from Its Clocking

After more than half century's evolution, electronic system supported by semiconductor IC has reached to a very sophisticated stage. At chip design level, several key technical challenges lied in front of future electronic system designer are listed next. All these issues are related to clocking in one way or another. Innovation in the area of clock generation is likely required to meet these challenges.

- Connect every electronic device to Internet (Internet of Things)
- Reduce system's power consumption
- Reduce system's EMI (electromagnetic interference)
- Improve network's time synchronization to *ns* range
- Make hardware more software-friendly
- Embed higher security into chip design
- Miniaturization challenge
- Interoperability challenge

Electronic system's power consumption becomes a concern of increasingly importance. This is because that more and more electronic systems are mobile or portable devices supported by battery, which has tight constraint on the current they can drain. For desktop computing and server farm, power reduction is also a crucial task. As clock frequency reaches GHz range, the heat generated from chip's switching activities is enormous. As a matter of fact, inside a chip, the clock network is usually the largest energy consumer and thus greatest heat generator (due to its highest toggle rate and largest fanout). As a result, expensive cooling system must be employed to keep system functional. Reducing power consumption can alleviate the cooling stress. Clocking can play an important role in solving this power-reduction problem.

Due to the ever-increasing sophistication of today's system, the demand for quick time-to-market and the wide variety of applications, software plays an increasingly important role in the overall system design. Consequently, hardware design trends to be progressively generic and product differentiation is preferably accomplished through software. To deliver improvement over all these areas of power, performance, programmability, and portability, computer industry is now establishing a new approach: heterogeneous system architecture (HAS). It provides a unified view of fundamental computing elements. It allows programmer to write applications that seamlessly integrate CPU (central processing unit, latency compute units) with GPU (graphic processing unit, throughput compute units), while benefiting from the best attributes of each [17]. In the development of this trend, clocking is inevitably a key issue to be investigated since the computation cores are driven by clocks.

For all electronic systems, a hostile effect of high clock frequency is the strong EMI caused by the energy radiation of the clock signal. The purer the clock spectrum is (less noisy clock), the severer is the problem. EMI is a problem of concern for almost all the consumer electronic systems operating today in our home, office, public place, etc. Manipulation on the spectrum of clock signal is an effective way of attacking this problem.

One of the popular demands for today's and future's electronic system is to connect it to the Internet (e.g., Internet of Things, or IoT). In this increasingly connected world, the requirement for the precision of time synchronization will become more stringent in future. Currently, the network synchronization precision is in the range of *millisecond* and is expected to reach *nanosecond* range in near term. The accuracy of time synchronization is closely tied to the quality of on-chip frequency source (clock generator).

Another fact in future designs is that, as more devices of all kinds of cost structures and physical characteristics being mobile and/or portable, wireless communication of reasonable data rate with cost-effective implementation becomes the key for gaining advantage in commercial competition. Clock is the most important design factor in data transmission. Innovative clock generation technique enables new approaches in facilitating data transmission.

Furthermore, as a chip is more likely to be connected to the Internet, the issue of chip security becomes more of a concern. Certain level of security must be incorporated in the chip design consideration. Clock signal can play a role in this regard.

System miniaturization is another trend in future electronic system design (e.g., for all kinds of wearable electronics). One of the possibilities for miniaturization is to eliminate the crystal frequency reference which is bulky for many applications. Crystal-less electronic system design also demands reconsideration of clock signal.

All these challenges and their solutions are closely related to the system's driving clock. A fixed frequency clock source, no matter how spectrally true is its spectrum, will not be efficient when addressing these difficult problems. Heterogeneous system usually includes many functional blocks that are running at difference clock frequencies. This requires an on-chip clock source that must be able to generate many frequencies with reasonable cost. More importantly, the clock source must be flexible so that its frequency can be changed quickly to accommodate the sophisticated communication activities occurring among the blocks. One of the most effective methods of reducing EMI is to spread the clock spectrum from a single frequency into a group of frequencies in a well-defined frequency range. This also asks for a flexible clock source. Since power usage is directly proportional to the clock frequency, a natural approach of reducing power is to use power-as-needed (or adjusting clock frequency based on loading). Dynamical voltage and frequency scaling (DVFS) is a popular method adopted today by many advanced chips. One of the requirements for implementing DVFS is a flexible clock source.

Improving network time synchronization to *ns* range demands small time granularity. In circuit implementation, this is translated to small frequency granularity. A rigorous clock source capable of only a few frequencies does not fit the need; a flexible clock source with strong frequency generation capability is a must-have. One of the solutions of incorporating security into chip design is through the interaction of pulse trains of complex frequencies. A clock source capable of generating many frequencies could be the enabler of this solution.

In summary, to meet today and future systems' requirements, a rigorous clock source such as PLL is no longer sufficient. We need a flexible clock source such that many frequencies can be generated from it and switching among frequencies can be accomplished quickly.

13.4 A Flexible Clock Generator: Time-Average-Frequency Direct Period Synthesis

One of the two fundamental physical properties used for describing an electrical signal, the voltage level, can be handled very flexibly by electrical circuit. Using basic devices such as transistor, capacitor, resistor, and diode, we can generate almost any desired voltage level (within a given range for a particular application). Also, we can switch a circuit's output voltage level from one to another in a very quick fashion (under reasonable loading). Therefore, it is fair to say that electrical circuit is *a flexible voltage source*. Similarly, it is desirable for us to pursue the same capability in handling the other property: time. As is well known, the flow-of-time inside electrical world is expressed through an electrical pulse train. The lapse-of-time of each pulse is used to gauge the speed of events happened or to be happened. The concept of clock frequency is created to quantify the difference in circuits' information processing rates. It is defined as the number of pulses existing within the time window of 1 s. In a more precise term, a flexible clock source can be described as that (1) its output clock frequency can be arbitrarily asked by user and be consequently generated by clock source; (2) its output clock frequency can be quickly changed from one to another upon user request.

This is, however, a much more challenging task than voltage handling. As discussed in Section 13.1, an electrical pulse train must be originated from some kind of electrical oscillation. When an oscillation circuit is not controlled (free run), its frequency varies greatly both in short and long terms. In order for it to be useful, the circuit must be controlled through comparing it to a good reference. This good reference is usually the quartz crystal that has the capability of preserving a stable and precise vibration when a voltage is applied to it. The frequency of the vibration is determined by its physical properties [18]. It is, however, not practical to mechanically reshape the quartz crystal every time a new frequency is requested. Hence, the dream of arbitrary frequency generation must be fulfilled through other means.

The difficulty associated with this challenge forces us to ask these questions: what is clock frequency? Do all the pulses in a clock pulse train have to be equal-in-length? Since clock pulse train is used just to mark other activities inside electronic system, *there is no fundamental restraint that all the pulses have to be equal-in-length*. This leads to the introduction of time-average-frequency (TAF) concept and theory [19]. This important concept is a crucial step toward the solution of this problem since it mathematically opens up other possibilities of creating a clock frequency. The other requirement of flexible clock source is the fast frequency switching. This task has to be accomplished by the approach of directly constructing each pulse in the pulse train. The feedback mechanism of "compare then correct" used in PLL is inherently incompatible with this requirement.

Following the philosophy adopted in previous discussion, an "unconventional" method of creating clock pulse train (frequency synthesis) is depicted in Figure 13.4. From two types of pulses, with length-in-time of T_A and T_B, respectively, a clock pulse train is generated by conjoining them together one after another. These two types of pulses are used in an interleaved fashion.

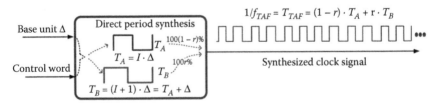

FIGURE 13.4
The architecture of TAF-DPS.

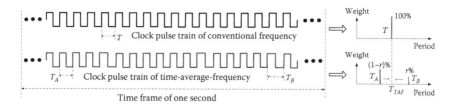

FIGURE 13.5
The clock pulse trains of convention frequency (top) and TAF (bottom).

The requirement is that the total number of pulses existing within the time frame of 1 s must be f_{TAF}, which is a number required by user for a particular application. By definition, f_{TAF} is the clock frequency [19]. When examined from a time window of 1 s , the possibility-of-occurrence of type T_B is represented by a fraction r (consequently, that of type T_A will be $1 - r$). A variable T_{TAF} can be defined from f_{TAF} as $T_{TAF} = 1/f_{TAF}$ and it thus can be calculated as $T_{TAF} = (1 - r) \cdot T_A + r \cdot T_B$. Unlike the case in conventional frequency, there are two (or more if desired) different types of pulses used. Therefore, this clock pulse train is termed TAF clock where f_{TAF} is the TAF and T_{TAF} is the average period. Figure 13.5 shows its time domain characteristic. The weighting factor between the groups of T_A and T_B is r, which is controlled by user for achieving a desired frequency [20].

In circuit practice, the T_A and T_B types of pulses are both created from a base unit Δ (e.g., 25 ps) as $T_A = I \cdot \Delta$ and $T_B = (I + 1) \cdot \Delta$, where I is an integer. Usually, I is a large number (in the range of tens, sometime hundreds). As a result, the difference between the sizes of T_A and T_B is kept within a small range so that its impact on serving as clock is not significant. The period (frequency) transfer function is expressed as $T_{TAF} = 1/f_{TAF} = FREQ \cdot \Delta = (I + r) \cdot \Delta$ where FREQ is termed as "frequency control word." The period transfer function is linear with respect to FREQ. When FREQ varies in small range, the frequency transfer function is almost linear [20].

There are two ways that the output frequency f_{TAF} can be changed. Small frequency tuning can be done by adjusting the r. Large step frequency change is achieved by adjusting the value of I. Since each pulse in the clock pulse train is directly constructed, the switching between frequencies can be accomplished almost instantly (in two clock cycles [20]). Most recent development in time-average-frequency direct period synthesis (TAF-DPS) can be found in References 21, 22. Table 13.1 makes the comparison on key characteristics of PLL and TAF-DPS. A typical example of TAF-DPS implementation is the Flying–Adder frequency synthesis architecture. Section VI.G and Table V of Reference 22 is the comparison between Flying–Adder clock generator and PLL. It can serve as the real case with numerical evidence to support Table 13.1.

TABLE 13.1

Comparison between PLL and TAF-DPS

	Reference Time Base	Source of Oscillation	Frequency Synthesis Method	Frequency Granularity	Frequency Switching	Output Characteristic
PLL	f_r	Ring or LC-tank	Feedback mechanism of compare-then-correct	f_r	Many cycles	One type of pulse
TAF-DPS	Δ	From the oscillation associated with Δ	Edge selection/combination	Usually below ppm[a]	Two cycles[b]	Two (or more) types of pulses

[a] $df = -(dF/F){\cdot}f$, where F is short for FREQ, dF is one LSB of the FREQ's fractional part. This frequency granularity is defined using TAF.

[b] In frequency fine-tune case where only the value of r is adjusted (the sizes of T_A and T_B are unchanged), it will take certain number of cycles before the waveform can show the impact of r change (small frequency change needs long observation time).

13.5 IC Clocking Is More Than PLL Design

Clock pulse train is the marker system to identify billions of events happened per second inside an electronic system. First, the clock pulse train has to be generated from somewhere in the chip. Second, it has to be delivered to every sequential cell in the chip (there could be millions of them in a large chip). Third, chip architect has to design sophisticated scheme of using clock to control the data flows among various functional blocks. Hence, IC clocking is an integrated issue. It must be investigated from the aforementioned three perspectives as a whole.

Traditionally, clock generation problem is handled by timing circuit designer (e.g., PLL designer) whose specialty is in transistor level circuit design. The task of using clock to drive functional blocks, however, is dealt with by system architect whose expertise lies in the areas of product definition and perhaps RTL level design. Further, the clock distribution problem belongs to the domain of physical design engineer. The latter two types of engineers often do not have knowledge on transistor level circuit design (or they simply do not have to care about it). As a result, there are significant skill and knowledge gaps among the engineers who are involved in IC clocking.

It is often considered by people that IC clocking is just PLL design. In other words, the IC clocking problem is solved if a good PLL is delivered from the PLL designer to system architect and physical implementation engineer. This is, however, far from the truth. Even when a PLL with very good jitter performance (e.g., sub-ps) is available to chip designer, the overall system performance will still be not satisfactory if the clock distribution network is not constructed in decent quality. In such cases, the large amount of resource spent on the high-quality PLL is easily negated by a lousy clock tree. In another front, if this ultra-low jitter PLL only can provide limited frequency choices and the switching between frequencies is slow, this high-dollar-price PLL might not be the best candidate for solving the complex data movement problems that chip architect struggles to address. Therefore, IC clocking is more than PLL design. It is a much bigger topic. It needs attention from engineers of several different disciplines.

When viewing from high level over all the chips developed today, the principal use of clock signal is to drive digital circuit (for computation and data communication). The tasks of using clock to drive data converters (ADC and DAC) and to function as RF carrier only account for a very small portion of the clock usage. This fact opens up the possibility for a new and innovative flexible clock source, such as TAF-DPS, to play a significant role in chip design. As will be evidenced from the following discussion, when driving digital circuit, a flexible clock source is more efficient than a spectrally pure but rigorous clock source. This is the next big opportunity for improving electronic system's information-processing efficiency. In Section 13.6, this statement will be justified with real examples.

13.6 TAF-DPS: A Circuit Level Enabler for System Level Innovation

Section 13.3 highlighted the key challenges in future electronic system design. For a complex chip design, these challenges present themselves to system architect as one or several real problems to be resolved, as illustrated in Figure 13.6. The solutions to these problems could be related to clock

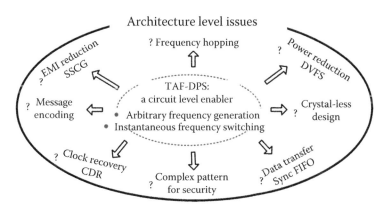

FIGURE 13.6
A flexible clock source TAF-DPS is the enabler for innovations in system level.

generator in one way or another. A flexible clock source can be an enabler for enabling various innovative approaches in architect level to better solve these problems. This section investigates this possibility for those issues in a brief way. Due to space limitation, the discussion will be in high level. The experiment data used in this section are collected from three chips: two in 55 nm and one in 65 nm. Some of the chip information can be found in Table VI of Reference 21 and Table II of Reference 22.

One of the very effective ways of reducing chip power is DVFS. Since power consumption is $P = C \cdot V^2 \cdot f$, lowering clock frequency can directly reduce power usage. A more significant fact, however, is that lowering clock frequency allows lower supply voltage to be used for supporting transistors' operations. This can lead to an even larger percentage drop on power usage. When PLL is used as the clock source for supporting DVFS, the full benefit of DVFS is difficult to be reached since the processors have to be paused when PLL switches its frequency (processor unavailable time). Although the frequency switching could be done quickly if a programmable frequency divider is used as the means of changing frequency, the resolution (frequency step) is coarse. This is not ideal for implementing DVFS. Thanks to the features of arbitrary frequency generation and instantaneous frequency switching, the TAF-DSP is squarely suitable for DVFS. In theory, the processor does not need to be paused at all since the clock waveform transits from one frequency to another seamlessly (glitch free).

The plot in Figure 13.7 is a transistor-level simulation showing the TAF-DPS's dynamical frequency switching capability. The clock waveform is shown in the bottom trace and its corresponding frequency measurement is shown at the top trace. The TAF-DPS clock generator's output period (frequency) is dynamically changed by a control circuit. The plot in Figure 13.8 is a measurement obtained from a chip. The waveform in the left is a clock pulse train whose frequency is dynamically switched among 200, 250, and 333 MHz. In the right-hand side, a zoom-in view is presented that shows the seamless waveform transition from 333 to 200 MHz, and then back to 333 MHz. These evidences support the claim that TAF-DPS clock source is a good tool for implementing DVFS.

The clock generator in a chip is the strongest source for electromagnetic radiation since clock signal has the highest toggle rate. And more significantly,

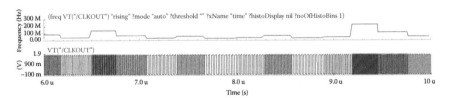

FIGURE 13.7
A transistor level simulation showing TAF-DPS for dynamical frequency switching: clock waveform (bottom) and its frequency measurement (top).

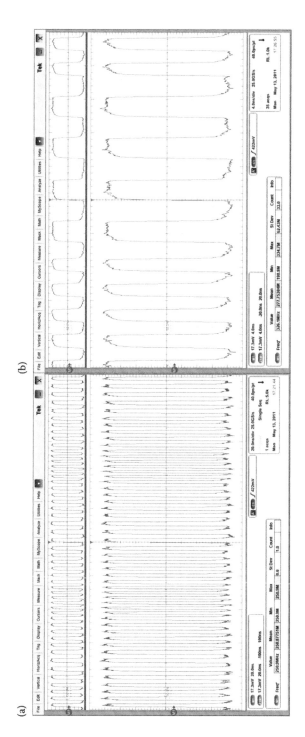

FIGURE 13.8

A measurement showing TAF-DPS for dynamical frequency switching among 200, 250, and 333 MHz (a) and zoom-in view from 333 to 200 MHz (b).

following the conventional wisdom that every pulse must have same length-in-time → a clock signal shows its presence at spectrum plot as a prefect single line → it radiates its energy at highest efficiency. To weaken its radiation strength, a natural approach is to transform this line into a group of lines of lower magnitudes (spread the spectrum). In circuit implementation, this can be achieved by adding noise into the clock signal (jittering the clock signal intentionally). When PLL is used for this purpose, this task is carried out by adding a controlled voltage pattern at the VCO's input, or by applying a desired division pattern at the loop divider. Both methods, however, produce its effect only indirectly through modulating the VCO. The accuracy of the spread on spectrum is hard to be controlled because modulating VCO is a nonlinear process. Worse yet, the clock's effectiveness in driving circuit is degraded in an uncontrolled fashion since the magnitude-of-change in clock pulses' lengths is not under direct control (in other words, the amount of added jitter is hard to be known to the designer).

On the other hand, the philosophy of TAF is naturally in harmony with the goal of spread spectrum. The two (or more) types of clock pulses automatically spread the clock energy from a concentrated frequency into a plurality of frequencies. Furthermore, from the equation of $T_{TAF} = (1 - r) \cdot T_A + r \cdot T_B$, it is seen that the output period (frequency) can be dynamically fine adjusted by slightly changing the value of r from time to time. This can produce a group of frequencies of lower magnitude centered at the unspread clock frequency (this is the center-spread; down-spread or up-spread can be easily achieved with TAF-DPS as well). There are multiple ways of changing the r value for achieving various spread results, such as random dithering, saw-tooth sweeping, or triangle sweeping. Figure 13.9 shows the measurement result from a case of triangular sweeping. There are three types of pulses produced by this clock generator. They correspond to frequencies of 888.89 MHz, 1 GHz, and 1.14 GHz. These three frequencies are observed in the experiment as evidenced in Figure 13.9. The weighting factors for these types of pulses (the fraction r) are swept triangularly at a fixed rate. The left plot in Figure 13.9 is the measurement of frequency versus time. As expected, three discrete frequencies are seen. On the right hand side, the same frequency versus time measurement after a divided-by-16 divider is shown. The characteristic of

(a) (b)

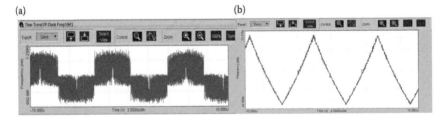

FIGURE 13.9
Measurement of triangularly sweeping weighting factor among three frequencies of 888.9 MHz, 1 GHz, and 1.14 GHz (a) and measurement after a divided-by-16 circuit (b).

triangular-sweep is clearly visible now since the divider has transformed the three frequencies into many lower frequencies. Figure 13.10 shows the clock spectrums. On the left is the unspread 888.9 MHz clock. In the middle and right, two spread spectrums of different sweeping rates are shown. Clearly, the clock signal's electromagnetic radiation strength is weakened (by ~15 and ~10 db, respectively). One important point worth mentioning is that there is no additional clock jitter added when using TAF-DPS-based spread spectrum method. This is because that the spectrum spread is achieved by only adjusting the weighting factor, or the fraction r. Another advantage of this approach is that the spread of the spectrum can be precisely calculated (and thus controlled). This is hard to be achieved in the case of PLL.

Frequency hopping spread spectrum is a method used in wireless communication. During the communication process, instead of using one fixed frequency, the carrier frequency is switched dynamically among a plurality of frequencies. Its key advantage includes resistant to narrowband interference and efficient use of bandwidth. Further, it is difficult to be intercepted. TAF-DPS constructs its output waveform directly. Thus, it is a good candidate for cost-effective implementation of frequency hopping. Figure 13.11 is a measurement of frequency versus time plot that shows the capability of frequency hopping. The TAF-DPS output hops among 800 MHz, 1 GHz, and 1.3 GHz quickly. In the top trace, the lapse-of-time between the two cursors is about 100 ns. This provides a reference for appreciating the quickness of this circuit's frequency hopping. The bottom trace shows the frequency hopping in a larger view.

Clock data recovery (CDR) is an important circuit technique that is often used in interblock or interchip digital data communication. In this approach, the clock signal is not explicitly transmitted with the data during the transmission. Instead, the clock information is embedded in the serially transmitted data stream. Therefore, the clock signal needs to be extracted from it at the receiver side. The left part in Figure 13.12 is the general architecture of conventional CDR. A detector is used to compare the transition edges embedded in the incoming data with the edges of locally generated clock (from VCO). The result, after being processed and filtered, is used to direct the VCO to generate the corresponding frequency. In most applications, the detector is digital type due to its operating advantage in high-speed environment. The VCO, however, requires an analog voltage to drive it. Thus, there is a need of digital-to-analog (D $\rightarrow$ A) process in between as illustrated in the figure. This D $\rightarrow$ A process slows the loop response speed. It is also resource intensive (in term of power, area, and design effort). On the right-hand side of Figure 13.12, the TAF-DPS-based CDR architecture (termed as TAF-CDR) is depicted. The TAF-DPS functions as the digital controlled oscillator (DCO) used in the loop. It takes digital value as its input. Thus, the D $\rightarrow$ A process is eliminated. Consequently, this TAF-CDR architecture is highly resource efficient.

The enabling component in this TAF-CDR architecture is the flexible clock source. Thanks to TAF-DPS's fast frequency switching, the loop has a very

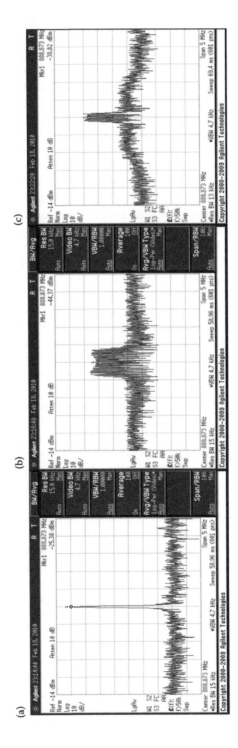

FIGURE 13.10
Clock spectrum measurement: unspread at 888.9 MHz (a), spread at one sweeping rate (b), and spread at another rate (c).

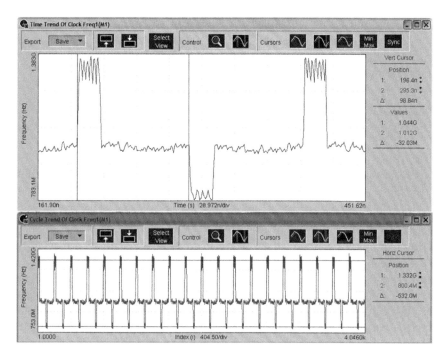

FIGURE 13.11
TAF-DPS output hops among three frequencies of 800 MHz, 1 GHz, and 1.3 GHz.

fast response speed. This results in a small value on loop latency. Further, this latency is also precisely calculable since this DCO's (TAF-DPS) response to control word change is known. Unlike the VCO used in conventional CDR, this DCO only generates three frequencies $f_{catch}, f_{hold}, f_{slow}$. The TAF-CDR loop uses them to perform three actions: *catch up*, *hold*, and *slow down*. By applying these actions appropriately, the loop is able to track the incoming data steam and extract the embedded clock frequency. Figure 13.13 shows an example of TAF-CDR with measured data. This circuit is designed for an ultra-low power 2 Mbps wireless communication chip (IEEE.802.15.4, 2 Mbps chip rate, 250 Kbps data rate). The results shown at the left plot, from top to bottom are the waveforms of incoming data, recovered data, and recovered clock. As

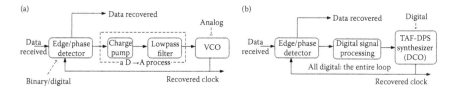

FIGURE 13.12
The conventional CDR architecture (a) and the time-average-frequency based clock data recovery (TAF-CDR) architecture (b).

(a)   (b)   (c)

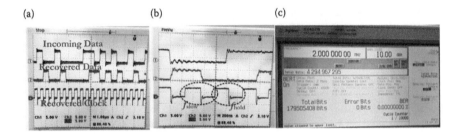

FIGURE 13.13
A TAF-CDR example: measured waveforms (a), zoom-in view of the waveforms (b), and the BERT result (c).

shown, the recovered data follows the incoming data (i.e., the data is correctly received by the receiver). Since the transmission rate is low at 2 Mbps (500 ns bit-time), the loop latency is designed as just one clock cycle. In other words, all the operations in the signal processing block of the left of Figure 13.12 can be finished in just one clock cycle. This is evidenced from the waveforms as one clock cycle delay between the incoming data and the recovered data is seen. The waveforms in the middle are the zoom-in version of that on the left. Selected two clock cycles corresponding to f_{slow} and f_{hold} are marked in the plot. As can be seen, the cycle of f_{slow} is longer than that of f_{hold}. The screenshot in the right shows the result of BERT (Bit Error Rate Test). The number of error bit is zero for a PN15 data stream (pseudo-random sequence) of $2^{32}-1$ bits.

Data transfer between blocks driven by clocks of different frequencies is a common task in digital design. This could happen in the scenario that the receiving block (domain B) processes its data in a continuous fashion while its incoming data (from transmitting block, domain A) is not a fixed-rate data stream. This can be caused by two reasons: (a) the clock frequency of the transmitting block varies from time to time or (b) the data is transmitted in an intermittent mode (some clock cycles do not have valid data). Overall the data rates at the two sides must be equal in a time average sense. Otherwise, valid data will be lost or new data will be created. Neither case is allowed in system operation. FIFO memory is often inserted in between the blocks to assist this type of data communication. Asynchronous FIFO is the straightforward approach but it requires several additional handshake signals. It also increases processing latency due to the time required for handshake handling. Synchronous FIFO eliminates the handshake signals but it demands better control of the clock signal on the receiving side. The receiving clock's frequency must be adjusted, based on the FIFO status, to accommodate the data rate variation associated with the incoming data stream. If PLL is used as the clock source, the FIFO size could be very large since (a) it takes long time for PLL to adjust its output frequency and (b) the choice of usable frequency is limited. For this type of application, TAF-DPS is the ideal clock engine as shown in Figure 13.14. Since the TAF-DPS can generate many frequencies and it can switch its frequency almost instantly, the FIFO

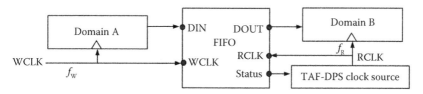

FIGURE 13.14
The architecture of using a TAF-DPS clock source and a synchronous FIFO for data transfer between blocks of different clock frequencies.

overflow and underflow problems can be avoided even when the FIFO size is reduced significantly (compared to the one used in PLL case). During operation, based on the FIFO status, the TAF-DPS can quickly adjust its frequency to move the read-pointer to the appropriate position.

One of the important trends in modern electronic system design is to replace crystal frequency reference source with crystal-less integrated solutions, such as MEMS oscillator, LC, or RC oscillator. The key advantages are lower overall system BOM (bill of material), reduced board space (to meet the miniaturization challenge of Internet of Things, for example), and potentially higher reliability. However, the biggest problem with crystal-less reference is its frequency accuracy and stability. Comparing to crystal-based reference, the crystal-less solution's frequency performance is inferior in several orders of magnitude. To make it operational, frequency trimming and temperature/voltage compensation must be employed. The principal idea is illustrated in the left part of Figure 13.15. The purpose of frequency trimming is to tune the oscillator and bring its frequency into the desired range. This is done by comparing it to an external frequency reference. The aim of frequency compensation is to counteract the frequency variation caused by environmental changes (supply voltage, temperature). In both the trimming and the compensation, the key circuit block is the one for implementing the frequency tuning mechanism. TAF-DPS is suitable for performing this task due to its capabilities of arbitrary frequency generation and instantaneous frequency

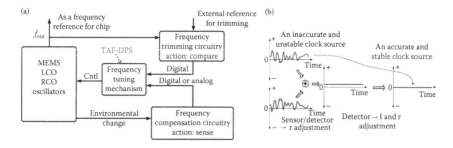

FIGURE 13.15
Using TAF-DPS for assisting crystal-less system design (a) and the scheme to stabilize the frequency (b).

switching. This scheme is illustrated in the right-hand side of Figure 13.15. Since TAF-DPS's frequency (or period) transfer is precisely known, it can be used to accurately cancel any frequency variation caused by environmental changes. This can be achieved by mostly adjusting the fractional part of the control word. For frequency trimming, which usually requires larger step of frequency change, the integer part of the frequency control word can be adjusted to bring the frequency into the range of designed value. As a result, an inaccuracy and unstable crystal-less source can be turned into a reliable one as illustrated in the figure. Comparing to other frequency tuning methods, such as adjusting capacitor array [23,24] or adjusting RC delay constant [25], the TAF-DPS solution is very cost effective [26,27].

13.7 Conclusion: The Philosophical Debate of Spectrally Pure Clock versus Flexible Clock

From all the discussion carried out in previous sections, we reach the stage that a comparison between spectrally pure clock and flexible clock can be made. The goal is to answer this important question: *Which one is more suitable for what application?* The term "spectrally pure clock" is referred to as the clock signal wherein all the cycles have exactly same length-in-time. In circuit practice, this type of clock signal usually comes directly from a crystal oscillator, or is generated from a PLL. The term "flexible clock" is used to describe a clock signal whose choice-of-frequency is ample and whose frequency can be changed quickly. Implementation-wise, the clock signal is generated from circuitry that uses direct synthesis method, such as TAF-DPS. Table 13.2 is the high-level comparison between the two types of clock signals.

Since the first day that clock signal is introduced into circuit design several decades ago, generating spectrally pure clock signal is always the aim of timing circuit designer. This fact has a side effect: it impedes people from looking at the clock signal from other perspectives. More seriously, it precludes people from investigating a more fundamental issue: What does

TABLE 13.2

Comparison of Spectrally Pure Clock and Flexible Clock

Type of Clock	Point of Focus	Clock Structure	Preferred Application
Spectrally pure clock (e.g., PLL)	A clean line in spectrum; clock waveform of high regularity	All cycles have same length-in-time (mathematically intended)	RF carrier for wireless communication; driving data converters
Flexible clock (e.g., TAF-DPS)	Ample frequencies; fast frequency switching	Intentionally, two or more types of cycles might be used	Driving digital circuit for computation, for moving data between blocks

TABLE 13.3

Clock Signals and Their Applications in Several Key IC Design Issues

Design Problems	Spectrally Pure Clock	Flexible Clock
Computation (logic and arithmetic operations)	Good; Best clock waveform regularity for driving digital circuit	Excellent, operating speed can be easily adjusted; Setup constraint needs to be tighten, the shortest cycle is the setup constraint
Digital data communication between entities	Not ideal; Requires larger size FIFO due to the rigorousness of the clock source	Desirable; The clock frequencies of both TX/RX can be adjusted dynamically to smooth data transfer
Power reduction	Not ideal; Clock frequency is hard to be adjusted for adapting to loading change	Desirable; Clock frequency can be dynamically adjusted to accommodate different loadings
EMI reduction	Bad; The cleaner the clock spectrum is, the stronger is the EMI radiation	In harmony; Using various types of clock cycles naturally spreads the clock energy → less radiation
Clock data recovery	Not ideal; Slow RX response due to the rigorousness of the clock source	Desirable; RX clock can track the TX data rate better due to its capability of adjusting itself quickly
Suitableness for mating crystal-less frequency source	Not ideal; Hard to handle frequency error due to the rigorousness of the clock source	Desirable; The capability of small frequency granularity is ideal for compensating the frequency error
Network time synchronization	Not ideal; Due to large frequency granularity	Desirable; Small frequency granularity is better for small time granularity
Frequency hopping	Not ideal; Take long time to adjust frequency due to the rigorousness of the clock source	Desirable; The approach of direct waveform construction is squarely fit for fast frequency change
RF carrier	Excellent; Lowest phase noise → good for signal recognition	Might not be as good; The direct waveform construction might introduce spurious tones (depend on circuit layout)
Drive data converter (ADC and DAC)	Excellent; Lowest phase noise → high SNR	Might not be as good; The direct waveform construction might introduce spurious tones (depend on circuit layout)

clock frequency really mean when clock is used to drive circuit? As is well known, engineering work is all about trade-offs. In this world, *no tool is perfect for every job*. This is very true in the case of using clock-to-drive-circuit. History has proven it many times that, when a new tool is available for better handling of a particular type of job, it is the opportunity for improving work efficiency and making better products. It is the author's belief that, for driving future electronic system, now is the time for reinvestigating this long-lasting presupposition that "all clock pulses must have same length-in-time." By breaking this deadlock, it is possible for us to develop new tools, to attack emerging problems with higher efficiency. The conclusion to this philosophical debate of "spectrally pure clock vs. flexible clock, which one is more efficient?" is as follows: it depends. The pros and cons of using these clock signals in dealing with several key design problems are summarized in Table 13.3. It is the author's desire to share this attitude of "take another look" with PLL designer, system architect, SoC integration engineer, application engineer, test/product engineers, and all other people who have something to do with IC design. It is author's further desire to encourage people to "step out of box" when dealing with emerging problems in future electronic system design. It is a paradigm shift in the design of electronic system [28].

References

1. Poore, R. *Overview on Phase Noise and Jitter,* Agilent Technologies, 2001, http://cp.literature.agilent.com/litweb/pdf/5990-3108EN.pdf accessed date: May 2013.
2. Tektronix. *Understanding and Characterizing Timing Jitter.* Application note, 2010.
3. Hajimiri, A., Lee, T.H. General theory of phase noise in electrical oscillators. *IEEE J Solid-State Circuits,* 33(2), 179–194, 1998.
4. Lee, T.H., Hajimiri, A. Oscillator phase noise: A tutorial. *IEEE J Solid-State Circuits,* 35(3), 326–336, 2000.
5. Abidi, A.A. Phase noise and jitter in CMOS ring oscillators. *IEEE J Solid-State Circuits,* 41(8), 1803–1816, 2006.
6. Mair, H., Xiu, L. An architecture of high-performance frequency and phase synthesis. *IEEE J Solid-State Circuits,* 35, 835–846, 2000.
7. Calbaza, D.E., Savaria, Y. A direct digital periodic synthesis circuit. *IEEE J Solid-State Circuits,* 37(8), 1039–1045, 2002.
8. Xiu, L., You, Z. A flying–adder architecture of frequency and phase synthesis with scalability. *IEEE Trans VLSI,* 10, 637–649, 2002.
9. Kroupa, V.F. *Direct Digital Frequency Synthesis,* IEEE Press, 1998.
10. Sotiriadis, P., Galanopoulos, K. Direct all-digital frequency synthesis techniques, spurs suppression, and deterministic jitter correction, *IEEE Trans. Circuit System I,* 59(5), 958–968, 2012.
11. Gardner, F.M. *Phase Lock Techniques,* 3rd edition, Wiley-Interscience, New York, 2005.

12. Egan, W.F. *Phase-Lock Basics*, 2nd edition, Wiley-IEEE Press, New York, 2007.
13. Kundert, K.S. *Predicting the Phase Noise and Jitter of PLL-Based Frequency Synthesizers*, www.designers-guide.com.
14. Helal, B.M., Hsu, C.M., Perrott, M.H. Low jitter programmable clock multiplier based on a pulse injection-locked oscillator with a highly-digital tuning loop. *IEEE J Solid-State Circuits*, 44(5), 1391–1400, 2009.
15. Wang, B., Ngoya, E. Integer-N PLLs verification methodology: Large signal steady state and noise analysis. *IEEE Trans. Circuit System I*, 59(11), 2738–2748, 2012.
16. Park, D., Cho, S.H. A 14.2 mW 2.55-to-3 GHz cascaded PLL with reference injection and 800 MHz Delta-Sigma modulator in 0.13 m CMOS. *IEEE J Solid-State Circuits*, 47(12), 2989–2998, 2012.
17. Kyriazis, G. "Heterogeneous System Architecture: A Technical Review," AMD, 2012. http://developer.amd.com/wordpress/media/2012/10/hsa10.pdf
18. Vig, J.R. *Quartz Crystal Resonators and Oscillators, For Frequency Control and Timing Applications—A Tutorial*, March, 2004. https://escies.org/download/webDocumentFile?id=62209
19. Xiu, L. The concept of time-average-frequency and mathematical analysis of flying–adder frequency synthesis architecture. *IEEE Circuit System Magazine*, 3rd quarter, pp. 27–51, 2008.
20. Xiu, L. *Nanometer Frequency Synthesis Beyond Phase Locked Loop*. John Wiley IEEE Press, Hoboken, New Jersey, 2012.
21. Xiu, L., Lin, K.-H., Ling, M. The impact of input-mismatch on flying–adder direct period synthesizer. *IEEE Trans. Circuit Syst. I*, 59, 1942–1951, 2012.
22. Xiu, L., Lin, W.T., Lee, K. A flying–adder fractional-divider based integer-N PLL: The 2nd generation flying–adder PLL as clock generator for SoC. *IEEE J Solid-State Circuits*, 48, 441–455, 2013.
23. McCorquodale, M.S., O'Day, J.D., Pernia, S.M. et al. A monolithic and self-referenced RF LC clock generator compliant with USB 2.0. *IEEE J Solid-State Circuits*, 42(2), 385–399, 2007.
24. McCorquodale, M.S., Carichner, G.A., O'Day, J.D. et al. A 25-MHz self-referenced solid-state frequency source suitable for XO-replacement. *IEEE J Solid-State Circuits*, 943–956, 56(5), 2009.
25. Sundaresan, K., Allen, P.E., Ayazi, F. Process and temperature compensation in a 7-MHz CMOS clock oscillator. *IEEE J Solid-State Circuits*, 433–442, 41(2), 2006.
26. Xiu, L. Circuit and method of using time-average-frequency direct period synthesizer for Improving crystal-less frequency generator's frequency stability, US patent pending, Serial No.: 14/140016.
27. Xiu, L. Direct period synthesis for achieving sub-PPM frequency resolution through time-average-frequency: The principle, the experimental demonstration and its application in digital communication. *IEEE Trans. VLSI*, accepted in June 2014, to be published soon.
28. Xiu, L. *From Frequency to Time-Average-Frequency: A Paradigm Shift in the Design of Electronic System*, John Wiley IEEE Press, Hoboken, New Jersey, 2015.

14

Machine Learning-Based BIST in Analog/RF ICs

Dzmitry Maliuk, Haralampos-G. Stratigopoulos, and Yiorgos Makris

CONTENTS

14.1 Introduction

The seemingly ever-increasing data rates and complexity of modern analog and RF integrated circuits (ICs) intensify the test effort that must be spent to guarantee the correct functioning of each manufactured part. Pushing device geometries to their physical limits has made them more susceptible to manufacturing process variations and defects. To ensure the compliance of each manufactured part to its specifications, hundreds of sequential tests are carried out in practice. This procedure demands sophisticated test equipment and creates a major bottleneck in the manufacturing throughput, thereby escalating significantly the overall manufacturing cost. As RF devices become a ubiquitous part of our everyday lives, there is a great incentive to reduce the implicated test costs, while maintaining the highest possible quality and reliability of the parts that pass the test.

Furthermore, the postproduction guarantees of fault-free functionality can no longer be made after the parts are deployed in the field of operation. The effects that cause MOS transistors to degrade with time, such as hot-carrier injection (HCI) and negative bias temperature instability (NBTI), have become more prominent in advanced process nodes. The results of transistor wear and tear is a gradual decline in performance or an abrupt failure due to catastrophic faults (such as dielectric breakdown). The consequences of such failures depend on particular applications, but are usually detrimental for safety-critical applications (e.g., avionics, medicine, nuclear reactors), sensitive environments (e.g., space operations), and remote-controlled systems. Understanding the aging mechanisms to increase the life span of ICs is the subject of failure analysis. Equally important, however, is the detection and, possibly, prevention of aging-related faults in ICs after their deployment in the field of operation. Commonly know as a *built-in self-test* (BIST) and typically used in digital circuits, this topic remains an elusive yet highly desirable feature of analog and RF circuits.

In this work, we discuss a machine learning-based BIST which offers a promising solution to the mentioned problems. A BIST-enabled IC will examine its own functional health using exclusively on-chip resources. The benefits of this feature are twofold. First, in postproduction test, each manufactured part could be tested via BIST to identify failing devices that either exceed their specification limits or contain defects. By simply avoiding the use of expensive test equipment and eliminating lengthy test times, the associated test cost could be significantly reduced. Second, BIST empowers ICs with self-test capabilities after their deployment in the field of operation. In such a scenario, BIST is periodically executed to ensure the correct functionality of the IC and report if any of its performance are out of the respective and desired design specification metrics to inform an operator of a potential system failure.

14.2 Machine Learning-Based BIST Architecture

Traditionally, RF testing has relied on expensive automated test equipment (ATE) to explicitly measure performance parameters (such as bit error rate,

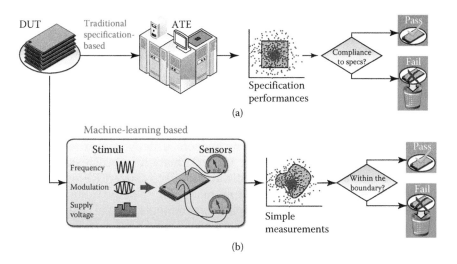

FIGURE 14.1
Traditional specification-based versus machine learning-based methodologies for testing analog/RF circuits.

noise figure, gain, etc.) to decide if a device under test (DUT) is functional or not (Figure 14.1a). Clearly, it is economically impossible to integrate the functionality of ATE on chip. However, an increased focus of the past decades on alternative test strategies has created the necessary prerequisites for a truly stand-alone BIST. These alternative test strategies can be broadly categorized as structural test, built-off test (BOT), built-in test (BIT), and machine learning-based test.

Structural tests aim to generate signatures that discriminate between faulty and functional circuits [1,2]. They are devised specifically to detect a list of faults that result from a prior inductive fault analysis. This approach presents a high potential to reduce test costs, but further research is required on fault modeling. BOT relies on relatively simple circuits to convert RF signals to DC signals [3,4]. These circuits are placed on the device load-board, thus allowing the interface of the DUT to an inexpensive tester. BIT is a more aggressive approach. It consists of building a miniature tester on-chip [5,6], reconfiguring the DUT in an easily testable form [7], and extracting on-chip informative measurements based on sensors [8–12]. Machine learning test is a generic technique which aims to infer the outcome of the standard tests based on low-cost measurements. This inference relies on statistical models that are built off-line using a representative training set of instances of the DUT. Different types of statistical models can be used, namely regression functions, in which case we can predict individually the high-level performances (this approach is known as alternate test [13–16]), or classifiers, in which case we can make directly a pass/fail test decision [17].

The machine learning-based approach represents a great candidate for a stand-alone BIST for two reasons. Firstly, it relies on measurements that can be obtained inexpensively using exclusively on-chip resources. Secondly, these alternate measurements along with performance parameters are subject to the same process variations, thus with appropriate information-rich alternative measurements we can track variations of performances. The test can be performed as follows (Figure 14.1b): The stimuli generators excite the DUT with predefined test stimuli (e.g., single tone RF input, modulated input, supply voltage modulation, etc.). The sensors usually comprise low-cost circuits extracting low-frequency or DC information from RF signals. For example, peak detectors are frequently employed to convert RF signals into DC voltages proportional to the maximum amplitude of the input. The sensor outputs comprise a vector of simple measurements which is used to infer a pass/fail label. This is equivalent to constructing a boundary in the space of simple measurements and classifying the DUT according to which side of the boundary the response footprint falls onto. Unlike in the traditional test, this boundary represents a highly nonlinear and complex hypersurface reflecting the fact that there is no simple relation between sensor outputs and performance parameters. Thus, test decisions cannot be derived analytically; rather they have to be learned from a representative set of instances of the DUT. An additional nuisance of the space of simple measurements is the overlap between the populations of good and bad devices. This creates a nonzero classification error as a result of inferring the test outcome from simple measurements instead of performing the specification-based test. Test escape and yield loss are often used as test metrics quantifying the quality of the machine learning-based test.

The aforementioned ideas can be incorporated to obtain a truly stand-alone BIST (Figure 14.2). Its architecture mainly consists of three components [18]: a stimulus generator, measurement acquisition sensors, and a measurement processing mechanism. When instructed to test itself, the DUT is connected to the stimulus generator which supplies a test stimulus to the DUT. The on-chip sensors monitor the DUT and provide the measurement pattern which is presented to the neural network. Finally, the neural network classifier maps on-chip sensor measurements into a one-bit output, which simply indicates whether the DUT complies to its specifications or not. The mapping is derived in an off-line learning phase on a sample set of fabricated chips. The learning phase results in an appropriate topology for the neural network and it also determines the weights which are stored in a local memory. During the test phase, the neural network classifies the DUT by processing the measurement pattern and examining its footprint with respect to the learned classification boundary.

Among all the components comprising the BIST architecture, the on-chip classifier plays a special role. It is not only the central component enabling the machine learning BIST but also the most challenging component to design. While the stimuli generators and sensors do require careful analysis

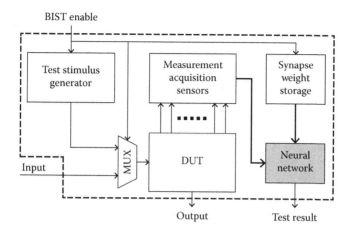

BIST enable

Input

MUX

Output

Test result

FIGURE 14.2
Machine learning-based BIST architecture. On "BIST Enable" command, the DUT is excited by the an on-chip test stimulus. Multiple on-chip sensors collect simple measurements which are processed by an on-chip neural network that produces a test result. (Maliuk, D. et al. Analog neural network design for RF built-in self-test. *Proceedings of the IEEE International Test Conference* (ITC), Austin, TX, pp. 23.2.1–23.2.10. © 2010. IEEE.)

to produce meaningful measurement, their hardware implementation aligns well with common analog/RF circuit design techniques. The neural classifier, on the other hand, is originally conceived as a mathematical model which finds its best expression in a software domain. The functional complexity of the neural classifier is also considerably higher comprising tens or hundreds of multiply-accumulate (MAC) operations. In addition, it requires pretuning, or training, of its weight coefficients to achieve the desired functionality along with an on-chip memory to store this learned functionality. In view of these challenges, the hardware implementation of a neural classifier is recognized as a key factor enabling a fully stand-alone BIST and represents the main focus of this chapter.

14.3 Cost-Efficient Implementation of Neural Classifier

In applications such as BIST, the additional power and area overhead incurred by test circuits is crucial. Since a neural network is functionally the largest component within the BIST architecture, it is highly probable that its hardware implementation will consume the major portion of resources allocated to BIST. We prefer an analog implementation because it delivers both exceptional power efficiency during run time and a compact size. Indeed, when high precision is not required, analog computation can be as much as 1000× more energy efficient than digital [19]. The area requirements are modest as well; a four-quadrant multiplier can be implemented by as few as four transistors operating below threshold [19]. The summation can be attained for free by connecting wires together and utilizing Kirchhoff's current law. Many other mathematical functions (such as square root, power, hyperbolic tangent, etc.) could be built with just a few transistors using the translinear principle [20]. An entire system could be compiled from these primitive circuits for a fraction of the price of a digital implementation. However, building such systems requires high level of expertise to address many of the concerns inherent to low-power analog design, such as process variation, mismatch, temperature, and noise. As an example, a simple current mirror circuit can introduce an error of up to 100% when copying an input current to an output branch [21]. High sensitivity of circuit parameters and the lack of programmability have restricted the use of low-precision analog VLSI circuits to adaptive, biologically inspired computational architectures, such as artificial neural networks [22], silicon retina [23], silicon cochlea [24], vector matrix multipliers [19], support vector machines [25–27], transform imagers [28], etc. Neural models are an example of adaptive, distributed processing networks requiring only two building blocks, namely a synapse and a neuron circuit, for their construction. The following reasons warrant the selection of the analog domain as a medium for the hardware neural networks implementation:

- It delivers superior power efficiency
- It requires small silicon real estate
- It can be directly interfaced with sensor outputs, thereby eliminating the use of analog-to-digital converters
- It leverages the floating gate (FG) technology for high-precision non-volatile weight storage

Despite the benefits, there are additional constraints imposed by the analog implementation, such as limited dynamic range of signals and weights, limited resolution of weight values programming, nonlinearities, and offsets in synaptic multiplications. These limitations could be successfully circumvented by taking advantage of the highly adaptive nature of neural models. In fact, a properly designed training strategy can learn around these nonidealities by treating a neural network as a "black box," that is, without precise knowledge of its analytical model. One implication of this approach is that we cannot preprogram the functionality of a neural network without taking into account its physical characteristics (i.e., off-chip training). In other words, each instance of the analog network has to be trained *in situ* by either an on-chip or chip-in-the-loop training strategy. We prefer the latter for its minimal on-chip support, which is in line with the strict area and power budget imposed by BIST. Transforming the aforementioned ideas into analog circuits and demonstrating their learning potential is the subject of subsequent discussion.

To address the requirements imposed by the BIST architecture, we present the designs of two reconfigurable analog neural network experimentation platforms [29,30]. Each platform features a large number of components and serves as a prototyping tool for identifying appropriate classifier models and their parameters for a BIST-related classification problem. Thereby, the final classifier circuitry to be integrated on chip is optimized to deliver high classification accuracy while utilizing a small fraction of each platform's resources.

14.4 Analog Neural Network with Digital Weight Storage (ANNDW)

In the first design, we aimed at demonstrating the proof-of-concept learning ability of analog neural networks [18,29]. In other words, we set out to prove that relatively simple analog circuits could learn nonlinear classification boundaries in the space of low-cost measurements despite the nonidealities of the analog implementation. The experimentation platform fabricated in a 0.5-μm CMOS process represents an array of 100 synapses and 10 neurons that could be configured into multilayer perceptron (MLP) models with a variable number of hidden neurons. The weight coefficients are implemented as digital words stored in 6-bit SRAM cells next to each synapse circuit. Analog operation domain of synapses and neurons in conjunction with the digital weight storage allow fast computational time and rapid training cycle. The network is trained in the chip-in-the-loop fashion with the simulated annealing-based parallel weight perturbation training algorithm.

14.4.1 Supported Neural Network Model

The first platform is designed to model a widely used artificial neural network model, namely a multilayer perceptron (MLP), which is capable of learning complex nonlinear classification problems due to the presence of hidden layers in its topology [31]. The block diagram of an MLP is illustrated in Figure 14.3a. The network is feed-forward; it does not contain feedback loops and each layer receives connections only from inputs or previous layers. The first layer (a.k.a. hidden) has M neurons which receive the inputs X_1, X_2, ..., X_N and a constant $X_0 = 1$. The second layer (a.k.a. output) contains K neurons (where $K = 1$ for binary classification) receiving the outputs $Z_1, Z_2, ...,$ Z_M of the first layer and a constant $Z_0 = 1$. The strength of connections is controlled by synapses which act as multipliers of input signals and their local weight values. The sum of synaptic products is passed through a nonlinear activation function of a neuron. The number of input and output neurons is

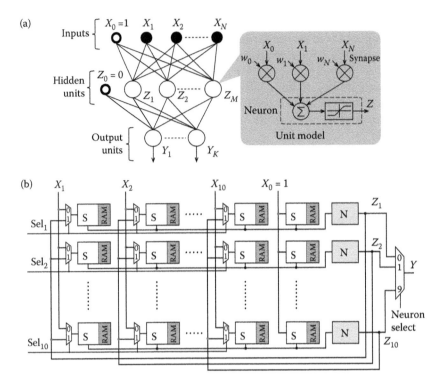

FIGURE 14.3
(a) Connection topology of a multilayer perceptron composed of one hidden and one output layer. (b) Block-level schematic consisting of a matrix of synapses (S) and neurons (N). A unit is represented by a row of synapses terminated with a neuron. The role of each unit is defined by Sel_j: hidden units ($Sel_j = 0$) receive primary inputs $X_{\{1,\ldots,10\}}$, while the output units ($Sel_j = 1$) receive the outputs of the hidden units $Z_{\{1,\ldots,10\}}$.

usually defined by the classification problem itself, while the number of hidden neurons reflects the learning capacity of the model.

14.4.2 System Architecture

Figure 14.3b illustrates a block-level schematic of a circuit implementation of the mentioned neural model. The circuit consists of a matrix of synaptic blocks (S) and neurons (N). The synapses represent mixed-signal devices, in the sense that they conduct all computations in analog form while their weights are implemented as digital words stored in a local RAM memory. Multiplexers before each synapse are used to program the source of its input: either the primary input (for the hidden layer) or the output of a hidden unit (for the output layer). The products of synapse multiplication are fed to the corresponding neuron, which performs summation and nonlinear transformation and produces an output either to the next layer or the primary output. The architecture is very modular and can be easily expanded to any number of neurons and inputs within the available silicon area. The signal encoding takes different forms: the outputs of the neurons are voltages, while the outputs of the synapses are currents. In addition, all signals are in differential form, thereby increasing the input range and improving noise resilience. The presented design can be configured into any one-hidden-layer topology within the given number of inputs and neurons.

14.4.3 Synapse and Neuron Circuits

The synapse circuit chosen for this design is a simple multiplying DAC [32], which represents a differential pair with programmable tail current, as shown in Figure 14.4b. The differential input voltage $\{V_{in} +, V_{in} -\}$ splits the tail current I_{tail} to produce the differential output current $\{I_{out} +, I_{out} -\}$. Current summation of different synaptic products takes place at the output nodes which are shared among all synapses connected to the same neuron. The core of the circuit is a differential pair N10–N11 performing a two-quadrant multiplication, while the four switching transistors P0–P3 controlled by bit B5 steer the current between the two summing nodes, thus defining the sign of the multiplication. The tail current is digitally controlled by the five switch transistors N5–N9 connecting the corresponding binary weighted current sources N0–N4 to the tail node. Thus, the tail current as a function of a digital weight word (bits B4–B0) can be represented by

$$I_{tail} = \sum_{i=0}^{4} B_i \cdot I_i = \sum_{i=0}^{4} B_i \cdot (I_0 \cdot 2^i) = I_0 \cdot W \tag{14.1}$$

where W is the weight value, B_i is the ith bit of the weight value, I_i is the current corresponding to the ith bit, and I_0 is the LSB current. Linear multiplication is only valid for a narrow range of input voltages which depends on both the tail current and the transconductance coefficient. The increase in this range was achieved by selecting long channel transistors (14 µm), while keeping the tail current low (<1.5 µA). As shown in the measured characteristic of Figure 14.4c, the input range depends on the tail current, which is set by the weight value, and is about 800 mV for the maximum weight. In order to accommodate a large gate–source voltage of the differential pair and a minimum drain voltage of the binary current sources the input voltages must be held relatively high. This translates into a high common-mode input voltage requirement that must be met by the neuron circuits and sensors supplying the inputs.

   A neuron converts the sum of differential currents from the connected synapses into a differential voltage which propagates to the next layer (for hidden neurons) or serves as a network output (for output neurons). The design of the neuron circuit pursues two goals. First, the output voltage of the neuron should be compatible with the input voltage requirements of the

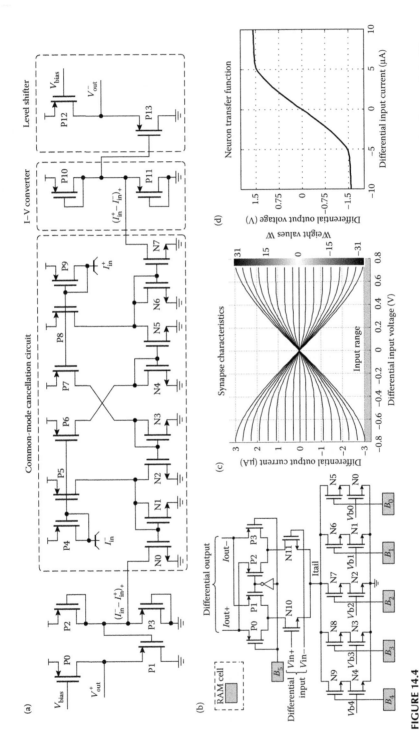

FIGURE 14.4

Key circuits of the ANNDW platform: (a) neuron circuit schematic; (b) synapse circuit schematic; (c) measured synapse transfer characteristic; (d) measured neuron transfer characteristic. (Maliuk, D. et al. Analog neural network design for RF built-in self-test. *Proceedings of the IEEE International Test Conference (ITC)*, Austin, TX, pp. 23.2.1–23.2.10. © 2010. IEEE.)

synapses, that is, the common-mode component should be high. Second, the circuit should handle relatively large dynamic range of input currents. While the useful information is contained in the difference, the common-mode current may vary significantly depending on the number of connected synapses, as well as on their weight values. For example, in our design the common-mode current ranges from 90 nA (one synapse, minimum weight value) to 30 μA (10 synapses, maximum weight values).

A circuit satisfying these requirements is shown in Figure 14.4a. The central part of the circuit is responsible for common-mode cancellation by subtracting the input currents from each other and producing a positive difference. The subtraction occurs at the drains of $N5$ and $N2$ and the positive difference flows into $N6$ and $N1$. The positive and negative components of the difference are $\max\{0, (I_{in}^- - I_{in}^+)\}$ and $\max\{0, (I_{in}^+ - I_{in}^-)\}$, respectively. Note that only one of these nodes carries nonzero current. The second stage is a simple current-to-voltage converter composed of two p-channel MOSFETs. A current extracted from the source of $P11$ reduces the source voltage. This conversion is linear when two transistors are identical and is given by

$$V = \frac{V_{dd}}{2} - \frac{I}{2K_P(V_{dd} - 2V_{TP})} \tag{14.2}$$

where V is the source voltage, I is the extracted current, K_P is the transconductance coefficient of two identical transistors, V_{TP} is the threshold voltage, and V_{dd} is the supply voltage. The circuit also provides a limiting function when the extracted current exceeds the internal current flowing through the circuit, thus introducing nonlinearity to the neuron characteristic. Notice from the formula above that the slope of the characteristic depends on the K_P, which is set at the design stage by specifying transistor sizes. Finally, the output of the converter is shifted upward to meet the requirements of the high common-mode input voltage for the synapses in the following layer. This level shifter is a simple source follower circuit where the amount of shift is controlled by V_{bias}. A shift of 1 V is used in this design. Figure 14.4d shows the measured transfer characteristic of the entire circuit which serves as a good approximation to the sigmoid activation function of the software neurons.

14.4.4 Hardware-Friendly Training Algorithm

The algorithm we employed for training our neural network represents a combination of the parallel weight perturbation [33] and the simulated annealing techniques. During each iteration a vector of weights **w** is perturbed by a random vector **pert** and an error E_{pert} is evaluated on the entire training set **X**. If the new error decreases as compared to the current error E, the new vector of weights is accepted. Otherwise, it is accepted with probability $e^{-|E_{pert} - E|/T}$, where T is a parameter called temperature. The ability to accept weight changes that result in an increase in the error is instrumental in avoiding local minima and is known as the simulated annealing technique. Higher temperatures at the initial stages favor the exploration of the whole search space, whereas low temperatures at the final stages facilitate convergence to the found minimum. The corresponding cooling schedule is given by

$$T(n + 1) = \alpha \times T(n) \tag{14.3}$$

where α is the scaling factor ($\alpha < 1$), and n is the iteration number. The error term is calculated using a vector of class labels $\widehat{\mathbf{Y}}$ returned by the neural classifier. It is found by passing the entire training set through the analog neural network and measuring its output response. Engaging the hardware during training is a distinct feature of the chip-in-the-loop learning strategy. The error between $\widehat{\mathbf{Y}}$ and a vector of true labels $\mathbf{Y}$ is obtained by

$$Err(\widehat{\mathbf{Y}}, \mathbf{Y}) = \sum_i \left(\frac{1}{1 + e^{\gamma \cdot \widehat{Y}_i \cdot Y_i}} + \max\{-\widehat{Y}_i \cdot Y_i, 0\} \right) \tag{14.4}$$

where $Y_i \in \{-1,1\}$ is a true label of the ith pattern, γ is the scaling parameter, and the sum is applied over the entire training set. This form differs from the standard squared-error function in that it does not penalize patterns that were classified correctly with large positive margin ($\widehat{Y}_i \cdot Y_i \gg 0$) and grows linearly for patterns with large negative margin ($\widehat{Y}_i \cdot Y_i \ll 0$). The latter property makes this error function more robust in presence of outliers.

14.5 Floating Gate-Based Analog Neural Platform

While the first neural network chip (ANNDW) serves as a proof-of-concept implementation demonstrating an adequate learning ability for the test-related classification problems, the issues of cost-efficient implementation are tackled with the second platform [30]. As mentioned before, practical implementation of BIST hinges on our ability to meet stringent area and power constraints of the circuits dedicated to neural networks, which, however, should not compromise their ability to learn fast and retain functionality throughout their lifecycle. In an effort to address these issues, we designed and fabricated a floating gate-based neural network (FGNN) platform which represents an array of synapses and neurons reconfigurable into two learning structures: a multilayer perceptron and an ontogenic neural network (ONN). With this platform, we intend to address the key cost-efficiency issues: a fully analog implementation built with current-mode subthreshold circuits, a dual-mode weight memory for fast training and long-term nonvolatile storage, and a learning ability of the proposed architecture. The dynamic weight storage mode is engaged during training due to its high speed, while the nonvolatile storage makes use of the FG technology for permanent weight storage during operation or standby.

14.5.1 Ontogenic Neural Network Model

Unlike multilayer perceptron, the ONN learns the boundary by both adjusting the weights and expanding its topology [17]. It has a similar structure, including a layer of inputs, several hidden neurons, and an output neuron (Figure 14.5a). However, a single neuron receives connections from both the inputs and the outputs of preceding neurons. A decision boundary is constructed by successively adding hidden neurons: each hidden neuron augments the feature space of the original inputs with the intention of making the derived space linearly separable. Trained by cascade-correlation, the ONN avoids the "moving target" problem inherent in back propagation [34], thereby resulting in more compact network sizes for a given classification problem. In the remainder of the chapter, the topologies of both neural models have an X–Y–Z designation, where X, Y, and Z are the number of inputs, hidden neurons, and output neurons, respectively.

14.5.2 Chip Architecture

The block diagram of the FGNN platform is presented in Figure 14.5b. A 30×20 array of synapses (S) and neurons (N) is arranged so that the neurons are aligned along the main diagonal of the upper matrix and along the right edge for the bottom part. Each row of synapses is locally connected to a corresponding neuron, forming a single unit. Global connectivity is programmable by means of multiplexors inserted between rows. The basic neural network operations, such as multiplication and nonlinear activation functions, are implemented via the translinear principle [35]. The signals and weights are represented by balanced differential currents for increased robustness and four-quadrant multiplication. As a result, a single weight value requires two current sources for differential current storage. For this purpose, we designed a current storage cell (CSC) featuring two modes of weight storage: dynamic, for rapid bidirectional programming, and nonvolatile, for long-term storage of learned weights. The dynamic mode is engaged during training, which requires thousands of weight update operations to be completed in a short period of time. Upon completion of training, the learned weights are stored permanently using FGTs.

The peripheral circuits provide support for fast programming and interfacing with the external world. The differential transconductors GM convert voltage-encoded input signals into balanced differential currents, as required by the core. The digitally controlled current source IDAC generates

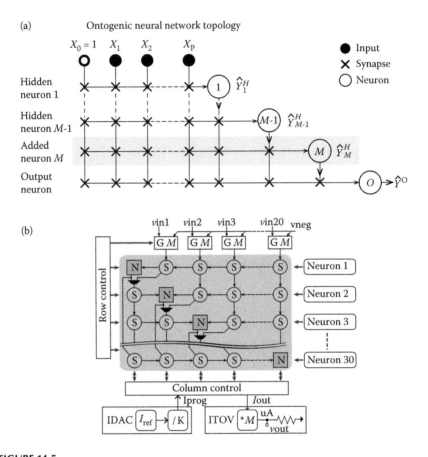

FIGURE 14.5

(a) Ontogenic neural network learning model. (b) System architecture of the floating gate-based neural platform.

target currents from an on-chip reference for dynamic programming of the CSCs. Finally, the current-to-voltage converter ITOV facilitates the reading of internal currents by converting them to voltages that can be sampled by an external ADC. Each of these blocks requires characterization prior to their first use to achieve a desired accuracy.

14.5.3 Weight Storage

The principle of weight storage is illustrated in Figure 14.6b. We use a multiple-input FG transistor (FGT) P1 to store the drain current I_w representing one of the weight value components. The drain current is modulated by the voltage on the FG node, which is itself determined by the FG node charge and the voltages on two control gates. The global voltage *vgate*1 of the first control gate is shared among all FGTs, while *vgate*2 is stored locally in the dynamic

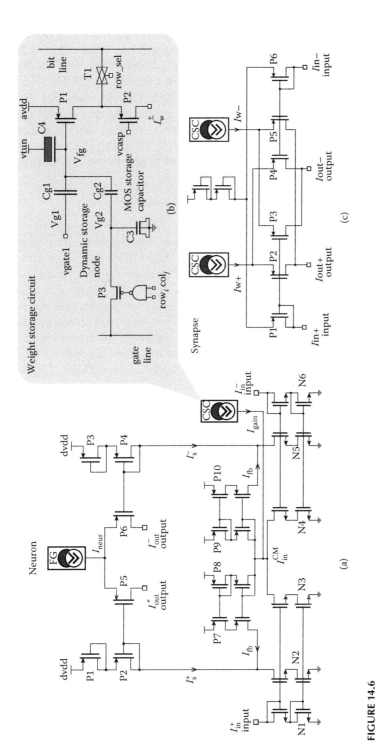

FIGURE 14.6

Key circuits of the FGNN platform: (a) neuron circuit schematic; (b) weight storage circuit; (c) synapse circuit schematic.

sample-and-hold (S/H) circuit which consists of the switch transistor P3 and the MOS capacitor C3. The low-coupling capacitor C2 makes I_w much less sensitive to charge leakage and other parasitic effects of the S/H circuit. The tunneling capacitor C4 is implemented as a minimum size PMOS transistor with its source, drain and well terminals connected to *vtun*. Hot-electron injection is used to add electrons to the FG, thus, lowering its voltage and increasing the drain current. Conversely, Fowler–Nordheim (FN) tunneling is used to remove electrons from the FG. The tunneling is used for global erase only, while the injection allows us to individually program drain currents of each CSC with high accuracy ($\approx$8 bits).

14.5.4 Synapse and Neuron Circuits

The synapse circuit, illustrated in Figure 14.6c, implements a four-quadrant multiplication of a differential input current $\{I_{in}^+, I_{in}^-\}$ by a differential weight current $\{I_w^+, I_w^-\}$. The circuit features two CSC cells for differential weight component storage and a six-transistor core P1–P6. The neuron circuit, illustrated in Figure 14.6a, implements a nonlinear activation function of the sum of the outputs of the connected synapses. This nonlinear transformation is completed in two stages. The first stage, represented by the bottom part of the circuit, controls the slope of the activation function. The slope is adjusted by programming the I_{gain} current, which is stored in a local CSC. The second stage, implemented by the top part of the circuit (P1–P6), performs nonlinear transformation of the normalized input current. The common-mode signal I_{neur} of the output current is set by a separate FGT.

14.5.5 ONN Training Algorithm

For training ONN network topologies, we employ a cascade-correlation (CC) algorithm [36]. Training starts with a minimum size topology including only an output neuron. Each subsequent neuron adds on to the learning capacity of the model, improving the error on the training set. Each iteration of the training algorithm consists of two steps. Suppose our current topology has $M-1$ hidden neurons (see Figure 14.5a for reference). Let $Y_j^H(X_i)$ represent the output of the jth hidden neuron for the training vector X_i and $Y_j^O(X_i)$ be the network output when it has k hidden neurons. In the first step, an Mth hidden neuron is added at the bottom so that it receives the primary inputs $X_1, \ldots, X_p$ as well as the outputs of all preexisting neurons $Y_1^H, \ldots, Y_{M-1}^H$. This neuron is trained to maximize the covariance between its output $Y_M^H(X_i)$ and the residual training error of the previous iteration $Y_{M-1}^E(X_i) = Y_i - Y_{M-1}^O(X_i)$. This covariance is given by

$$C = \sum_i \left(Y_M^H(X_i) - \overline{Y_M^H} \right) \left(Y_{M-1}^E(X_i) - \overline{Y_{M-1}^E} \right) \qquad (14.5)$$

where $\overline{Y_M^H}$ and $\overline{Y_{M-1}^E}$ represent averaged quantities over the entire training set. The covariance maximization is equivalent to the minimization of $-C$, which can be accomplished by the RPROP algorithm with the error function $E = -C$. Once the covariance is maximized, the weights of this neuron become permanent and the output layer is retrained to minimize the error on the training set. The second step is again completed by the RPROP algorithm. Note that in each iteration, only the weights of the neuron being added undergo modification, followed by the weights of the output neuron, while the other weights are kept unchanged. This feature greatly simplifies the gradient estimation by the hardware and leads to stable performance even for large-sized topologies. Hidden neurons are added until one of the following two events is achieved, depending on the classification task: a zero-error on the training set or a target-error on a validation set.

14.6 Case Study I: Parametric Faults (LNA Circuit)

14.6.1 Problem Definition and DUT

In our first case study, we demonstrate the effectiveness of training our neural network platforms to separate faulty from functional LNAs using a vector of BIT measurements and compare it with the results of equivalent training performed in software. For our experiment, we selected the DUT to be a standard source-degenerated LNA shown in Figure 14.7a. The initial design targeted the set of nominal performances listed in Table 14.1. The LNA is integrated along with two RF amplitude detectors that are placed at its input and output ports. The amplitude detectors provide DC signals proportional to the RF power seen at their inputs. These DC signals comprise the BIT measurement pattern that is finally presented to the classifier. The objective is to have very simple sensor structures with small area and power overhead, such that they qualify for a BIST application. The schematic of the selected amplitude detector is presented in Figure 14.7b and its key features are summarized in Table 14.2. The circuits were designed in the TSMC 0.18 μm CMOS process and were integrated into a single die as shown in Figure 14.7c.

14.6.2 Data Source

A set of 1000 instances reflecting manufacturing process variations is generated through a Monte Carlo postlayout simulation (Figure 14.7d). The parameter distribution and mismatch statistics were provided by the design kit. The amplitude detectors undergo the same process variations as the LNA. Next, each instance is fully characterized using standard test benches to

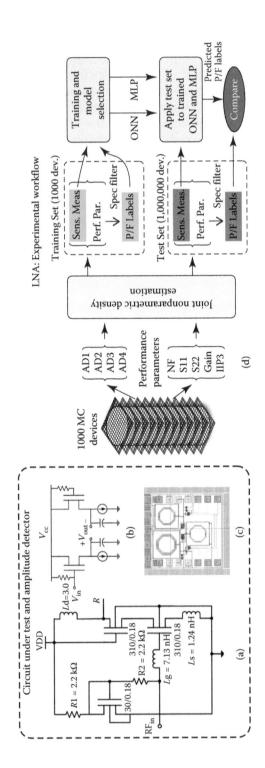

FIGURE 14.7

Case study I: (a) LNA circuit schematic; (b) peak detector; (c) layout of LNA and peak detector; (d) course of experiments. (Maliuk, D. et al. Analog neural network design for RF built-in self-test. *Proceedings of the IEEE International Test Conference (ITC)*, Austin, TX, pp. 23.2.1–23.2.10. © 2010. IEEE.)

TABLE 14.1

LNA Features

Performance	Nominal
Central frequency (in GHz)	1.575
NF (in dB)	2
S11 (in dB)	−10
Gain (in dB)	15
IIP3 (in dBm)	−5
S22 (in dB)	−10
Power consumption (in mW)	18

obtain the set of performance parameters. In addition, we obtain a four-dimensional BIT measurement pattern (e.g., the test signature) by applying to the LNA two test stimuli of powers −10 and 0 dBm consisting of a single tone at 1.575 GHz. It should be noted that these power levels might be higher than those of typical signals expected during a real application. This is because the input amplitude detector does not have enough gain to detect small incoming signals, in response to which its output is likely to be hidden below the noise level.

For the original sample of fully characterized devices we set the specification limits to Mean $\pm 3 \cdot$ StdDev (where + or − depends on whether the performance has a lower or upper limit). The Mean and StdDev of individual performances are computed across the set of 1000 devices. Our choice of specifications resulted in eight faulty devices out of 1000. Figure 14.8a illustrates the distribution of good and faulty devices projected onto a 2D subspace of the measurement pattern. In particular, the "×1" and "×2" axes correspond to the normalized input and output amplitude detector readings when excited by the −10 dBm signal. The eight faulty devices fall in the tail of the distribution which is sparsely populated. For training the classifier, however, it is preferable to have a balanced dataset, such that one class does not overshadow the other. Furthermore, sparsely populated areas produce randomness in the curvature of the separation boundary and, thereby, can deteriorate the classification. Therefore, the training set must be populated with many marginal devices whose footprint lies in close proximity to the boundary. Similarly, we would like to evaluate the classification rate using a larger validation set. A technique to add the desired features to the original dataset is discussed next.

TABLE 14.2

Power Detector Features

Central frequency (in GHz)	1.575
Dynamic range (in dB)	50
Power consumption (in μW)	6.66
Area overhead (in μm²)	4 2 × 80

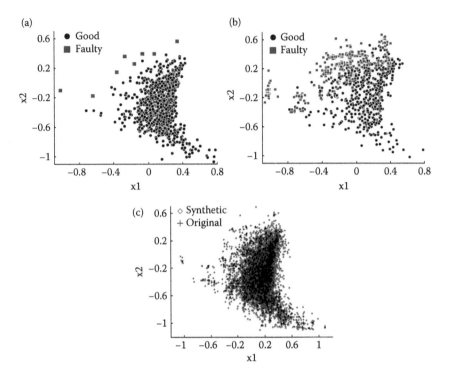

FIGURE 14.8
LNA datasets (2D projections): (a) original dataset consisting of 1000 devices generated through Monte Carlo simulation; (b) synthetic training set derived through resampling of the original distribution; (c) original and synthetic devices shown together. One million of synthetic devices comprise the validation set. (Maliuk, D. et al. Analog neural network design for RF built-in self-test. *Proceedings of the IEEE International Test Conference (ITC)*, Austin, TX, pp. 23.2.1–23.2.10. © 2010. IEEE.)

14.6.3 Dataset Enhancement

A Monte Carlo simulation samples with priority the statistically likely cases. Therefore, it is typically required to go through many passes so as to assemble a training set with an adequate number of marginal devices. Notice that a few simulations targeting only process and environmental corners do not suffice because they do not account for the actual statistics. To alleviate the computational effort, we adopt the technique in Reference 37.

The main idea is to use the available set of 1000 devices to estimate the joint probability density function of the performances and the measurement pattern. Subsequently, the estimated density can be sampled to obtain new observations that correspond to synthetic devices following the original distribution. Sampling can be performed very fast, in particular 1 million observations can be obtained in 25 min.

To build the training set, we sample iteratively devices from the density with the aim to generate three equal sets, a set of faulty devices, that is, devices with at least one performance falling outside the $3 \cdot$ StdDev specification limit, a set of marginally in-the-specifications devices, that is, devices that have at least one of their performances falling between the $2 \cdot$ StdDev and $3 \cdot$ StdDev specification limits, and a set of functional devices whose performances fall within the $2 \cdot$ StdDev specification limit. The generated training set is illustrated in Figure 14.8b, which shows a marked contrast to Figure 14.8a that contains the original devices. The generated training set contains 900 devices, of which 1/3 are faulty, 1/3 are marginally functional, and 1/3 are functional close to the nominal design. Now the boundary can be better approximated since the area around it is populated with many samples.

Notice that this training set is "biased" in the sense that it is enhanced with a large number of faulty devices. In contrast, the trained classifier is validated using a large validation set of devices (e.g., 1 million) that is produced by "natural" sampling of the density. In this way, the validation set emulates a large set of devices that will be seen in production. As an example, Figure 14.8c illustrates the original set of 1000 devices together with 10^4 randomly generated synthetic devices. This technique allows us to decompose the classification error into test escape (e.g., probability of a faulty device passing the test) and yield loss (e.g., probability of a functional device failing the test) and express these test metrics with PPM accuracy.

14.6.4 Experimental Results

In this section, we put to test each of the presented neural platforms and compare their performance with software neural networks of equivalent topologies. In the first set of experiments, the ANNDW platform is configured into three MLP topologies, each consisting of a single hidden layer with 2, 4, and 8 hidden neurons. The training is repeated five times for each network configuration to average out the randomness due to stochastic nature of the training algorithm. For comparison purposes, the same experiments are repeated with software neural networks of identical topologies using the MATLAB® neural networks toolbox. The software training is carried out using the resilient backpropagation algorithm.

The results on the training and validation sets are presented in Table 14.3. A large discrepancy between the training and validation errors is the result of having a "biased" training set with many marginal devices and a

TABLE 14.3

LNA Training Results on ANNDW Platform

Number of Hidden Neurons	Software Network			Hardware Network		
	2	4	8	2	4	8
Training error (%)	5.82	4.91	4.88	6.82	5.53	5.75
Validation error (%)	0.566	0.548	0.581	0.727	0.435	0.491

"natural" validation set where the majority of devices is distributed around the nominal point. In terms of training error, the software classifier consistently outperforms the hardware classifier by about 1%. However, the validation errors, representing the true accuracy of classification, are similar for both networks (the difference is <0.2%). The hardware version achieves even smaller error for the models with more than two hidden neurons. In fact, the best performance is shown by the hardware network with four hidden neurons resulting in the error of 0.435% or equivalently 4350 misclassified PPM.

The validation error can be further decomposed into test escapes T_E (percentage of faulty devices passing the test) and yield loss Y_L (percentage of good devices not passing the test). Since T_E is more undesirable than Y_L, the decision boundary can be biased to favor Y_L over T_E, that is, to accept fewer faulty devices at the expense of rejecting more functional devices. To this end, the hardware network with four hidden neurons is repeatedly trained for different values of a parameter that regulates the positioning of the boundary. Larger values of the parameter correspond to higher penalties imposed on T_E and cause the boundary to shift toward the functional class. The results on the training set are illustrated in Figure 14.9a, which shows the reduction of misclassification of faulty devices as we gradually increase the value of the parameter (the value 1 corresponds to an "unbiased" boundary, see Table 14.3). Figure 14.9b illustrates the trade-off between T_E and Y_L obtained on the validation set. When the parameter value equals 3 we obtain $Y_L \approx 10 \cdot T_E$. This trade-off point corresponds to the rule-of-ten, which suggests that it is 10 times more expensive to accept a faulty device than rejecting a functional one. The absolute numbers for T_E and Y_L are 1062 and 10063 PPM, respectively. While these numbers are rather large, they do demonstrate the equivalence of the hardware classifier to its software counterpart. More discriminative measurements or a larger measurement pattern are needed to reduce this error to industry acceptable levels.

The second set of experiments employs the FGNN platform configured into both the MLP and ONN topologies along with equivalent software models for comparison purposes. Besides fitting the weights, each of the presented classifiers has a model complexity parameter—the number of hidden neurons—that needs to be selected. We employ a popular technique called cross validation (CV), whereby a small portion of the original training set is reserved as validation data to test a model trained on the remaining data. This step is repeated multiple times for each number of hidden neurons with individual CV errors aggregated into a graph such as the one shown in Figure 14.10. Also shown are the mean and the standard error bars of CV errors. The fact that the CV error does not improve after a few hidden neurons suggests that the optimal boundary is fairly simple. We select the best model according to the one-standard-error rule as the most parsimonious model whose score is within one standard error of the best score. For the MLP and ONN classifiers, these models contain three and two hidden neurons, respectively. The resulting classifiers are retrained on the entire training set and evaluated

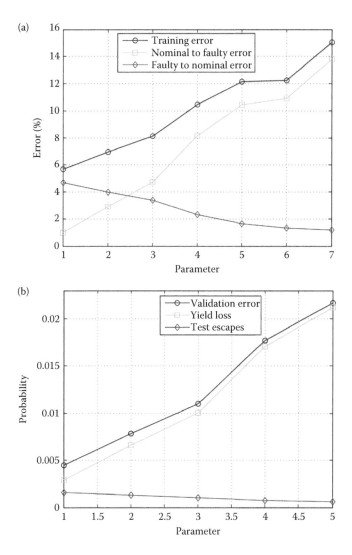

FIGURE 14.9
(a) Training error decomposed into false positives and false negatives for biased boundaries. Larger values of the parameter shift the boundary toward the functional class resulting in lower "faulty-to-nominal" error. (b) Test escape and yield loss error for biased boundaries obtained on the validation set. (Maliuk, D. et al. Analog neural network design for RF built-in self-test. *Proceedings of the IEEE International Test Conference (ITC)*, Austin, TX, pp. 23.2.1–23.2.10. © 2010. IEEE.)

on the test set with the results shown in Table 14.4. For comparison purposes, the table also reports the test error as obtained by training a software MLP with two and four hidden neurons. Although the classification accuracy is similar across various implementations, the software MLP with four hidden neurons and the analog ONN achieve the lowest test error. The average

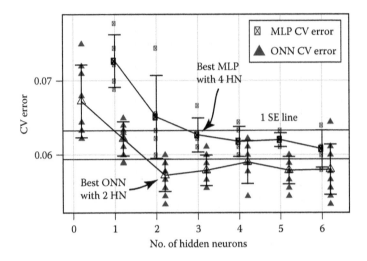

FIGURE 14.10
CV error versus the number of hidden neurons for MLP and ONN learning models trained on the FGNN platform.

TABLE 14.4

LNA Training Results on FGNN Platform

	Analog		Software	
	ONN	MLP	MLP	MLP
Model size (HN)	2	3	2	4
Test error (%)	0.48	0.58	0.73	0.435
Die area (mm²)	0.126	0.138	N/A	N/A
Power (μW)	1.58	1.63	N/A	N/A

power consumption of the analog ONN with two hidden neurons is 2.6 μW. Finally, we copied the learned weights onto the FGTs and verified that the test error remained consistent within several days.

14.7 Case Study II: Defect Filter (RF Front-End)

14.7.1 Problem Definition and DUT

Another interesting class of problems that the presented BIST framework is designed to solve is the identification of manufacturing defects and catastrophic faults (which can happen during the lifetime of ICs). Unlike the previous case, a boundary trained to separate functional from parametric faults does not perform well on defects which do not follow the distribution of devices with process variation and appear as outliers in the space of sensor measurements. Moreover, while it is straightforward to obtain devices as affected by process variation by performing Monte Carlo simulation, there are no widespread models for defect generation. Thus, relying on devices affected by process variation only, the problem of training a defect filter can be solved by constructing a boundary around the class of instances with process variation (considered as functional) serving to protect them against any defects, which appear as outliers.

14.7.2 Data Source

For the purpose of this case study, we used the data from an RF front-end chip presented in Reference 38. The reader is referred to the original publication for details on the DUT and sensors. The front-end consists of an LNA, a mixer and a number of on-chip sensors serving as input features for the classification task. In particular, we consider the readings of three sensors: two envelope detectors and a DC probe. These sensors produce voltage outputs and can be conveniently interfaced with the neural classifier's inputs. The sensor measurements are obtained for 1000 devices generated through postlayout MC simulation and 79 devices with defects. The list of defects includes all possible open- and short-circuits injected one at a time at the layout level.

14.7.3 Experimental Results

The course of experiments is shown in Figure 14.11. The original dataset is split into a training set consisting of 500 randomly selected devices generated with process variation and a test set comprising the remaining 500 devices and all defects. Since training a neural classifier requires two classes,

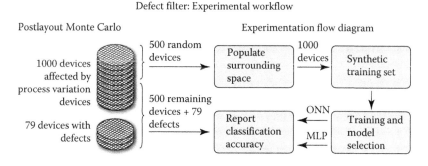

FIGURE 14.11
Defect filter course of experiments. The synthetic training set is used for training and model selection. The test set is used to report classification performance.

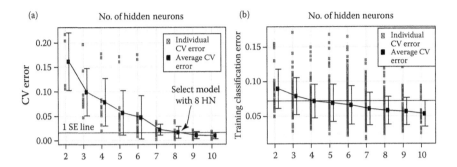

FIGURE 14.12

Cross validation versus the number of hidden neurons for training (a) the MLP and (b) the ONN classifiers.

we generate a second class by uniformly populating the space around the 500 devices from the training set with artificial data. The objective is to leave a small gap between the classes for the separation boundary. Next, both analog classifiers are trained on the synthetic training set using the cross-validation technique. Figure 14.12 illustrates training results in terms of CV errors versus the model complexity for the MLP and ONN classifiers. The CV errors beyond 10 neurons remain at the same level and are not shown. It should be pointed out that absolute values of the CV errors mean little due to the artificial nature of the training set and are used only for model selection.

The one-standard-error line indicates that the best models for the ONN and MLP classifiers contain 4 and 8 neurons, respectively. These models are retrained on the entire synthetic set and applied to the test set with the results shown in Table 14.5. Note that the MLP classifier's error of 1.2% is considerably lower which, however, is achieved at a larger model size. Also note that yield loss accounts for almost the entire test error for both classifiers with only a single defect being misclassified as a functional device by MLP (0.17% test escape). Figure 14.13 illustrates 579 devices from the test set and a decision surface produced by the trained MLP classifier. The devices affected by process variation (blue) are located inside the cavity formed by the decision surface while the defective devices (red) are located outside. Note that only

TABLE 14.5

Training Results on Defect Filter

	Test Error (%)	Test Escapes (%)	Yield Loss (%)	Model Size (HN)	Die Area (mm²)	Power (µW)
Analog ONN	3.79	0	3.79	4	0.21	2.64
Analog MLP	1.2	0.17	1.03	8	0.282	3.45

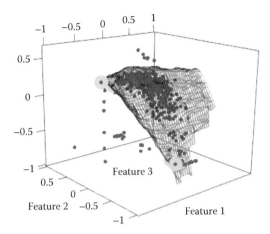

FIGURE 14.13
Decision surface of the MLP classifier as obtained by measuring the response on a fine 3D gird of inputs and interpolating a surface over points where the response crosses a zero threshold. Misclassified devices are highlighted in gray.

those few devices with process variation lying in the tail of distribution fall outside the enclosing boundary and, as a result, are misclassified.

14.8 Conclusions

In this chapter, we discussed how machine learning can be used to make a fully stand-alone BIST in analog/RF circuits possible. In particular, we proposed BIST architecture that incorporates an on-chip neural classifier that can be trained to predict the results of specification-based testing without reliance on any external test equipment. Despite the benefits of reducing postproduction test costs and endowing ICs with self-test capabilities, integrating BIST on chip inevitably raises practicality concerns expressed in terms of added area, energy consumption, and overall complexity. Toward addressing these and other constrains imposed by BIST, we fabricated several experimental platforms consisting of programmable arrays of synapse and neuron circuits, developed strategies to train them, and evaluated their performance on real data from BIST applications. The results presented in this chapter laid the groundwork for high-level integration of a neural classifier, stimuli generators, and sensors into a fully stand-alone BIST. The developed neural platforms allow prototyping neural networks of various sizes and topologies to determine the most suitable one for a given classification problem. This is extremely helpful during the design stage facilitating resource allocation for various components of BIST including the neural classifier. Among other things, the developed platforms could be used to explore fault tolerance properties of analog neural networks and develop techniques to enhance their resilience to faults and aging effects. At the end, the neural classifier must retain correct functionality beyond the projected lifetime of a DUT. On a similar note, the inherent redundancy of neural networks could be leveraged to combat the sensitivity of synaptic weights on temperature and voltage variations. Addressing these questions opens avenues for future research.

References

1. Silva, E., Pineda de Gyvez, J., Gronthoud, G. Functional vs. multi-VDD testing of RF circuits. *IEEE International Test Conference*, Austin, TX, 2005, pp. 17.2.1–17.2.9.
2. Acar, E., Ozev, S. Defect-oriented testing of RF circuits. *IEEE Trans. Computer-Aided Design Integr. Circuits Syst.*, 27(5), 920–931, 2008.
3. Ferrario, J., Wolf, R., Moss, S., Slamani, M. A low-cost test solution for wireless phone RFICs. *IEEE Commun. Mag.*, 41(9), 82–88, 2003.
4. Bhattacharya, S., Chatterjee, A. A DFT approach for testing embedded systems using DC sensors. *IEEE Design Test Comput.*, 23(6), 464–475, 2006.
5. Hafed, M. M., Abaskharoun, N., Roberts, G. W. A 4-GHz effective sample rate integrated test core for analog and mixed-signal circuits. *IEEE J. Solid-State Circuits*, 37(4), 499–514, 2002.
6. Rye, J.-Y., Kim, B. C. Low-cost testing of 5 GHz low noise amplifiers using new RF BIST circuit. *J. Electron. Testing: Theory Applications*, 21(6), 571–581, 2005.
7. Ramzan, R., Zou, L., Dabrowski, J. LNA design for on-chip RF test. *IEEE International Symposium on Circuits and Systems*, Kos, Greece, 2006, pp. 4236–4239.
8. Gopalan, A., Margala, M., Mukund, P. R. A current based self-test methodology for RF front-end circuits. *Microelectron. J.*, 36(12), 1091–1102, 2005.
9. Cimino, M., Lapuyade, H., De Matos, M., Taris, T., Deval, Y., Bégueret, J. B. A robust 130 nm-CMOS built-in current sensor dedicated to RF applications. *IEEE European Test Symposium*, Southampton, UK, 2006, pp. 151–158.
10. Huang, Y.-C., Hsieh, H.-H., Lu, L.-H. A low-noise amplifier with integrated current and power sensors for RF BIST applications. *IEEE VLSI Test Symposium*, Berkeley, CA, 2007, pp. 401–408.
11. Valdes-Garcia, A., Venkatasubramanian, R., Silva-Martinez, J., Sánchez-Sinencio, E. A broadband CMOS amplitude detector for on-chip RF measurements. *IEEE Trans. Instrum. Meas.*, 57(7), 1470–1477, 2008.
12. Abdallah, L., Stratigopoulos, H.-G., Kelma, C., Mir, S. Sensors for built-in alternate RF test. *15th IEEE European Test Symposium (ETS)*, Prague, Czech Republic, 2010, pp. 49–54.
13. Akbay, S. S., Torres, J. L., Rumer, J. M., Chatterjee, A., Amtsfield, J. Alternate test of RF front ends with IP constraints: Frequency domain test generation and validation. *IEEE International Test Conference*, Santa Clara, CA, 2006, pp. 4.4.1–4.4.10.

14. Ellouz, S., Gamand, P., Kelma, C., Vandewiele, B., Allard, B. Combining internal probing with artficial neural networks for optimal RFIC testing. *IEEE International Test Conference*, Santa Clara, CA, 2006, pp. 4.3.1–4.3.9.

15. Voorakaranam, R., Akbay, S. S., Bhattacharya, S., Cherubal, S., Chatterjee, A. Signature testing of analog and RF circuits: Algorithms and methodology. *IEEE Trans. Circuit Syst. I*, 54(5), 1018–1031, 2007.

16. Stratigopoulos, H.-G., Mir, S., Acar, E., Ozev, S. Defect filter for alternate RF test. *IEEE European Test Symposium*, Sevilla, Spain, 2009, pp. 101–106.

17. Stratigopoulos, H.-G., Makris, Y. Error moderation in low-cost machine learning-based analog/RF testing. *IEEE Trans. Comput.-Aided Design Integr. Circuits Syst.*, 27(2), 339–351, 2008.

18. Maliuk, D., Stratigopoulos, H.-G., He, H., Makris, Y. Analog neural network design for RF built-in self-test. *Proceedings of the IEEE International Test Conference (ITC)*, Austin, TX, 2010, pp. 23.2.1–23.2.10.

19. Schlottmann, C. R., Hasler, P. E. A highly dense, low power, programmable analog vector-matrix multiplier: The FPAA implementation. *IEEE J. Emerg. Select. Topics Circuits Syst.*, 1(3), 403–411, 2011.

20. Minch, B. A. Analysis and synthesis of static translinear circuits. Tech. Rep. No. CSL-TR-2000-1002, Cornell University, 2000. http://citeseerx.ist.psu.edu/viewdoc/download?doi=10.1.1.141.1901&rep=rep1&type=pdf

21. Barranco, B. L., Gotarredona, T. S. On the design and characterization of femtoampere current-mode circuits. *IEEE J. Solid-State Circuits*, 38(8), 1353–1363, 2003.

22. Tam, S., Holler, M. An electrically trainable artificial neural network (ETANN) with 10240 floating gate synapses. *Procdings of International Joint Conference on Neural Networks*, Tampa, FL, 1989, pp. 191–196.

23. Mead, C. *Analog VLSI and Neural Systems*. Addison-Wesley, USA, 1989.

24. Lyon, R., Mead, C. An analog electronic cochlea. *IEEE Trans. Acoust. Speech Signal Proc.*, 36(7), 1119–1134, 1988.

25. Kang, K., Shibata, T. An on-chip-trainable gaussian-kernel analog support vector machine. *IEEE Trans. Circuits Syst. I, Reg. Papers*, 57(7), 1513–1524, 2010.

26. Kucher, P., Chakrabartty, S. An energy-scalable margin propagation-based analog vlsi support vector machine. *IEEE International Symposium on Circuits and Systems*, New Orleans, LA, 2007, pp. 1289–1292.

27. Chakrabartty, S., Cauwenberghs, G. Sub-microwatt analog VLSI trainable pattern classifier. *IEEE J. Solid-State Circuits*, 42(5), 1169–1179, 2007.

28. Robucci, R., Gray, J., Abramson, D., Hasler, P. E. A 256 × 256 separable transform cmos imager. *IEEE International Symposium on Circuits and Systems*, Seattle, WA, 2008, pp. 1420–1423.

29. Maliuk, D., Stratigopoulos, H.-G., Makris, Y. An analog VLSI multilayer perceptron and its application towards built-in self-test in analog circuits. *IEEE International On-Line Test Symposium (IOLTS)*, Corfu, Greece, 2010, pp. 71–76.

30. Maliuk, D., Makris, Y. A dual-mode weight storage analog neural network platform for on-chip applications. *IEEE International Symposium on Circuits and Systems*, Seoul, South Korea, 2012, pp. 2889–2892.

31. Haykin, S. *Neural Networks: A Comprehensive Foundation*, 2nd edition. Prentice-Hall, USA, 1998.

32. Koosh, V. F., Goodmanr, R. M. Analog VLSI neural network with digital perturbative learning. *IEEE Trans. Circuit Syst. II*, 49(5), 359–368, 2002.

33. Jabri, M., Flower, B. Weight perturbation: An optimal architecture and learning technique for analog VLSI feedforward and recurrent neural networks. *IEEE Transac. Neural Netw.*, 3(1), 154–157, 1992.

34. Honavar, V., Uhr, L. Generative learning structures and processes for generalized connectionist networks. *Inf. Sci.*, 70, 75–108, 1993.

35. Serrano-Gotarredona, T., Linares-Barranco, B., Andreou, A. G. A general translinear principle for subthreshold MOS transistors. *IEEE Transac. I, Fundam. Theory Appl.*, 46(5), 607–616, 1999.

36. Fahlman, S. E., Lebiere, C. The cascade-correlation learning architecture. *Proc. Advanc. Neural Inform. Process. Syst.*, 2, 524–532, 1990.

37. Stratigopoulos, H.-G., Mir, S., Makris, Y. Enrichment of Limited training sets in machine-learning-based analog/RF testing. *Design, Automation and Test in Europe Conference*, Nice, France, April 2009, pp. 1668–1673.

38. Abdallah, L., Stratigopoulos, H.-G., Mir, S., Kelma, C. RF front-end test using built-in sensors. *IEEE Design Test Comp.*, 28(6), 76–84, 2011.

15

Closed-Loop Spatial Audio Coding

Ikhwana Elfitri

CONTENTS

Having in mind the growing demand for the reliable delivery of high-quality multichannel audio in various multimedia applications such as home entertainment, digital audio broadcasting, computer games, music streaming services as well as teleconferencing, efficient coding techniques [1] have become paramount in the audio processing arena. The traditional approach for compressing multichannel audio has been to encode each audio channel using a mono audio coder, such as Dolby AC-3 and MPEG advanced audio coder (AAC) [2]. However, for the majority of coders adopting this method, the number of bits to be transmitted tends to increase linearly with the number of channels.

Recently, a new concept for encoding multichannel audio signals has been proposed. It comprises the extraction of the spatial cues and the downmixing of multiple audio channels into a mono or stereo audio signal. The downmix signals are subsequently compressed by an existing audio encoder and then transmitted, accompanied by the spatial cues coded as spatial parameters. Any receiver system that cannot handle multichannel audio can simply remove this side information and just render the downmix signals. This provides the coder with backward compatibility, which is important for implementation in various legacy systems. In addition, by utilizing the spatial parameters, the downmix signals can be directly upmixed at the decoder side into a multichannel configuration that may be different from the one used at the encoder side. This technique is known as spatial audio coding (SAC).

Various SAC techniques, such as binaural cue coding (BCC) [3] and MPEG 1/2 layer 3 (MP3) Surround [4], have been proposed. Interchannel level difference (ICLD), interchannel time difference (ICTD), and interchannel coherence (ICC) are extracted as spatial parameters that are based on human spatial hearing cues. Techniques such as parametric stereo (PS) [5] and MPEG surround (MPS) [6,7] may also utilize signal processing techniques, such as decorrelation. The great benefit of these perceptual-based coders is that they can achieve bitrates as low as 3 kb/s for transmitting spatial parameters, as in the case of MPS.

As one can observe, each of these coding techniques has its unique advantages and disadvantages. However, they can all be classified as open-loop systems, where the encoders of BCC, MP3 Surround, and MPS do not consider the decoding process during encoding. The major drawback of an open-loop system is that there is no mechanism employed to reduce the error introduced by quantizing the spatial parameters and coding the downmix signals. In this chapter, an analysis by synthesis spatial audio coding (AbS-SAC) technique is presented, which provides the advantages of a closed-loop system in order to improve the quality of multichannel audio reproduction. We believe that the AbS-SAC technique can be implemented in any of the recent SAC schemes, even though in this work the AbS-SAC is applied solely in the context of the MPS architecture.

The rest of this chapter is organized as follows. Section 15.1 provides an overview of MPS. Subsequently, the closed-loop R-OTT module and the simplified AbS algorithm are presented in Sections 15.2 and 15.3, respectively. Experiments and results achieved by the proposed method are given in Section 15.4 followed by conclusions in the last section.

15.1 Overview of MPEG Surround

To provide a meaningful illustration, a basic block diagram of MPS is shown in Figure 15.1. An analysis quadrature mirror filterbank (A-QMF) is used to decompose the audio signal in each channel into subband signals, while a

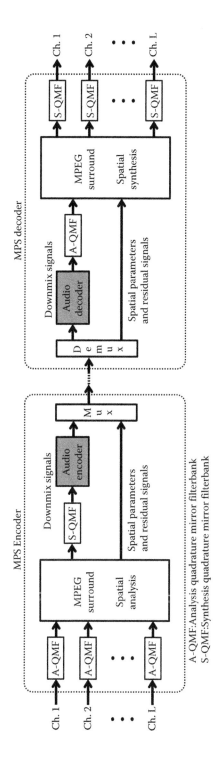

FIGURE 15.1
Block diagram of MPEG surround.

synthesis QMF (S-QMF) is used to transform them back into a time-domain signal. Considering the importance of reducing the number of audio channels, multiple audio signals at the encoder side are typically downmixed into a mono or stereo signal, which can be further encoded by any kind of audio coder such as MP3 and high-efficiency advanced audio coder (HE-AAC). The major benefit of downmixing audio channels is a possibility to employ an existing, legacy audio coder, allowing backward compatibility. In order to be capable of creating back all audio channels at the decoder side, channel level differences (CLDs), ICCs, and channel prediction coefficients (CPCs) must be extracted as spatial parameters and transmitted as side information of the downmix signal. Furthermore, residual signal can be computed as error compensation due to downmixing process and transmitted to the decoder for enabling high-quality audio reconstruction. Interestingly, when operating at lower bitrates the residual signal can be ignored and replaced in the decoder by a synthetic signal constructed using a decorrelator.

15.1.1 Spatial Analysis and Synthesis

The MPS system comprises two pairs of elementary building blocks for channel conversion and the reverse process: one-to-two (OTT) and two-to-three (TTT) modules. The OTT module is used to convert a single channel to two channels while the TTT module is used to convert two channels to three channels. The reverse conversions are done by the reverse OTT (R-OTT) module and the reverse TTT (R-TTT) module. CLDs, ICCs, and residual signal are extracted from the R-OTT module, whereas CPCs, ICCs, and residual signal are calculated from the R-TTT module. The whole process in the encoder and decoder is built up by combining several OTT and TTT modules in a tree structure. This section simply describes the extraction of CLD, ICC as well as residual signal, as they are implemented within the proposed framework. For further details on CPC the readers can refer to Reference 8.

The schematics of the OTT and R-OTT modules are depicted in Figure 15.2. The R-OTT converts two input channels into one output channel and then extracts CLD and ICC as spatial parameters. Conversely, the OTT resynthesizes two channels from one channel, utilizing the spatial parameters.

The first spatial parameter, CLD denoted as C, relates energies of the audio signals in the first and second channels which can be written as

$$C = \frac{e_{x_1}}{e_{x_2}} = \frac{\sum_n x_1[n]x_1^*[n]}{\sum_n x_2[n]x_2^*[n]} \tag{15.1}$$

where the $x_1^*[n]$ and $x_2^*[n]$ represent the complex conjugate of $x_1[n]$ and $x_2[n]$, respectively. For transmission, the quantized logarithmic values of the CLDs are conveyed. The second spatial parameter, ICC denoted as I, reflects

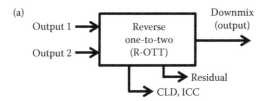

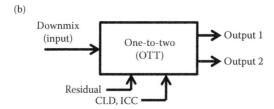

FIGURE 15.2
Block diagram of (a) the OTT module and (b) the R-OTT module as used in MPS.

the degree of correlation between both input channels which can be determined by

$$I = \text{Re}\left\{ \frac{\sum_n x_1[n]x_2^*[n]}{\sqrt{e_{x_1}e_{x_2}}} \right\}$$ (15.2)

The downmix signal $y[n]$ is a scaled sum of the input signals. One possible representation of the downmix signal can be written as

$$y[n] = \frac{x_1[n] + x_2[n]}{a + b}$$ (15.3)

where the energy constants a and b are calculated as

$$(a + b)^2 = \frac{e_{x_1} + e_{x_2} + 2I\sqrt{e_{x_1}e_{x_2}}}{e_{x_1} + e_{x_2}}$$ (15.4)

representing the energy preservation constraint [8].
Furthermore, the residual signal $r[n]$ is determined from the following decomposition:

$$x_1[n] = ay[n] + r[n]$$ (15.5)

$$x_2[n] = by[n] - r[n]$$ (15.6)

which produces a single residual signal for reconstructing both $x_1[n]$ and $x_2[n]$.

At the decoder side, both audio signals are recreated by estimating a and b as follows:

$$\hat{a} = X \cos(A + B) \tag{15.7}$$

$$\hat{b} = Y \cos(A - B) \tag{15.8}$$

where the X, Y, A, and B variables given as

$$X = \sqrt{\frac{\hat{C}}{1 + \hat{C}}} \tag{15.9}$$

$$Y = \sqrt{\frac{1}{1 + \hat{C}}} \tag{15.10}$$

$$A = \frac{1}{2}\arccos(\hat{I}) \tag{15.11}$$

$$B = \tan\left[-\left(\frac{X - Y}{X + Y}\right)\arctan(A)\right] \tag{15.12}$$

are determined from the quantized values of CLD, $\hat{C}$, and the quantized values of ICC, $\hat{I}$. Hence, both signals can be reconstructed as

$$\hat{x}_1[n] = \hat{a}\hat{y}[n] + \hat{r}[n] \tag{15.13}$$

$$\hat{x}_2[n] = \hat{b}\hat{y}[n] - \hat{r}[n] \tag{15.14}$$

which are similar to Equations 15.5 and 15.6 but use the decoded downmix and residual signals, $\hat{y}[n]$ and $\hat{r}[n]$, respectively.

15.1.2 Quantization and Coding

The logarithmic value of the extracted CLD, calculated as $10 \cdot \log_{10}(C)$, is represented using one of the following nonuniform quantization values:

$$CLD = [-150, -45, -40, -35, -30, -25, -22, -19, -16, -13, -10, -8,$$
$$-6, -4, -2, 0, 2, 4, 6, 8, 10, 13, 16, 19, 22, 25, 30, 35, 40, 45, 150]$$

where five bits are allocated to send the index of this quantized CLD. Additionally, the extracted ICC, *I*, is represented by one of the following nonuniform quantization values:

$$ICC = [-0.99, -0.589, 0, 0.36764, 0.60092, 0.84118, 0.937, 1]$$

where three bits are allocated for transmitting the index of the quantized ICC.

The residual signal is encoded in the same way as in LC-AAC. The MPS standard specifies the transformation of the residual signal from the subband domain to the spectral coefficients of the MDCT transform. A frame, comprised of 1024 spectral coefficients, is segmented as scale factor bands, whereas many as 49 scale factor bands are used.

For each band, a scale factor is determined and the spectral coefficients are quantized as follows:

$$ix[k] = \text{sign}(r[k]) \cdot \text{nint}\left[\left(\frac{|r[k]|}{\sqrt[4]{2^{S_F}}}\right)^{0.75}\right] \tag{15.15}$$

where $ix[k]$ is the quantized spectral coefficient with its value limited from -8191 to $+8191$, $r[k]$ is the spectral coefficient of the residual signal, and S_F is the scale factor. Consequently, 14 bits are required to represent the index of the quantized value. The quantizer may utilize a psychoacoustic model to compute the maximum allowed distortion while an analysis by synthesis procedure is carried out to select the best scale factor and quantized spectral coefficient, resulting in a minimal error.

Huffman coding is then carried out for further compression where several Huffman codebooks, designed for different sets of spectral data, are provided. The quantized spectral coefficients of one or more scale factor bands are grouped and then encoded with an appropriate codebook depending on the maximum absolute value of the quantized spectral coefficients within the group. Groups that have all of their quantized spectral coefficients at zero values are associated to codebook 0, where all of the quantized spectral coefficients within this group do not need to be transmitted. Particular attention is given to the last codebook which is provided for groups with the maximum absolute values greater than or equal to 16 when a special escape sequence is used. A set of *n*-tuples of quantized spectral coefficients within a group, consisting of either two or four coefficients, are then represented as Huffman codewords. To keep their size small, most codebooks are given unsigned values. Thus, the sign of each nonzero coefficient is represented as an additional bit appended to the unsigned codeword.

15.2 Closed-Loop R-OTT Module

15.2.1 Analysis-by-Synthesis Framework

Analysis-by-synthesis (AbS) technique [9] is a generic method that has already been implemented in many areas, such as estimation and identification. Several decades ago, this concept was proposed as a framework for encoding speech signals and determining the excitation signal in a linear predictive coding (LPC)-based speech coder. Since then, many other speech coders have been proposed within this framework, such as the most popular code-excited linear prediction (CELP), which is currently specified as one of the tools in the MPEG-4 Audio standard. CELP is also currently adopted in the development of the MPEG standard, ISO/IEC 23003-3/FDIS, Unified Speech and Audio Coding (USAC). The AbS technique is currently applied in many applications, including the quantization of spectral coefficients of the MPEG-AAC audio codec.

An AbS system is able to synthesize a signal by a set of parameters where the values of these parameters are usually made variable in order to produce the best-matched synthesized signal. The difference between the observed signal and the synthesized signal, called the error signal, is utilized in an error minimization block. A set of parameters which produce minimum error signal are selected as optimal parameters and sent to the decoder.

The framework of the AbS-SAC [10] is given in Figure 15.3. A signal upmixing module from spatial synthesis block, similar to that performed at the decoder side, is embedded within the AbS-SAC encoder as a model for reconstructing multichannel audio signals. Assuming that there is no channel error, the audio signals synthesized by the model in the encoder will be exactly the same as the reconstructed audio signals at the decoder side. The error minimization block is used to compare the input signals with the reconstructed signals based on a suitable criterion such as mean squared-error (MSE) or other perceptual relevant criterion. The resultant downmix

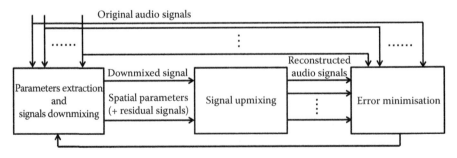

FIGURE 15.3
Framework of analysis-by-synthesis spatial audio coding (AbS-SAC).

and residual signals as well as the optimal spatial parameters are then transmitted to the decoder.

Based on this framework, various implementations are possible. They are listed as follows:

- Any approach of spatial analysis and synthesis can be implemented.
- Different numbers of input and output channels can also be used.
- Various types of suitable error criterion can be utilized.
- For taking into consideration the error introduced in the communication channel, a block modeling the channel error can be inserted between the spatial analysis and synthesis block.
- The AbS-SAC approach can be implemented for only a single parameter recalculation without multiple iterations. This can be considered as a modification to the original open-loop spatial analysis block.
- The implementation can be intended to find the optimal synthesized signals by performing the trial and error procedure. Either the original blocks or the adapted version of the spatial analysis and synthesis can be applied.

15.2.2 R-OTT Module within AbS-SAC Framework

The OTT and R-OTT modules, as used in MPS, can be implemented within the AbS-SAC framework, where two channels of the original audio signals are fed to an R-OTT module as the spatial analysis block. On the other hand, the OTT module is performed as the spatial synthesis block for reconstructing two channels of synthesized audio signals, so that Equation 15.8 becomes the formula of the model. In this chapter, this is referred to as a closed-loop R-OTT module where a new optimized downmix signal can be approximated as

$$y_{new}[n] = \frac{x_1[n] + x_2[n]}{\hat{a} + \hat{b}} \qquad (15.16)$$

Moreover, based on the new optimized downmix signal Equation 15.5 can be used to obtain the expression for the new optimized residual signal as

$$r_{new}[n] = x_1[n] - \hat{a}y_{new}[n] = \hat{b}y_{new}[n] - x_2[n] \qquad (15.17)$$

where either $x_1[n] - \hat{a}y_{new}[n]$ or $\hat{b}y_{new}[n] - x_2[n]$ can be used to determined $r_{new}[n]$. If both input signals have the exact same magnitude but opposite phases (i.e., $x_1[n] = -x_2[n]$), then the downmix signal has all-zero values, $y_{new}[n] = 0$. Consequently, the residual signal can be determined as

$r_{new}[n] = x_1[n] = -x_2[n]$, and a specific information has to be transmitted to the decoder conveying this information.

The proposed closed-loop R-OTT algorithm relies on an approximation in Equation 15.16 when recalculating the downmix signal as a new, optimized signal, $y_{new}[n]$. Therefore, the signal distortion reduction process is based on how to create the new optimized downmix signal on the encoder side, such that the synthesized downmix signal on the decoder side fulfils the desired criteria. In practice, this is achieved by ensuring that the approximation error, which is the difference between the synthesized downmix signal, $\hat{y}[n]$, and the new optimized downmix signal, $y_{new}[n]$, is minimized. In order to obtain the minimum approximation error, both the synthesized and the approximated signals should be synchronized and compared.

The closed-loop R-OTT method is ideally capable of considerably minimize the error introduced by the quantization process of the spatial parameters. This is because the new optimized downmix and residual signals, $y_{new}[n]$ and $r_{new}[n]$, are computed based on estimated energy constants, $\hat{a}$ and $\hat{b}$, so that the quantization errors of CLD and ICC are now compensated for through the newly optimized signals. Consequently, the quantization errors of CLD and ICC no longer affect the overall distortion of the synthesized audio signals.

15.3 Simplified AbS Algorithm

15.3.1 Full Search AbS Optimization

Referring to the proposed AbS-SAC framework for SAC as given in Figure 15.3, a case of the simplest AbS implementation to encode 2-channel audio signals can be illustrated. In order to find the optimal downmix and residual signals, as well as the optimal parameters, a full search AbS optimization should be applied. An OTT module is used as a model for reconstructing two channels of audio signals. As the full searching procedure is performed, there is no need for applying the spatial analysis block.

An AbS optimization procedure can be carried out in such a way that the inputs of the optimization procedure are the quantization values of the downmix and residual signals, as well as the spatial parameters. All of these inputs can be varied to reconstruct various forms of audio signals. The purpose of the AbS optimization procedure is to examine all possible outcomes obtained by combining all inputs in every possible way, that is, all possible combinations of every variable. For each combination, the OTT module reconstructs audio signals and the error minimization block then computes signal distortion. Any combination that obtains minimum error is chosen as the optimal one.

The quantization values of the spectral coefficient quantizer of the downmix and residual signals that are becoming the inputs of the AbS optimization procedure range from −8191 to +8191, meaning that there are 16,383 quantization values for each spectral coefficient of the downmix signals. On the other hand, 31 and 8 quantization values of the CLD and ICC, respectively, are also available, provided that the MPS's quantizers are used. As a result of combining all those quantization values, for each index of the spectral coefficient the number of available combinations can be computed as $16,383 \times 16,383 \times 31 \times 8 = 6.6564 \times 10^{10}$. Note that for simplifying the calculation, the spatial parameters are assumed to be calculated for every spectral coefficient. The scale factor band is ignored.

15.3.2 An Approach for Algorithm Simplification

A simplified trial and error procedure can be applied in order to find suboptimal signals and parameters as a solution for the impractical implementation requirements of the full search AbS procedure. Three steps of simplifications are applied to make the algorithm simple. First, the number of parameters and spectral coefficients involved in the searching procedure are significantly reduced. For instance, rather than finding an optimal spectral coefficient from all quantization values, a suboptimal coefficient is simply chosen from a limited number of quantization values which are assigned based on decoded spectral coefficients. Considering the trade-off between the complexity and the degree of suboptimality, the number of coefficients and parameters involved in the searching procedure can be made variable.

Second, the main AbS-SAC algorithm is performed as a sequential process in that the suboptimal signals and parameters are not selected at the same time but one after the other. The suboptimal spectral coefficients of the downmix and residual signals can be determined first. Once the suboptimal downmix and residual signals are found, the suboptimal CLDs and ICCs can be selected. Alternatively, suboptimal spatial parameters are selected first followed by choosing suboptimal spectral coefficients. Performing the searching algorithm in a sequential process will significantly reduce the number of possible combinations to be examined.

Finally, the sequential process is performed iteratively until an insignificant error reduction is achieved. The reason for performing the iteration process is that the suboptimal spatial parameters are found based on the selected suboptimal downmix and residual signals. Additionally, the suboptimal spectral coefficients of the downmix and residual signals are selected based on the chosen suboptimal spatial parameters. Consequently, it is possible to reoptimize the downmix and residual signals after determining suboptimal spatial parameters. In contrast, it is also possible to reselect new suboptimal spatial parameters once suboptimal downmix and residual signals are found. It is expected that undertaking the iteration process will gradually reduce signal distortion.

The limited number of quantization values, where the suboptimal spectral coefficient is selected from, can be determined as follows: the spectral coefficients decoded by the spectral decoder become the inputs to the algorithm. A number of quantization values defined as the candidates for suboptimal spectral coefficient named as predetermined vector, $ix_p[k]$, can then be assigned based on the decoded spectral coefficients, $ix[k]$, as

$$ix_p[k] = [ix[k] - v, ix[k] - v + 1, \ldots, ix[k] - 1, ix[k],$$
$$ix[k] + 1, \ldots, ix[k] + v - 1, ix[k] + v] \tag{15.18}$$

where $ix[k]$ is the decoded spectral coefficient as in Equation 15.15, k is the index of spectral coefficient, and $2v + 1$ is the size of the predetermined vector, $ix_p[k]$, with v an integer number reflecting the computational complexity of the searching procedure.

A limited number of quantization values of spatial parameters are determined in a similar way. The CLDs and ICCs obtained from the quantizer are used as the initial values. A set of predetermined values of CLDs, C_p, and a set of predetermined values of inter channel coherences, I_p, are determined using

$$P_p = \begin{cases} [P(1), P(2), \ldots, P(p-1), P(p), P(p+1), \ldots, P(2w), P(2w+1)] \\ \text{if } P(p) \leq w + 1; \\ [P(p-w), P(p-w+1), \ldots, P(p-1), P(p), P(p+1), \ldots, P(p+w-1), \\ P(p+w)] \quad \text{if } w + 1 < P(p) \leq P_{max} - w; \\ [P(P_m - 2w), P(P_m - 2w + 1), \ldots, P(p-1), P(p), P(p+1), \ldots, P(P_m - 1), \\ P(P_m)] \quad \text{if } P_m - w < P(p); \end{cases} \tag{15.19}$$

where $P(p)$ is the initial decoded parameter, P_m is the size of the codebook, P is the spatial parameter which is either C, or I, P is the index of the initial decoded parameter, and w is an integer number reflecting the complexity of the procedure. For CLD $w \leq 15$ and for ICC $w \leq 3$.

15.3.3 Basic Scheme of the Encoder

The suboptimal AbS optimization is performed based on the proposed MDCT-based closed-loop R-OTT module. The downmix and residual signals, as well as the spatial parameters extracted from the closed-loop R-OTT module, are used as the initial input of the suboptimal AbS optimization. Practically, the AbS-SAC encoder is implemented, as shown in Figure 15.4. The audio signal in each channel is transformed to spectral coefficients by means of MDCT. A tree scheme employing the closed-loop R-OTT modules

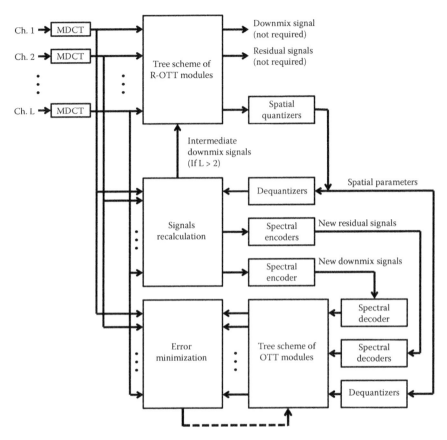

FIGURE 15.4
Block diagram of the AbS-SAC encoder.

is then performed in order to extract the spatial parameters. The new optimized downmix and residual signals are calculated based on the quantized spatial parameters. The downmix signal, as well as the residual signals, are then encoded by the spectral encoders, which actually consist of spectral coefficient quantization and the noiseless coding scheme. Prior to being supplied to the tree of OTT modules, all signals are decoded back by the spectral decoders. The decoded downmix and residual signals, as well as the quantized spatial parameters, are then used by the tree of OTT modules to upmix the audio signals. The error minimization block compares the spectral coefficients of the reproduced audio signals with those of the original signals and then computes the errors.

A closed-loop optimization procedure, utilizing the error minimization block and the tree of OTT modules, can be carried out. The inputs of the optimization loop are the decoded spectral coefficients of the downmixed and residual signals, as well as the decoded spatial parameters. The purpose of the closed-loop approach is to examine all possible outcomes obtained by combining all quantization values of the spectral coefficient and spatial parameter quantizer in every possible way.

15.3.4 Suboptimal Algorithm

The operation of the algorithm can be explained as follows. For each index of the spectral coefficient, the candidates for suboptimal spectral coefficients of the downmix and residual signals are combined. Then, all resulting possible combinations are examined to jointly choose a suboptimal coefficient, which provides the smallest error among the other tested coefficients, for the downmix signal, as well as a suboptimal coefficient for each residual signal. In performing this task, the tree of OTT modules takes the decoded spatial parameters, given by the dequantizer, as inputs. However, the process of selecting the suboptimal spatial parameters is not performed at this stage.

Note that this task has to be carefully completed by considering the Huffman encoding process. As explained previously (see Section 15.1.2), there is a case where all spectral coefficients within a group have zero values. For such a case there is no need to transmit the magnitude of the spectral coefficients. Modifying one or more spectral coefficients within that group causes the Huffman encoding process to be associated with another codebook. As a result, the spectral coefficients within the group need to be transmitted, which causes an increase in the transmitted bitrate. However, the error reduction achieved by modifying those spectral coefficients may not provide a worthy advantage, due to an increase in the transmitted bitrate.

For this reason, the search for the suboptimal spectral coefficients, particularly downmix and residual signals, is not performed if the maximum absolute value of a group is zero. The process of choosing the suboptimal spectral coefficients is then followed by the selection of the suboptimal spatial parameters. For each parameter band, predetermined values of CLDs and ICCs are combined to form all possible combinations. A set of CLDs and ICCs is then selected as the suboptimal spatial parameters. At this stage, the downmix and the residual signal optimization is not performed.

The whole sequential process of searching for suboptimal signals and parameters is repeated as an iteration process. For the second and subsequent iterations, the chosen suboptimal signals and parameters should be used as inputs of sequential process rather than the ones from the spectral decoder and the spatial dequantizer. The iteration process is terminated when error reduction below a given threshold has been reached. The quality of the reconstructed audio is expected to improve with every iteration, however, the amount of improvement may reduce. Even if further iterations are executed the resultant values may not converge to the optimal values of both the spectral coefficients and the spatial parameters. Thus, the goal of this simplified AbS algorithm is not to provide the optimal or near-optimal signals and parameters. Instead, it is intended to provide a solution for impractical implementation of the full search AbS procedure while minimize the signal distortion.

The whole sequential process of searching for suboptimal signals and parameters is repeated as an iteration process. For the second and subsequent iterations, the chosen suboptimal signals and parameters should be used as inputs of the sequential process rather than the ones from the spectral decoder and the spatial dequantizer. The iteration process is terminated when an error reduction below a given threshold has been reached. The quality of the reconstructed audio is expected to improve with every iteration, however, the amount of improvement may reduce. Even if further iterations are executed, the resultant values may not converge to the optimal values of both the spectral coefficients and the spatial parameters. Thus, the goal of this suboptimal algorithm is not to provide the optimal or near-optimal signals and parameters. Instead, it is intended to provide a solution for the practical implementation of the optimal searching procedure while minimizing the signal distortion.

15.3.5 Complexity of Suboptimal Algorithm

The proposed simplified AbS algorithm is scalable, and its algorithmic complexity depends mainly on the number of loop procedures that have to be performed in finding the suboptimal spectral coefficients and spatial parameters, as well as the number of R-OTT modules whose signals and parameters are optimized. In order to reduce the complexity of the algorithm, the number of loop procedures and the involved R-OTT modules can be decreased. Based on equations from 15.7 through 15.14, for each spectral coefficient an R-OTT module performs a number of operations, N_{ott}, consisting of 264 additions/multiplications. Hence, the number of operations, N_{op}, for each index of spectral coefficient required by the simplified AbS algorithm can be determined as

$$N_{op} = N_{ott} \times N_{loop} \tag{15.20}$$

where N_{ott} is the number of operations performed by an R-OTT module and N_{loop} is the number of loop procedures need to be performed. As an illustration, the number of loop procedures, N_{loop}, that have to be executed for encoding 5-channel audio signals is given in Table 15.1.

15.4 Results

In order to evaluate the proposed system, a number of experiments designed to assess the encoding of 5 and 10 audio channels were conducted. The audio

TABLE 15.1

Complexity of Simplified AbS Algorithm

v	Size of Predetermined Vector	Number of Loops
0	1	0
1	3	243
2	5	3125
3	7	16,807
w	Size of Predetermined Vector	Number of Loops
0	1	0
1	3	6561
2	5	390,625

TABLE 15.2

List of Audio Excerpts for Experiments

Excerpt Name	Description
Applause	Clapping hands of hundreds of people
Drum	Drum and male vocal with guitar as background
Laughter	Sound of hundreds of people laughing
Talk	Male and female speech with music background
Vivaldi	Classical music with vocal

excerpts, sampled at 48 kHz, listed in Table 15.2 were prepared for the experiments. They were selectively chosen from a broad range of long sequence 5.1 audio signals ranging from speeches, pop, and classical music, as well as specific sounds such as clapping hands. For each audio sequence, a limited 12-s audio excerpt was selected based on the possibility of more transient events. All of the 10-channel audio signals were produced by upmixing the 5-channel signals using a simple amplitude panning technique. The tree scheme of R-OTT modules for downmixing five channels into a mono downmix, as given in Figure 15.5, was used in the experiments. However, the low-frequency

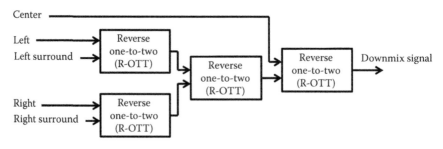

FIGURE 15.5
A tree scheme of R-OTT modules used in the experiments.

enhancement (LFE) channel was excluded for simplicity. Each channel of audio signals was segmented into 2048 time-domain samples with 50% overlap. For calculating CLDs and ICCs 20 parameter bands were used. The 20 parameter bands for the MDCT-based R-OTT were determined by mapping the 20 parameter bands of the CM-QMF to the 49 scale factor bands of the spectral coefficients. The downmix signal was encoded by AAC. The AAC multichannel codec, implemented as FAAC 1.28 and FAAD2 2.7, was used for benchmarking to demonstrate the usefulness of the proposed AbS-SAC approach, even though it is not the best implementation of the AAC standard.

15.4.1 Evaluation of Closed-Loop R-OTT Module

The evaluation of the closed-loop R-OTT module is aimed at demonstrating that the closed-loop approach can both improve the segSNR compared to the open-loop and perform much better in the MDCT domain. Table 15.3 shows the results of the experiment comparing segSNRs of 5-channel audio encoders employing open-loop and closed-loop R-OTT modules in both the CM-QMF-based and MDCT-based scenarios. All encoders operate at a bitrate of 160 kb/s per audio channel. Here all encoders are optimized to provide maximum segSNR performance.

The results show that the closed-loop R-OTT method improves average segSNR of all the tested audio excerpts. It clearly indicates that the closed-loop R-OTT method is capable of minimizing signal distortion, resulting in segSNR improvement. Furthermore, the MDCT-based closed-loop R-OTT module outperforms, in terms of segSNR, the CM-QMF-based closed-loop R-OTT module for all tested audio excerpts other than Applause. For this Applause audio excerpt, both CM-QMF and MDCT schemes are competitive. It is an indication that the closed-loop R-OTT module generally performs better in the MDCT domain when compared with the CM-QMF domain. Moreover, the results also show that the open-loop R-OTT module has smaller segSNRs in the MDCT domain rather than in the CM-QMF domain. As expected, it indicates that the open-loop R-OTT module is basically not an appropriate method to be applied in the MDCT domain. However,

TABLE 15.3

SegSNRs (dB) of Open- and Closed-Loop R-OTT Modules

Audio Excerpt	CM-QMF Open Loop	CM-QMF Closed Loop	MDCT Open Loop	MDCT Closed Loop
Applause	24.42	30.53	21.33	30.04
Drum	23.46	32.82	21.31	39.72
Laughter	23.92	32.87	21.31	37.05
Talk	24.34	31.13	20.65	37.82
Vivaldi	21.91	27.35	19.20	44.35
Mean	23.61	30.94	20.76	37.80

employing a closed-loop approach indicates that a significant improvement is achieved, even better than the CM-QMF-based method.

15.4.2 Evaluation of Suboptimal Algorithm

The performance of the suboptimal algorithm is assessed by investigating the SNR of the reconstructed audio signals against the complexity of the algorithm. The suboptimal algorithm is tested for various computational complexities by assigning $v = 0$, 1, 2, 3 and $w = 0$, 1, 2. Assigning $v = 0$ and $w = 0$ means that the suboptimal algorithm is not performed. Moreover, $v = 0$ and $w \neq 0$ means that the suboptimal algorithm chooses the suboptimal spatial parameter but does not find the suboptimal spectral coefficients of the downmix and residual signals. On the other hand, defining $v \neq 0$ and $w = 0$ instructs the algorithm to select the suboptimal spectral coefficients without selecting suboptimal spatial parameters. The algorithm is terminated when the segSNR improvement is less than 10^{-4}. The average segSNRs achieved by the AbS-SAC, operating at three different bitrates, 42.8, 65.4, and 83.2 kb/s per audio channel, are given in Table 15.4.

As can be seen, the encoder employing suboptimal algorithm (i.e., $v \neq 0$ and $w \neq 0$), for various values of v and w, can improve the segSNR although the improvement is different for each complexity level. The results suggest that the best configuration to perform the proposed suboptimal algorithm, in terms of the segSNR improvement with the least complexity, is achieved by setting $v = 1$ and $w = 1$. However, to lower the complexity to a more reasonable level one can set $v = 1$ and $w = 0$ and still achieve considerable segSNR improvement.

15.4.3 Objective Evaluation

The goal of this experiment is to objectively assess perceptual quality of the proposed AbS-SAC for various operating bitrates. To our knowledge, no

TABLE 15.4

Average SegSNRs (dB) for Various Complexity Levels

Bitrate/Channel		$v = 0$	$v = 1$	$v = 2$	$v = 3$
	$w = 0$	21.82	24.31	24.45	24.50
42.8 kb/s	$w = 1$	22.38	25.12	25.16	25.18
	$w = 2$	22.41	25.19	25.19	25.22
	$w = 0$	27.82	30.41	30.45	30.46
64.4 kb/s	$w = 1$	28.33	30.77	30.79	30.79
	$w = 2$	28.34	30.79	30.81	30.82
	$w = 0$	30.66	33.51	33.54	33.72
83.2 kb/s	$w = 1$	31.12	33.72	33.74	33.74
	$w = 2$	31.13	33.73	33.75	33.75

objective perceptual test is currently available for high-quality multichannel audio signals. Thus, we have adapted the perceptual evaluation of audio quality (PEAQ), an ITU-R BS.1387-1 recommendation for assessing a mono audio signal, and currently under standardization process to include multichannel audio assessment, for evaluating multichannel audio signals. The objective difference grade (ODG), that has five grades: 0 (imperceptible), −1 (perceptible but not annoying), −2 (slightly annoying), −3 (annoying), and −4 (very annoying), was first measured for each channel of audio signal. The average values of the ODG scores over all channels are then presented as the final results for multichannel audio. A software developed by McGill University is used for calculating ODG score. Moreover, the experiments also include encoding of 10-channel audio signals. This is intended to show that, for larger channels at the given bitrate per audio channel, the performance improvement is even higher. Considering the complexity of the AbS-SAC encoder, for encoding 5 audio channels, the suboptimal algorithm is assigned with $v = 1$ and $w = 1$ while, for encoding 10 audio channels, parameters are set to $v = 1$ and $w = 0$.

The results of the experiments for encoding 5-channel and 10-channel audio signals are given in Figure 15.6. For simplicity, the results of AAC 10-channel are not shown, as they are almost identical to those achieved on AAC 5-channel. The overall ODG, averaged over all audio excerpts as shown in the lowest right plot, shows that the AbS-SAC, applied to both 5 and 10 channels, significantly outperforms, in terms of PEAQ, the tested AAC multichannel for all operating bitrates from 40 to 96 kb/s per audio

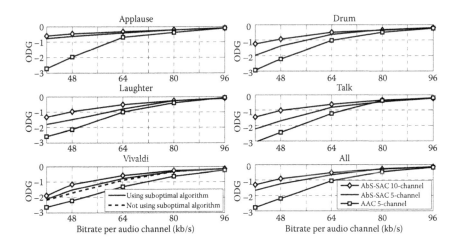

FIGURE 15.6
Objective difference grade (ODG) of the AbS-SAC for various bitrates in comparison with the tested AAC multichannel. The ODG scores of the tested AAC multichannel, for encoding 10-channel audio signals, are not plotted as they are similar to the ODG scores of the tested AAC multichannel for encoding 5-channel audio signals.

channel. However, the performance increase is greater when encoding 10 channels. It can be seen that an improvement of more than 2 points of ODG grade is achieved on the Applause audio excerpt at a bitrate of 40 kb/s per audio channel. Moreover, up to 1 point of ODG improvement is achieved on every tested audio excerpt. The results indicate that the proposed AbS-SAC technique significantly improves encoding performance for a wide range of tested audio materials.

15.4.4 Subjective Evaluation

The proposed AbS-SAC approach, for encoding 5-channel audio signals, has also been evaluated using listening tests. The subjective assessment of small impairments in the audio system, as recommended in the ITU-R BS.1116-1 using the "double-blind triple stimulus with a hidden reference" method, is used. The subjective difference grade (SDG), having five grades that are similar to ODG, is used. Three codecs were taken under test: AbS-SAC, AAC multichannel, and HE-AAC multichannel. In order to reduce the difficulty, that the listeners would experience in scoring the tested audio excerpts because of too small impairment, a low but still realistic bitrate should be chosen. In the experiments, a bitrate of 51.2 kb/s per audio channel, equal to an overall 5-channel bitrates of 256 kb/s, is chosen for both the AbS-SAC and AAC multichannel. Below this bitrate, the proposed AbS-SAC cannot provide a significant segSNR improvement. On the other hand, operating both coders above the chosen bitrate would increase the difficulty for the listeners in assessing the tested audio excerpts. In addition, it is still in the range of the normal operation bitrates of the AAC multichannel which is used as a benchmark. Moreover, the HE-AAC multichannel operates at its maximum typical bitrate of 32 kb/s per audio channel, which is equal to 160 kb/s for all five audio channels. Operating the HE-AAC above this bitrate is not useful in terms of coding efficiency, which means that the HE-AAC multichannel may not achieve a better performance.

A total of 20 listeners participated in this listening test. As specified in the expertise of the listeners are evaluated by averaging their SDG scores over all audio excerpts. Based on this average SDG score, a postscreening method was applied. Three listeners with an average SDG score greater than zero are assumed to be unable to correctly distinguish between the hidden reference and the tested audio object. Thus, the data from those three listeners was discarded. Only the SDG scores from the other 17 listeners were used for the results.

Figure 15.7a presents the average SDG score of each tested audio codec averaged over all audio excerpts. The error bars show the 95% confidence intervals of the mean scores. The results show that the SDG scores of all the tested codecs are competitive and very close to a grade of imperceptible. However, the proposed AbS-SAC approach achieves the highest SDG score. Furthermore, Figure 15.7b shows the SDG score of every tested audio excerpt

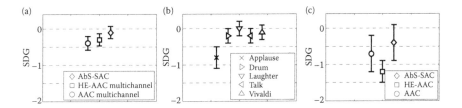

FIGURE 15.7
The results of the subjective test, to compare performance of the proposed AbS-SAC, AAC multichannel, and HE-AAC multichannel: (a) SDG scores of the tested audio codecs averaged over all audio excerpts, (b) SDG scores of all audio excerpts averaged over the tested audio codecs, (c) SDG scores of the tested audio codec for the Applause audio excerpt.

averaged over all tested audio codecs, where the Applause audio excerpt has the lowest SDG score. As expected, it suggests that the Applause audio excerpt is the most critical item among the tested audio excerpts. For this Applause audio excerpt the proposed AbS-SAC approach also achieves the highest SDG score as shown in Figure 15.7c.

15.4.5 Complexity Assessment

In order to assess the performance of the suboptimal algorithm with regards to complexity, the bottom-left graph in Figure 15.6 shows the ODGs of two variants of the proposed AbS-SAC technique, with two extremely different complexity scales, for encoding five channels of the Vivaldi audio excerpt. The first one is the AbS-SAC codec using the suboptimal algorithm where $v = 1$ and $w = 0$ and all spectral coefficients at every OTT module are optimized, and the other does not use the suboptimal algorithm. The results clearly demonstrate that without the suboptimal algorithm the proposed codec is still able to achieve significant quality improvement while the suboptimal algorithm improves the performance further.

15.5 Conclusions

This chapter proposes a new AbS-SAC technique where the AbS concept is applied when choosing the suboptimal downmix signal and the spatial parameters so as to minimize the encoded signal distortion. It is demonstrated that the closed-loop R-OTT algorithm significantly reduces the error introduced by the spatial parameter quantization process resulting in significant segSNR improvement. In addition, it is shown that the frequency domain parameterization is more suitable for the AbS-SAC method instead of the sub-band domain as applied in MPS for encoding error reduction.

Additionally, the MDCT allows simplification of the coding structure by removing the transformation of the residual signals from the sub-band domain to spectral coefficients of the MDCT for the purpose of quantization. Moreover, a simplified AbS-SAC search algorithm has also demonstrated its ability to find suboptimal signals and parameters with significantly lower complexity to address the practicality of implementation of the optimal searching procedure. Subjective tests show that the AbS-SAC method outperforms, in terms of SDG score, the tested AAC multichannel, at a bitrate of 51.2 kb/s per audio channel. In addition, the AbS-SAC method has consistently higher PEAQ ODG scores than the tested AAC multichannel, for bitrates ranging from 40 to 96 kb/s per audio channel.

References

1. Brandenburg, K., Faller, C., Herre, J., Johnston, J. D., Kleijn, W. B. Perceptual coding of high-quality digital audio. *Proc. IEEE*, 101(9), 1905–1919, 2014.
2. Herre, J., Dietz, M. MPEG-4 high-efficiency AAC coding. *IEEE Signal Process. Mag.*, 25(3), 137–142, 2008.
3. Baumgarte, F., Faller, C. Binaural cue coding—Part I: Psychoacoustic fundamentals and design principles. *IEEE Trans. Speech, Audio, Lang. Process.*, 11(6), 509–519, 2003.
4. Moon, H. A low-complexity design for an MP3 multichannel audio decoding system. *IEEE Trans. Speech, Audio, Lang. Process.*, 20(1), 314–321, 2012.
5. Breebaart, J., van de Par, S., Kohlrausch, A., Schuijers, E. Parametric coding of stereo audio *EURASIP J. Appl. Signal Process.*, 1305–1322, 2005.
6. Hilpert, J., Disch, S. The MPEG surround audio coding standard [Standards in a nutshell]. *IEEE Signal Process. Mag.*, 26(1), 148–152, 2009.
7. Herre, J., Kjorlings, K., Breebaart, J., Faller, C., Disch, S., Purnhagen, H., Koppens, J. et al. MPEG Surround—The ISO/MPEG standard for efficient and compatible multichannel audio coding. *J. Audio Eng.Soc.*, 56(11), 932–955, 2008.
8. Breebaart, J., Hotho, G., Koppens, J., Schuijers, E., Oomen, W., de Par, S. V. Background, concepts, and architecture for the recent MPEG Surround standard on multichannel audio compression. *J. Audio Eng. Soc.*, 55(5), 331–351, 2007.
9. Kondoz, A. *Digital Speech: Coding for Low Bit Rate Communication Systems.* London: John Wiley Ltd, 2004.
10. Elfitri, I., Gunel, B., Kondoz, A. Multichannel audio coding based on analysis by synthesis. *Proc. IEEE*, 99(4), 657–670, 2011.

Index

A

A-QMF, *see* Analysis quadrature mirror filterbank (A-QMF)

AAC, *see* Advanced audio coder (AAC)

AbS-SAC technique, *see* Analysis by synthesis spatial audio coding technique (AbS-SAC technique)

AbS, *see* Analysis-by-synthesis (AbS)

AC analysis, *see* Frequency-domain analysis

Access-based methods, 283
  DFT methods, 280
  IEEE Std. 1149.4, 284–285
  systematic method, 283

ADC, *see* Analog-to-digital converters (ADC)

ADPLL, *see* All-digital PLL (ADPLL)

ADS™, *see* Agilent Advanced Design System (ADS™)

Advanced audio coder (AAC), 379

AGC circuits, *see* Automatic gain control circuits (AGC circuits)

Agilent Advanced Design System (ADS™), 186, 189

Aging mechanisms, 111

AHI, *see* Anode hole injection (AHI)

All-digital PLL (ADPLL), 274

Alternate built-in test
  accuracy of parameter computation, 313
  built-in test configuration, 309
  convergence of genetic algorithm, 311
  FFT, 314
  loopback test, 314
  MATLAB®-based nonlinear equation solver, 311
  model parameter computation, 310
  using sensors, 309
  signal generators, 314
  transceiver test case study, 311–312
  transmitter parameters, 310

Alternate test, 351

Alternate testing paradigm, 302
  basis, 302
  low noise RF amplifier, 307–309
  multiple DUT specifications, 303
  precision Op-Amp, 304–307
  in production, 304
  training set of devices, 303

AMD, *see* Approximate minimum degree (AMD)

Analog-to-digital converters (ADC), 31, 177, 262, 290, 302
  6-bit 3.5-GS/s Flash, 45, 47
  transfer function, 33

Analog filters, 290–291

Analog front-end processing stage, 3

Analog/mixed-signal testing, 280

Analog Neural Network with Digital Weight Storage (ANNDW), 355
  hardware-friendly training algorithm, 359–360
  neuron circuits, 357–359
  supporting neural network model, 355–356
  synapse, 357–359
  system architecture, 356–357

Analysis-by-synthesis (AbS), 386–387
  algorithm, 388–390
  complexity of simplified, 394
  encoder, 390–392
  optimization, 388–389
  suboptimal algorithm, 392–393

Analysis by synthesis spatial audio coding technique (AbS-SAC technique), 380; *see also* Closed-loop spatial audio coding
  encoder, 391
  framework, 386
  R-OTT module within, 387–388

nd by CPI Group (UK) Ltd, Croydon, CR0 4YY

24/10/2024

01778307-0016

Printed and bound by CPI Group (UK) Ltd, Croydon, CR0 4YY

24/10/2024

01778307-0016